Simulation and Modelling of Continuous Systems

Prentice Hall International
Series in Systems and Control Engineering

M. J. Grimble, Series Editor

BANKS, S. P., *Control Systems Engineering: Modelling and simulation control theory and microprocessor implementation*

BANKS, S. P., *Mathematical Theories of Nonlinear Systems*

BENNETT, S., *Real-time Computer Control: An introduction*

BITMEAD, R. R., GEVERS, M. and WERTZ, V., *Adaptive Optimal Control*

BUTLER, H., *Model Reference Adaptive Control: From theory to practice*

CEGRELL, T., *Power Systems Control*

COOK, P. A., *Nonlinear Dynamical Systems*

ISERMANN, R., LACHMANN, K. H. and MATKO, D., *Adaptive Control Systems*

KUCERA, V., *Analysis and Design of Discrete Linear Control Systems*

LUNZE, J., *Feedback Control of Large-Scale Systems*

LUNZE, J., *Robust Multivariable Feedback Control*

MATKO, D., ZUPANČIČ, B. and KARBA, R., *Simulation and Modelling of Continuous Systems: A case study approach*

MCLEAN, D., *Automatic Flight Control Systems*

OLSSON, G. and PIANI, G., *Computer Systems for Automation and Control*

PARKS, P. C. and HAHN, V., *Stability Theory*

PATTON, R., CLARK, R. N. and FRANK, P. M. (editors), *Fault Diagnosis in Dynamic Systems*

PETKOV, P. H., CHRISTOV, N. D. and KONSTANTINOV, M. M., *Computational Methods for Linear Control Systems*

SÖDERSTROM, T. and STOICA, P., *System Identification*

SOETERBOEK, A. R. M., *Predictive Control: A unified approach*

STOORVOGEL, A., *The H∞ Control Problem*

WATANABE, K., *Adaptive Estimation and Control*

WILLIAMSON, D., *Digital Control and Instrumentation*

# Simulation and Modelling of Continuous Systems

## A Case Study Approach

**Drago Matko**

**Borut Zupančič**

**Rihard Karba**

*Faculty of Electrical and Computer Engineering*
*University of Ljubljana, Slovenia*

**Prentice Hall**
New York London Toronto Sydney Tokyo Singapore

First published 1992 by
Prentice Hall International (UK) Ltd
Campus 400, Maylands Avenue
Hemel Hempstead
Hertfordshire, HP2 7EZ
A division of
Simon & Schuster International Group



Typeset in 10/12pt Times
by P & R Typesetters Ltd, Salisbury, UK

Printed and bound in Great Britain
by Dotesios Ltd, Trowbridge, Wiltshire

Library of Congress Cataloging-in-Publication Data

Matko, Drago.
Simulation and modelling of continuous systems: a case-study approach / Drago Matko, Borut Zupančič, Rihard Karba.
p. cm. — (Prentice Hall International series in systems and control engineering)
Includes bibliographic references and index.
ISBN 0-13-808064-X (pbk.)
1. Computer simulation. I. Zupančič, Borut. II. Karba, Rihard. III. Title. IV. Series: Prentice-Hall International series in systems and control engineering.
QA76.9.C65M383 1992
003'.8—dc20

92-12881
CIP

British Library Cataloguing in Publication Data

A catalogue record for this book is available from the British Library

ISBN 0-13-808064-X (pbk)

1 2 3 4 5 96 95 94 93 92

To Professor Bremšak,
who filled us with enthusiasm for modelling and simulation

# Contents

# Preface

Simulation, meaning the imitation or reproduction of certain conditions, is probably older than is commonly thought. The reproduction, which is usually on a different scale or representation, but has, or at least pretends to have, the same characteristics as the object being simulated, is described by the model; and modelling is the procedure for its design. Models actually play the role of real objects, with the intention of representing some of their characteristics but without having to perform experiments, which could be expensive, dangerous, slow or even physically impossible. Thus modelling and simulation are inseparably linked and, as such, frame the scope of this book.

The principles of modelling and simulation are met in various areas of life and science, such as malingering in medicine, empathy in psychology, mimicry in biology, etc., which indicates that various kinds of models do exist. However, in this book only systems which can be described by differential or difference equations, i.e. the mathematical models, will be treated. The latter are used in all fields which are based on systems approaches such as physics, mathematics, control systems, biology, chemistry, medicine, pharmacokinetics, and even in economics and the social and political sciences.

Modelling and simulation usually represent one of the basic methodologies in the treatment of interdisciplinary projects and are directly or indirectly included in all modern methods of analysis and design. While research and theory abound, the best results are obtained through interactive work on computers, where efficient simulation languages, combined with the user's experiences, play an important role.

Although modelling and simulation approaches have been known for a long time, until relatively recently they were not used to a great extent because the simulation tools (analog computers or simulation languages on large mainframes) were extremely expensive and were therefore used only as the last possible method. The great advances in computer technology, especially the appearance of the personal computer (PC), quite literally brought the tools for solving mathematical problems into our homes and thus encouraged interest in modelling and simulation. Complex and well-tested languages from the preceding decades were transferred to PCs without serious limitations. So modelling and simulation became cost-effective methods and simulation tools were available to almost everybody: developers, researchers, students, professors, etc. Extensive use of simulation stimulated interest in many areas of system theory, such as nonlinear systems (modelling and simulation of chaotic systems represent the main topic of this field).

The roots of this book are in more than twenty years of simulation and modelling tradition at the Laboratory of Analog–hybrid Computation and Automatic Control at the Faculty of Electrical and Computer Engineering in Ljubljana. Over these years we obtained a great deal of experience, not only with various modelling approaches, simulation tools and practical projects, but also in the field of education. However, in the last few years we have met with some problems due to the lack of appropriate modern literature. As modelling and simulation approaches have an interdisciplinary character, a large amount of literature exists on both subjects, but the vast majority of works deal with specific problems and a global view on the entire modelling and simulation cycle, from the model's development and validation to its implementation, is rarely attainable in one work.

Our intention here is to analyse the whole modelling and simulation cycle in a clear and simple way, exploring those essential aspects of the area which are rarely mentioned in other works. This is done without complex theoretical work, through practical advice for efficient problem solving. Although modelling and simulation are inseparable techniques which are tightly interlaced through the whole cycle, modelling is more problem oriented while simulation is relatively independent of the problem being studied. Therefore, continuous system simulation methods, tools and specifics are stressed to a greater extent here than in other works, which mainly deal with discrete event simulation. The modelling and simulation approach is illustrated by numerous case studies and solved problems, which are discussed in detail and correspondingly documented.

In our opinion the book can be used as teaching and learning material for both undergraduate and graduate students in a variety of technical and nontechnical areas. However, we believe that it will also be of benefit to anybody who is interested in modelling a particular problem, and obtaining the solution, on his home PC. Finally, we hope that the book will also provide scientists from technical and biomedical, as well as social and behavioural, backgrounds with some ideas about the possibilities of the modelling and simulation approach. It is assumed that the reader has some basic knowledge of mathematics, particularly differential equations, physics and computer programming.

Chapter 1 analyses the modelling and simulation cycle. Some basic concepts of systems, system approach, system dynamics, modelling and simulation are given initially. Then types of models and their classification, as well as theoretical and experimental modelling approaches, are discussed. Following this, the iterative and interactive procedure of modelling and simulation is analysed, stressing the role of model validation. Finally, the term 'case study approach to modelling and simulation' is discussed. The chapter concludes with solved problems which serve as examples through the whole book.

In the first part of Chapter 2, the indirect approach to the solving of differential equations is introduced as the essence of continuous systems simulation. This, along with the direct and implicit approaches, is supported by tool-independent simulation schemes. Two methods of transfer function simulation are also analysed in detail. In the second part of Chapter 2, two types of discrete systems are discussed briefly with the respect to the nature of event appearance, namely the discrete systems described by difference equations and the discrete event systems.

Chapter 3 gives the basic features and classification of simulation systems, with the accent on the simulation languages. The concept of digital continuous simulation systems that enable the user to simulate his problem with the aid of general purpose programming languages is described. The chapter concludes with a description of the software structure of modern simulation languages.

Chapter 4 deals with the tools for the simulation of predominantly continuous systems. In an historical overview, emphasis is placed on those languages that have introduced some new features to simulation tools. The CSSL standard, which significantly influenced the development of simulation languages, is described briefly. Then three conceptually different languages are presented. An understanding of this chapter is particularly important for a thorough grasp of the solved problems and case studies throughout. Although the majority of problems are solved using our own simulation language SIMCOS, it must be noted that the programs presented can be run with or without slight modification on every CSSL standard simulation language (e.g. ACSL, CSSL IV). Finally, the basic concepts of analog–hybrid computation are included, together with procedures for amplitude and time scaling, as well as techniques for the estimation of maximum values of variables.

Chapter 5 describes numerical integration methods, which represent the heart of each digital simulation system. Emphasis is not placed on algorithms, but rather on those features that help the user to obtain more reliable results; thus, aspects of numerical stability, error evaluation, integration algorithm choice, and calculation interval choice are discussed. The problem of algebraic loop solving is also described briefly.

The first part of Chapter 6 discusses modern trends which are influenced by software engineering. Some features which are included only in the simulation languages of the new generation, such as modular simulation model definition, characteristics of combined simulation languages, separation of modelling and experiment, and user interfacing are presented. The second part of this chapter discusses the influence of modern technology on the development of special purpose simulation concepts. Basic parallelism concepts are examined and some modern systems based on these concepts are presented.

The final chapter includes particular case studies from various fields. The first examines the case of discrete cascade control of hydraulic plant; modelling aspects are discussed and two digital simulation languages are compared. The second case study deals with the modelling and simulation of distillation columns of various types including evaluation of the designed control of this nonlinear multivariable problem. A case of drug pharmacokinetics under conditions of haemodialysis is discussed in the third case, which compares the advantages of analog simulation with the use of digital simulation language. The fourth case study deals with the static and dynamic solution of the cantilever beam. Some specific approaches to simulation, such as solution of boundary value problems and partial differential equations, are given. The fifth case study examines model reference adaptive control of a positioning system, with emphasis on procedures for simulation scheme development and testing. The final study develops the simplified model of a diesel engine. The optimization procedure for maximum torque with respect to the injection angle is then applied to the simulated model and a parameter study is used to obtain the rpm–power characteristic of the motor.

Although the book is intended to provide a complete view of modelling and simulation, we feel that it will be helpful to readers already familiar with specific aspects of treated areas. Thus, for example, those who wish to become familiar with modelling and simulation cycles only are advised to read Chapter 1, Section 2.1 and Chapters 3, 4 and 7, while readers interested only in simulation principles and tools should read Section 2.1 and Chapter 4. The reader whose main interest is in the basic concepts of digital simulation systems, a knowledge of which is not required for the use of a particular simulation tool, will be advised to read Chapter 3; but if only the simulation of discrete systems is his of her main focus of interest, this can be found in Sections 2.2 and 3.3. Finally, if the reader is already familiar with modelling and simulation to some extent and wants either to increase the reliability of simulation results from a numerical point of view, or to learn some specific approaches, such as optimization, boundary value problems and the solving of partial differential equations, or wishes to gain some new information about the influence of software engineering and modern technology on simulation, then he or she should read Chapters 5 and 6 respectively. However, undergraduate students will find Chapters 1, 2, 3 and 4, together with some of the case studies presented in the last chapter, of most interest.

Several people have influenced this work, and we thank them all. We would particularly like to express our sincere gratitude to our colleagues M. Atanasijević-Kunc for providing the materials for one case study and to J. Kocijan for the careful technical preparation of the draft manuscript. Finally, many thanks to our families for their endurance throughout the preparation of the book.

Ljubljana, December 1991

Drago Matko
Borut Zupančič
Rihard Karba

# 1

# Modelling and Simulation

Modelling and simulation are inseparable procedures which include the complex activities associated with the construction of models representing real processes, and experimentation with the models to obtain data on the behaviour of the system being modelled (Koskossidis and Brennan, 1984). Here, modelling deals primarily with the relationships between actual dynamic processes and models; simulation refers above all to the relationships between the model and the simulation tool, while the model time responses, being the outputs of the simulation tool, are again evaluated in connection with the process being studied, which completes the exercise. In recent times the modelling and simulation approach has become increasingly unavoidable for solving different kinds of practical problems. Mathematical models of dynamic systems and computer simulation find application in technical and nontechnical areas as diverse as engineering and hard sciences, economics, medicine and life sciences as well as ecology and some social sciences. The purpose of studying systems through the modelling and simulation approach is to achieve different goals without actually constructing or operating real processes (Neelamkavil, 1987). The aims are as follows:

- to increase understanding of some mechanisms in the studied process;
- to predict system behaviour in different situations where any level of predictive ability represents a benefit;
- to enable the design and evaluation of synthesized control systems;
- to estimate those process variables which are not directly measurable;
- to test the sensitivity of system parameters;
- to optimize system behaviour;
- to enable efficient fault diagnosis;
- to make possible the exploration of such situations which in the actual system would be hazardous, problematic or expensive to set up, and thus to achieve safe and inexpensive operator training;
- to verify models obtained in some other way.

Though a variety of modelling techniques and simulation tools exist, neither the computer nor the model can completely replace human decisions, judgement, intuition and experience which still play a significant role in determining the validity and usefulness of models for practical applications. Therefore, in this chapter a brief overview of basic concepts will be given to serve as initial problem statements prior to further steps towards the case study approach using modelling and simulation. However, it is by no means our intention to collect and comment on as many definitions as possible from the huge amount of literature available, but only to indicate some problems in the cyclical procedure of model generation and use which is intertwined throughout with computer simulation (Koskossidis and Brennan, 1984; Shearer and Kulakowski, 1990; Spriet and Vansteenkiste, 1982).

## 1.1 BASIC CONCEPTS

In this section we will eludicate briefly some essential concepts which in our opinion are closely connected to the modelling and simulation approach.

### 1.1.1 Systems

Increasingly, the word *system* is used, not only by academics, philosophers and professionals in various areas, but also in other sections of society, and even in politics (Shearer and Kulakowski, 1990; Neelamkavil, 1987). In spite of its common use, or perhaps because of it, the exact meaning of this term is often not understood correctly. A system can be defined as a combination of elements or components interrelated to each other and to the whole which act together to achieve a certain goal. It is obvious that the term 'interrelation' plays the most important role in this definition. A system may be composed of one or more subsystems which consist again of some subsubsystems, and so on. Elements, attributes (parameters and variables), interrelationships and activities are associated with systems. Any process which changes the attributes of an element represents an activity. Every system interacts with its environment through so-called inputs and outputs. Inputs have their origins outside systems and are not directly dependent on what happens in systems. Outputs, on the other hand, are generated by systems as these interact with their environments. The absolute minimum description of elements, attributes and activities at a particular point in time, which is necessary for predicting the future behaviour of the system, defines the states of the system. These may change as a result of internal or external activities that are described with the aid of the corresponding state variables (Neelamkavil, 1987). Note that the properties of particular components may not show the overall characteristics of the total system and that in general the views of different people on the same system tend to be biased; this must be taken into account in the system's description.

### 1.1.2 Systems Approach

The problems of the modern age can no longer be strictly divided into physics, chemistry, engineering, mathematics, medicine, and so on; rather, they have a more or less interdisciplinary character. As conventional analytical methods are not powerful enough to tackle such problems, *Systems theory*, together with some other branches such as computer sciences, simulation, control engineering, cybernetics, robotics, biomedical engineering, and many others, become very important. Though ancient in origin, systems theory has only become usable recently due to the fast development of technology, computers and communications; but nowadays the systems approach is essential for solving problems in nearly all human activities. *Methodology* can be defined as a set of systematic procedures, based upon knowledge accumulated over a number of years, for solving different classes of problems. Systems theory is a methodological science which developed as the result of experience accumulated in studying many different systems. It is based on properties common to every system, such as goals, states, constraints, stability, control, dynamic behaviour, etc. Systems theory is therefore also an interdisciplinary science which tends to combine existing experiences of different systems into a unique approach which takes some attributes common to the majority of systems into account. The purpose of the *systems approach* is to learn, design, change, preserve, and possibly also control, the behaviour of systems. In the variety of possible solutions, we are looking for the most acceptable solution, taking into account the totality of the problem (including environment) and the different constraints. Similarly, systems theory unifies existing knowledge of modelling by separating those characteristics common to the development and use of models, irrespective of the area where such methodology is used. Many common charactersistics exist and on the basis of these the methodology we call *systems approach to modelling* has been created. One of its properties, among others, is that it reminds us of essential steps in the modelling procedure which would very likely be forgotten, especially in the applicative problems.

### 1.1.3 System dynamics

All existing systems change with time, and when the rates of change are significant, systems are called dynamic systems (Shearer and Kulakowski, 1990). Their main feature is that their output at any instant depends on their history (they have memory) and not just on current input. Initially, the term *dynamics* represented only a part of general mechanics dealing with the movements of mass point in space under the influence of external forces. It was based on Newton's laws and was used increasingly in various areas, leading to the development of theories of hydrodynamics, aerodynamics and elasticity. Differential equations of various types describing dynamic systems appeared in other technical and nontechnical areas. Globally, the coordinate of the system is not only the position of mass point, but also includes all state indicators of the system. Movement means not only motion in space (coordinate change), but also every time dependent coordinate change in the global sense. Dynamics thus deal with time-dependent change of the process states. Change must be interpreted in the widest meaning of the word (for example, change of axis size, change in plant shape, change in relations between certain people, etc.). The concept of system dynamics, being represented either by process time response or by solution of the corresponding differential equation(s), which again is time-dependent, is thus very important in the modelling and simulation approach.

### 1.1.4 Models

Although more will be said about models and modelling in the next section, let us try in the following list to give some basic properties in order to summarize in some sense the information gleaned from the wealth of definitions in the current literature (Neelamkavil, 1987; Meyer, 1985; Aburdene, 1988):

- A model is an object or concept which is used to represent something else. That is, the reality is converted to a comprehensive form (Meyer, 1985).
- A model is a simplified representation of a system intended to enhance our ability to understand, explain, change, preserve, predict, and possibly control, the behaviour of a system (Neelamkavil, 1987).
- A model is a substitute of some concrete system or equipment.
- A model contains only essential aspects of an existing system (or a system which we want to build).
- A model must represent our knowledge about a system in a suitable form, enabling also the use of some other media (for example, paper or computer memory).
- A model must stress only those effects of factors in the system which are important from a modelling goal's point of view.
- A model must be as simple as possible because constructing universal models is impractical and uneconomic.

*Modelling* is thus the process of establishing interrelationships between important entities of a system, where models are represented in terms of goals, performance criteria and constraints. Modelling has an iterative (cyclic) character that is the consequence of many feedback loops from the results of every stage of the modelling procedure.

### 1.1.5 Simulation

As we did for the model, we also give a set of definitions for simulation and for the most common computer simulation, though not in the strictest sense. In our opinion this is the best way to present its essence:

- Simulation is the technique of constructing and running a model of a real system in order to study its behaviour without disrupting the environment of the real system (Koskossidis and Brennan, 1984).

- Simulation of dynamic processes is the iterative method which enables the study of a system's properties through experimentation with the corresponding model of real plant (Korn and Wait, 1978).
- Simulation is the process of imitating important aspects of the behaviour of a system in real time, compressed time or expanded time by constructing and experimenting with a model of the system (Neelamkavil, 1987).
- In comparison with analytical methods, simulation is more realistic and easily understandable but only where used correctly.
- Simulation enables the substitution of real world, complex experiments and pilot plants using the cheap and simple microcomputer, so that the experimentation is possible without any risk while the results are very illustrative.
- Simulation is a technique for conducting experiments on a model.
- Computer simulation means the running of a special program on a suitable type of computer which generates time responses of the model that imitate the behaviour of the process being studied.
- Simulation is the procedure of solving differential equations through integration (this will be discussed further in the next chapter).

It is obvious that nearly all definitions also include statements concerning modelling, which again indicates the interdependent nature of modelling and simulation.

## 1.2 MODELS AND MODELLING

In this section we give a brief survey of model types and classification. Furthermore, theoretical and experimental approaches will be discussed and, finally, modelling methodology and procedure will be analysed along with the role of model validation.

### 1.2.1 Types of Models

Models can be divided into many types. Let us briefly give one of the possible classifications (Neelamkavil, 1987), i.e. one can distinguish *physical*, *symbolic* and *mental* models.

#### Physical models

Physical models are representatives of physical systems; their construction is often expensive, time consuming or impractical. Physical models which have static character can be either *scale-models* (reduced-size models of cars, buildings, ships, etc.) or *imitation models* (molecular structures, dolls, cartoons, etc.). Those with dynamic character are divided into *analog models* – the modelled system being represented by the aid of corresponding analogy which is probably suitable for some reasons (although the electric analogy is often used here, the meaning of the term 'analogy' must be understood in the global sense in which, for instance, rats or monkeys represent a kind of analog model of humans in the testing of new drugs); and *prototypes* – reduced-size copies of the real systems, laboratory and pilot plants of different industrial processes, miniature railway systems, etc.

### Mental models

Mental models have heuristic or intuitive characteristics and exist only in the human mind. They are fuzzy, imprecise and problematic for communication. Accumulated human experiences represent the mental models which support the planning and decision-making processes. Personal views of an object or event can be based on a mental model as well as on the human ability to interpret operations, etc.

### Symbolic models

Symbolic models are less problematic to manipulate and build than are physical models. They can be further divided into *mathematical* and *nonmathematical* models. The latter can be either *linguistic* (verbal or written descriptions of events, experiences, scenes, etc.); *graphic* (paintings, pictures, graphs, drawings); or *schematic* (flow charts, maps, network diagrams, etc.). They have the common property that it is often very problematic to obtain precise information from them, especially from verbally expressed models. For many reasons *mathematical* models are the most important and the most widely used category of models. They are concise, unambiguous and uniquely interpretable, while their manipulation and the evaluation of alternatives are relatively inexpensive. A mathematical model can be defined as the mapping of relationships between the physical variables of the system to be modelled into corresponding mathematical structures (Fasol and Jörgl, 1979). When such relationships are given for the steady state only, the model has *static* character and is described with algebraic equation(s). On the other hand, *dynamic* mathematical models include the transient as well as the steady state behaviour of a system and are described by a system of differential equations (of various types) and by a set of boundary conditions.

## 1.2.2 Classification of Mathematical Models

Insight into the different possibilities of mathematical models, character and form can be obtained by classifying the models (Fasol and Jörgl, 1979; Shearer and Kulakowski, 1990; Norton, 1986). Among numerous possibilities we shall choose and briefly discuss only those categories that we use in subsequent chapters. Note here that one of the important divisions, namely *static* and *dynamic* models, has already been mentioned, while the others are as follows:

- *Linear* models are those which can be described using linear mathematical structures (linear differential equations) and which obey the principle of superposition, whereas in the opposite situation the models are *nonlinear*. Though it is known that every real system has more or less nonlinear character, this can be neglected in many cases and nonlinear models can be converted to linear models by the procedure of linearization around a suitable working point (in most cases the steady state).
- *Lumped parameter* models are described by ordinary linear or nonlinear differential equations where there is only one independent variable (in most cases time). For systems where the variables are significantly dependent on spatial coordinates at a certain moment in time, the *distributed parameter* models described by partial differential equations must be used. Note here that in many cases only one most important spatial coordinate is taken into account.
- *Stationary (time invariant) models* are those where the shapes of their outputs are independent of the moment of onset of their inputs or disturbances, while in the opposite situation the models are *time varying*. Again, we can state that systems, especially if observed over a long time period, nearly all exhibit time variable characteristics. Changes are caused by such phenomena as corrosion, ageing, damage, etc. However, for many modelling aims, such changes can be neglected except for cases where they are outstanding.
- *Continuous time* models have their dependent variables defined over a continuous range of independent variables, while *discrete time* models have their dependent variables defined only for distinct values of independent variables. The former models are described using differential equations and the latter using difference equations. Although the vast majority of processes have continuous characteristics, discretization is becoming more and more important (e.g. computer realization of controllers in modern automatization).
- *Deterministic* models are those in which the probability of events does not feature, whereas in *stochastic* models the relations between variables are given in terms of statistical values. In real processes, random noise and disturbances are always present. However, the intensity of their influence on the system behaviour determines whether the process is deterministically or stochastically modelled. Here the character, especially the spectral density, as well as the measurability of noise and disturbance, play a very important role. Deterministic models can be further divided into *parametric* (algebraic, differential equations where the parameters of such fixed structures must be determined) and *nonparametric* models (response obtained directly from the system or indirectly through experimental analysis). Parametric models can thus be obtained from nonparametric models with the aid of numerous and various identification methods, while conversion in the opposite direction is achieved by simulation.

### 1.2.3 Theoretical Modelling

The essence of theoretical modelling lies in the decomposition of the studied system into particular subsystems, which must be as simple as possible. The corresponding relations between chosen subsystems must then be determined on the basis of different equilibrium equations and theoretical laws for the area under investigation. In the case of technical systems modelling, the known mass, energy and momentum balances are most frequently used, which gives the overall model expressed by differential equations (partial or ordinary, linear or nonlinear), together with the algebraic relations determining the boundary conditions. The main problem in this type of modelling is in the searching for relations among subsystems. It is stressed especially in the so-called soft sciences where the problems may be ill-defined. As the model is a quantitatively formal description of the system it can be very complex and complicated in the first phase of modelling, which is the consequence of numerous relations between subsystems included in the model. The complexity may cause the model to be unusable. Therefore, the possible global simplifying assumptions, approximations and neglections must first be carefully considered, while the others can be taken into account in the further modelling procedure. Here linearization and approximation with lumped parameters are often used, as well as the model order reduction methods. So from the initial partial nonlinear differential equations, which frequently represent the first phase of modelling, one can obtain linear ordinary differential equations of low order. Of course, a corresponding compromise between accuracy and complexity of the model must be found. Here the modelling goals are very important. Theoretical modelling requires very thorough and comprehensive knowledge of the modelled process and often the collaboration of an expert from a corresponding area in model development procedure is indispensable. An important aspect in theoretical modelling is that once a model has been developed for a particular subsystem it can be used for other similar systems by adjusting the respective parameters. Finally, it must be noted that theoretical modelling can be performed during the design and planning stages of a modelled system or for systems where measurements are for some reason impossible (expensive, dangerous, etc.). The interested reader can find more about theoretical modelling in the literature, for example in Weber (1973), Wellstead (1979), Shearer and Kulakowski (1990), Meyer (1985), etc.

### 1.2.4 Experimental Modelling

The basic principle of this kind of modelling lies in the definition of system inputs and outputs and the measurement of input and output signals, which enables the corresponding mathematical model generation. Using this approach the structure and parameters of the model which give responses that are equal, or as alike as possible, to the measured outputs of the system, using the same input signals, must be determined. Only the input–output relations are interesting here and no information about the mechanisms which cause these relations can be obtained.

Contrary to theoretical modelling, where only some general principles of the approach are given, the methods of experimental modelling or identification are algorithmized, and very numerous and diverse (Norton, 1986). Identification methods form a special class in the system approach to modelling (Söderström and Stoica, 1989; Ljung, 1987). They are classified according to the various criteria as the model, input signals and criterion function types, as well as according to the need of simultaneous (on-line) identification. Sometimes the methods are also classified in regard to the method of the data acquisition or in respect of some numerical aspect. Experimental methods are relatively simple and usually provide linear models of low order, also taking into account the influences of the environment. Their application is possible even in very complex or ill-defined systems. However, no information about the physical background of the system can be obtained. The models are usable only as a whole. Each use of some parts of the model to draw an analogy with the actual subsystem may result in undesirable consequences. The problem of separation of the useful signal from the measured one, which also contains the part originating from the disturbances, nearly always appears. Also, the problem of the choice of corresponding input signals which would excite the system enough for successful identification is not negligible. It is problematic, especially due to the fact that they are not similar to the operating input signals, and it is often very difficult to obtain the permission to experiment with real plant. Here also, the so-called *identifiability* concept can be mentioned. This involves the question as to whether the parameters of an assumed model can be estimated uniquely from experimental measurement. If an infinite number of estimates fit the data the model is said to be 'unidentifiable' from the experiment (Godfrey and DiStefano, 1985). Further analysis distinguishes *structural* (*a priori*) identifiability which is based on ideal measurements, while *numerical* (*a posteriori*) identifiability represents an additional restriction on the accuracy with which the parameters must be estimated, taking into account the noise and sampled data available in reality. Identifiability becomes a critical aspect of the modelling process when parameters are analogous to some important attributes of the real system and the model is needed to quantify them. Although identifiability is an interesting theoretical aspect, it may not necessarily guarantee practical identification results (Eykhoff, 1985). Finally, let us stress that in cases where corresponding measurements are impossible the identification methods cannot be used.

### 1.2.5 Combined Modelling

The combination of theoretical modelling and identification is the most frequently used approach, simply due to the fact that it should exhibit the best properties of both procedures. Here the structure of the system is defined through theoretical modelling, while the values of some or all the parameters are determined by the use of experimental techniques. The main advantage of such an approach is in the fact that the basic functional connections of the system are included in the model, which enables better understanding of the system behaviour as well as more flexible use of the model. Let us mention again the wide variety of parameter estimation techniques available, which in practice are largely based on real data (Eykhoff, 1974; Norton, 1986; Ljung, 1987; Söderström and Stoica, 1989). One of the best-known procedures, called 'curve fitting', will be discussed in the case study (Section 7.3).

## 1.3 MODELLING AND SIMULATION APPROACH

In this section we analyse the procedure of model development and utilization and its close relationship with simulation throughout the whole course of the cycle. As has already been stated, modelling is an iterative and interactive procedure where cycling in the numerous local and global loops is often necessary. Such a procedure has evolutionary characteristics in the sense that it includes the course of model generation from speculations, hypotheses, general model forming to the final simplified specific model usage. Here the transformation of qualitative information to quantitative data must be made (Neelamkavil, 1987). The cyclical procedure of modelling and simulation performed by the aid of computer is shown in Fig. 1.1, where it can be seen that the computer is the modeller's main partner. It is used at every stage of the modelling process, not only for simulation purposes but also in the phases of data acquisition and collection, data reduction and signal processing, system analysis, identification, hypothesis testing, parameter sensitivity analysis, verification and validation and, using the corresponding software, also for optimization, matrix operations and statistical computations. Interactive computer facilities and computer graphics makes computer aided modelling more effective and accurate, as well as less time consuming. The whole cycle shown in Fig. 1.1, which passes from the real system through model building to the formal and computer models and then through model utilization back to the real system, can be basically divided into two main parts. The first includes the phases of model building. This is the formalization step which must give the model of the real system. It is scientifically orientated in the sense that we tend to obtain sufficient understanding of the phenomena being studied by constructing their abstract representation. In contrast to the first part on the right-hand half of Fig. 1.1, the second part on the left-hand side of the diagram has more engineering characteristics. Through the analyses, interpretation of the results and utilization of the formal model we wish to learn how the real system can be manipulated for our own purposes, which means the achievement of the modelling aims (Spriet and Vansteenkiste, 1982). Although only the main steps in the modelling and simulation procedure are shown in Fig. 1.1, it can be seen that many connections and loops exist. Let us comment on the phases of the procedure, stressing in particular the role of simulation and validation. In the process of model building, system analysis must initially be undertaken. Here a lot of decisions must be made which are crucial for further model development. These are as follows:

- Modelling aims must be stated as clearly and unambiguously as possible.
- Various kinds of constraints must be taken into account.

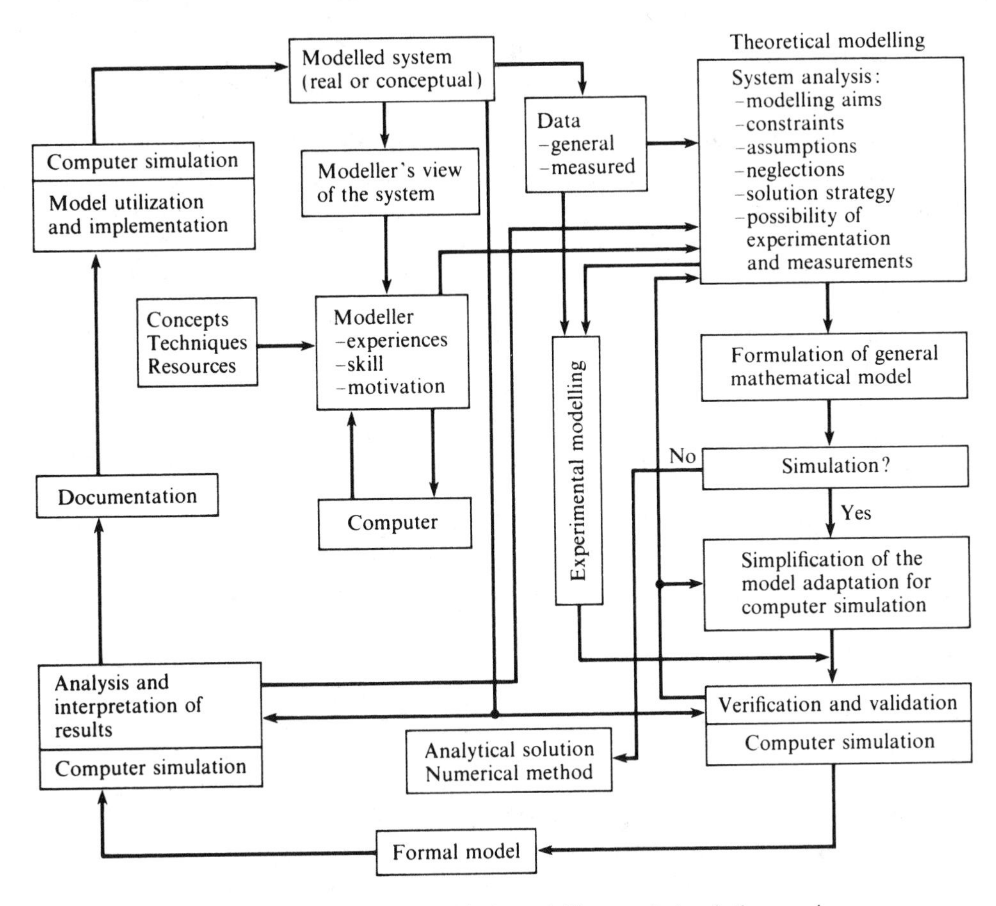


Figure 1.1 Computer aided modelling and simulation cycle.

- General assumptions and omissions must be carefully argued.
- The possibilities of measurements and experimentation on the system being studied must be investigated.
- A solution strategy must be chosen which includes the methods of collection and analysis of data, estimation of parameters, type of the model, available computational facilities, applicability and flexibility of simulation tools, generality of the solutions, possibility of model extension, etc.

Note here that even such basic decisions, as those mentioned above can be modified or totally changed in the iterative process of model development. After the formulation of the general mathematical model by the approach mentioned at the theoretical modelling stage, the question as to the necessity of computer simulation for solving and manipulating the model obtained is stated. In spite of the fact that analytical solutions rarely exist and that direct numerical methods are not often acceptable, we must note that simulation requires far more detailed input than analytic models. But, on the other hand, minor structural changes in the model (perhaps the addition of some nonlinearity) may mean that an entirely different approach must be chosen to obtain the analytic or numeric solution. The overall simulation strategy, however, remains the same, which shows the flexibility of simulation approach in solving mathematical models.

The next step is the simplification of the model because simplicity is the essential criterion for a good model (Neelamkavil, 1987). Here, further approximations and omissions are, of course, made again with corresponding arguments. Note that oversimplification of the model may lead to loss of accuracy and generality, while models which are too detailed and which tend to be universal may make the model more complex than the system being studied, which is rarely justifiable. Therefore, it is better to develop simple but more specific models, although it may occur that several different models representing various modelling goals are designed for the same system.

Also, adaptation of the mathematical model to the used simulation tool is sometimes necessary in order to obtain the simulation model.

Theoretical modelling procedure, from model formulation onwards, can also be replaced or combined (completed) with experimental modelling techniques (identification) which use measurements of the system inputs and outputs as the data base.

Prior to firm decisions about the final version of the formal or simulation model and before the beginning of experimentation with it, the procedures of model *verification* and model *validation* must be performed. The term 'validation' is concerned with demonstrating that the model is an adequate representation of reality, whereas the term 'verification' involves checking the design consistency (accuracy and correctness of modelling and solution methodologies, algorithms, computer programs, etc.). In short, we wish to prove that the model works as we proposed. As the real system is never completely known, and the model is never an exact representation of the real system, validation can be approached but never achieved (Neelamkavil, 1987). A model has no value until it is judged valid, except perhaps in so far as it enhances our understanding of the system being modelled. Far too little attention has been given to validation (Neelamkavil, 1987; Spriet and Vansteenkiste, 1982; Meyer, 1985), which is, in comparison with verification, more problematic. Let us try to explain the approaches to model validation for which no systematical procedure, or even algorithm, exists. A great majority of modellers are thus not interested in ultimate reality at any cost, rather, they wish to develop and validate practical models at reasonable cost and within acceptable time limits. Although there was little emphasis on validation, several relatively successful models were developed and implemented in the past (especially in the engineering sciences). So the objective of validation should slant more towards establishing the degree of confidence by examination to demonstrate how accurately the model represents the system. In this sense, the following can be considered (Neelamkavil, 1987):

- *Validity of concepts*. In the *rationalist approach*, one accepts that the model is a set of logical deductions from a series of theorems or axioms whose truths are unquestionable, and in this sense validation is reduced to the question of tracing the fundamental assumptions on which the model is based. On the other hand, the *empirical approach* refuses to accept any axioms or theorems, and validation involves the collection of empirical evidence to support the postulates or assumptions.

- *Validity of methodology.* Here justification of the methodology used in the formulation of the model and solution of the problem is examined (approximation of nonlinear problem by linear methods, representation of continuous systems by their discrete equivalents, inaccurate use of computational methods, etc.). It is obvious that the wrong methodology could lead to absurd solutions.
- *Validity of data.* The data may be of questionable value for several reasons and they must thus be carefully established before anything is concluded from them. Data can be defective as a consequence of observational errors, calibration errors, interpolation/extrapolation, inaccurate parameter estimation, etc.
- *Validity of results.* The degree of fit between the model response and the theoretical results or measured data plays a key role here. Only the predictive validity of the model is important. The degree of fit is obtained by using statistical methods such as analysis of variance, regression, factor analysis, spectral analysis, etc., which render the data useful for the interpretation of the results.
- *Validity of inference.* Inference is treated as valid when the conclusion of study drawn by several reasonable people are the same.

Modelling is not a precise science and hence the criteria for testing the robustness of scientific theories should not be strictly applied to the models. However, the need for study of retrospective (model responses fit past data), predictive (model responses should agree with future data) and structural (model reflects the internal behaviour of the real system faithfully) validity becomes increasingly obvious. Two extreme situations concerning validation may occur. The first is the case where the validity of the model is rejected in spite of the fact that the model is actually valid, which is termed the model builder's risk. In the second situation, by contrast, model validity is accepted when it is actually invalid, which can be termed the model user's risk. The first situation may incur unnecessary model modifications and associated costs, while the second may lead to disastrous consequences if such a model is implemented. Thus tight collaboration between model builders, model sponsors and model users must be ensured.

Along with the criteria mentioned above for model validation, it should be noted that models possess certain characteristics which may help in estimating their suitability (Meyer, 1985). These characteristics are as follows:

- *Accuracy.* A model is said to be accurate if its response is correct or very near to correct in respect of the system being modelled.

- *Descriptive realism.* A model is said to be descriptively realistic if it is based on assumptions which are correct, which means that it is deduced from a correct or, at least, believable description of mechanisms in the modelled system.
- *Precision.* A model is said to be precise if its predictions are definite numbers (or curves) and not a range of numbers (or a set of curves).
- *Robustness.* A model is said to be robust if it is relatively immune to errors in the input data.
- *Generality.* A model is said to be general if it applies to a wide variety of situations.
- *Fruitfulness.* A model is said to be fruitful either if its conclusions are useful or if it inspires or points the way to other good models.

It can be seen that decisions regarding model validity are some of the most problematic in the whole procedure of model development but at the same time are unavoidable if correct further use of the model is to be ensured. No systematic procedure can be advised except the principles discussed, which must be adapted correspondingly for particular examples. Also, some general instructions can be given, such as that the data used for model development cannot be used for model validation, etc. Its importance is proved by the fact that in the case of nonsatisfactory validity, the loop (Fig. 1.1) returns to the beginning of the model development procedure. In the modelling cycle the concept of parameter sensitivity analysis is often also included when dealing with the determination of the model parameters, which, when perturbed, cause the greatest changes in model response. This concept is closely connected with the robustness aspects. These properties play an important role in the model implementation phase (Lunze, 1988).

Among the further steps in the modelling and simulation cycle shown in Fig. 1.1, the corresponding interpretation of the results may also play a crucial role also in the phase of model implementation. Documentation, on the other hand, would have to become an integral part of the modelling process but until now it has rarely been accorded the importance that it deserves. Recently, the research efforts in the field of modelling has been mainly devoted to the development of hierarchical and object oriented goal driven automated generation of models using expert systems and corresponding knowledge bases. Interested readers will find out more about this topic in the excellent book by Cellier (Cellier, 1991).

Finally, let us give some situations in which the modelling and simulation approach could be beneficial, along with some of its limitations (Neelamkavil, 1987; Aburdene, 1988). Modelling and simulation should be used in the following situations:

- The real system does not exist and it is expensive, time consuming, hazardous or impossible to build and experiment with prototypes.
- Experimentation with the real system is expensive, dangerous, or likely to cause serious disruptions.

- There is a need to study the past, present or future behaviour of the system in real time, expanded time or compressed time.
- Some kinds of mathematical models have no analytical solutions.
- Satisfactory validation of simulation models and results is possible.
- Expected accuracy (results cannot be better than the input data) of simulation results is consistent with the rquirements of the particular problem.

Limitations of modelling and simulation are as follows:

- Expensive in terms of manpower and computer time.
- Generally yields suboptimal solutions.
- Validation difficulties.
- Results can be easily misinterpreted and it may be difficult to trace the source of the errors.

Let us note here that several authors have concluded that the modelling and simulation approach, being an iterative, experimental problem solving technique, is neither a science nor an art, but a combination of both.

## 1.4 MODELLING AND SIMULATION: A CASE STUDY APPROACH

In this short section we will define what, in our opinion, a case study in modelling and simulation means. From the preceding sections it is obvious that modelling, together with simulation, is a comparatively new area of activity involving the fusion of ideas from very different fields. There is thus a rather extensive literature with significantly interdisciplinary characteristics, which also includes the bibliography of bibliographies (Ören, 1974). As stated in the literature (Cellier, 1982), the only sure point concerning the attempts to collate the references in the field is their incompleteness. It can be noted that the first of the two 'partners', namely modelling, is area dependent in the sense that the problem oriented expert must perform or collaborate in the model development for a particular field. He or she must not only define the modelling task (and aims) but must also decide about the assumptions, modelling methodology, model validity, etc. The corresponding knowledge about the existing laws on the expert's specific area must be also provided from his or her side. Due to the fact that the development of universal models is not justifiable, as mentioned above, it would in our opinion be unwise to pay too much attention to the different methods and approaches to modelling for the various and diverse areas. It is clear that such models cannot be included in the framework of this book. On the other hand, simulation as an approach or as a tool is relatively area independent in the sense that the simulated model can perhaps influence the choice of the simulation tool, but more by virtue of its type and character than by the field in which the modelled process originates. Thus, the basic concepts of simulation approaches, simulation tools and associated special problems will be given in this book.

Both the preceding statements define the *case study approach to modelling and simulation*. This term should mean that when faced with a task a system engineer or an expert from some other discipline who undertakes the study should, as far as is possible be specific in regard to those aspects connected with the area from which the process is modelled (collaboration with a problem oriented expert is obligatory), while the engineer/expert should be as flexible and general as possible in regard to those aspects connected with the simulation concepts and simulation tools. It is, of course, important that the engineer has a global view of the total modelling and simulation cycle. Note that the term *case study* is used here in its widest sense. It must of course have a background rooted in a real problem, but it can be any kind of task defined on a level which allows a systems approach and the initiation of a modelling and simulation cycle. The numerous case studies outlined in this book will, hopefully, illustrate the essence of such approach, which, in our opinion, ensures successful realization of the given task in an acceptable time. For those people who will work in this interesting and demanding area, we would like to quote Confucius: 'I hear and I forget, I see and I remember, I do and I understand.'

## 1.5 PROBLEMS

In this section we will give some solved problems (stressing modelling phase) which will serve through the whole book as basic examples for the representation of different simulation aspects.

### Solved problems

#### Problem 1.1 Example of a car suspension system

Different variants of a simplified car suspension system model have been well known in the available literature for many years (Jackson, 1960; Giloi, 1975). In most cases this model is described by a system of two linear second order differential equations simulating the time courses of the car body and tyre displacement. Here a quarter of the chassis mass, car spring and shock absorber, as well as half of the axle mass with the mass of the wheels and tyre springs are taken into account. For our purposes one additional simplification will be introduced. Namely, we shall neglect the half of the axle mass with the mass of the wheels, which is small in comparison with a quarter of the car body mass. Besides, let us suppose that we are interested only in the car body displacement time profile. Bearing in mind this description of the problem, as well as the given assumptions, the mechanical system shown in Fig. 1.2 can be generated to approximate to the car suspension system. The basis for the modelling of the system shown in Fig. 1.2 is the momentum balance law. Its special case, known as Newton's second law of motion, is for the translational mechanical

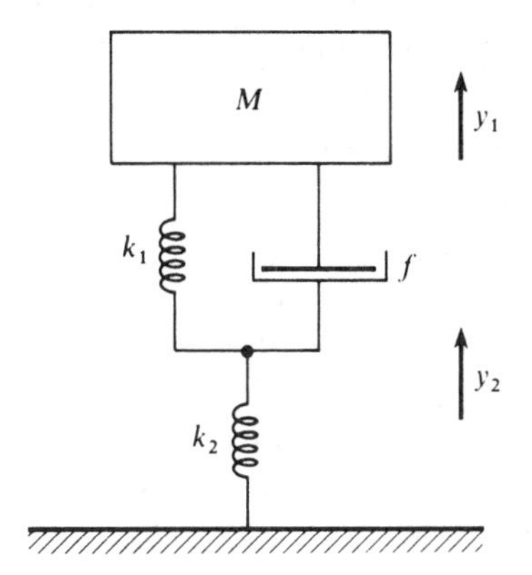


Figure 1.2 Simplified model of car suspension system where $M$ represents a quarter mass of the car body, $k_1$ represents the constant of car spring, $f$ represents the constant of car shock absorber, $k_2$ represents the constant of tyre spring, $y_1$ is car body displacement and $y_2$ is wheel displacement.

system given in the form

$$\sum F = Ma \tag{1.1}$$

where $M$ = mass (kg); $a$ = acceleration (m/s$^2$); and $F$ = force (N).

The mechanical spring force required to deflect a spring a distance $y$ from its free length is given by the relation (Hooke's law)

$$F_s = ky \tag{1.2}$$

where $k$ is the spring coefficient (N/m). The viscous damper force $F_D$ required to move one end of the dashpot at velocity $v$ relative to the other end can be expressed as

$$F_D = f_v = f\dot{y} \tag{1.3}$$

where $f$ is the damping coefficient (Ns/m) and $\dot{y}$ denotes the first time derivative of the displacement. Applying these relations to the system shown in Fig. 1.2, the following two differential equations describing both displacements are obtained:

$$M\ddot{y}_1 = -f(\dot{y}_1 - \dot{y}_2) - k_1(y_1 - y_2) \tag{1.4}$$

$$0 = -f(\dot{y}_2 - \dot{y}_1) - k_1(y_2 - y_1) - k_2 y_2 \tag{1.5}$$

To eliminate $y_2$ the sum of these differential equations is

$$M\ddot{y}_1 = k_2 y_2 \tag{1.6}$$

where $y_2$ is calculated as

$$y_2 = -\left(\frac{M}{k_2}\right)\ddot{y}_1 \tag{1.7}$$

By introducing $y_2$ in Eqn (1.4) we obtain

$$\frac{fM}{k_2}\dddot{y}_1 + M\left(1 + \frac{k_1}{k_2}\right)\ddot{y}_1 + f\dot{y}_1 + k_1 y_1 = 0 \tag{1.8}$$

and, finally,

$$\dddot{y}_1 + \frac{k_1 + k_2}{f}\ddot{y}_1 + \frac{k_2}{M}\dot{y}_1 + \frac{k_1 k_2}{Mf}y_1 = 0 \tag{1.9}$$

Eqn (1.9) is a linear third order ordinary differential equation with constant coefficients describing the dynamics of the car body in which we are interested. Car body displacement is, in our case, caused by some change in $y_1$ from its stationary value, which may occur, for instance, when the driver leaves the car, reflected by the following initial condition:

$$y_1(0) = -y_{10} \tag{1.10}$$

For the data $M = 500$, $k_1 = 7500$, $k_2 = 150\,000$, $f = 2250$ and $y_{10} = 0.05$, we obtain Eqn (1.9) in the following form:

$$\dddot{y}_1 + a\ddot{y}_1 + b\dot{y}_1 + cy = 0 \qquad y_1(0) = y_{10}$$

with the following parameters: $a = 70$, $b = 300$, $c = 1000$ and $y_{10} = -0.05$. Thus the model is prepared for further modelling and simulation procedures.

**Problem 1.2 Example of a heating control problem**

Warming of a room with an electrical heating device which is switched on and off by the thermostat is a classic and cheap kind of temperature control often used in everyday life. The effects of control depend above all on three main factors: the power of the heating device, the hysteresis of the thermostat controller and the delays which are included in the process. These factors influence the oscillations of temperature. As we desire low oscillations, the hysteresis must be relatively narrow. But, on the other hand, such hysteresis causes frequent switching in the heating device and this can also be undesirable. The power of the heating device, however, has the main influence on the time taken to reach the desired temperature.

The design of the heating device and its thermostat could be performed on site in the room. But the necessary measurements are very time consuming due to the fact that time constants of such systems are in the range of hours. The exact mathematical analysis is also problematic due to the hysteresis representing the nonlinearity which causes, even for closed loop systems with simple process models, very complicated dynamic behaviour. Here, simulation offers a very good possibility of solution to such a problem. If a suitable model of the process exists, the whole problem can be simulated (using the chosen simulation tool) very quickly (much faster than for the real system), giving the information desired about the closed loop system behaviour.

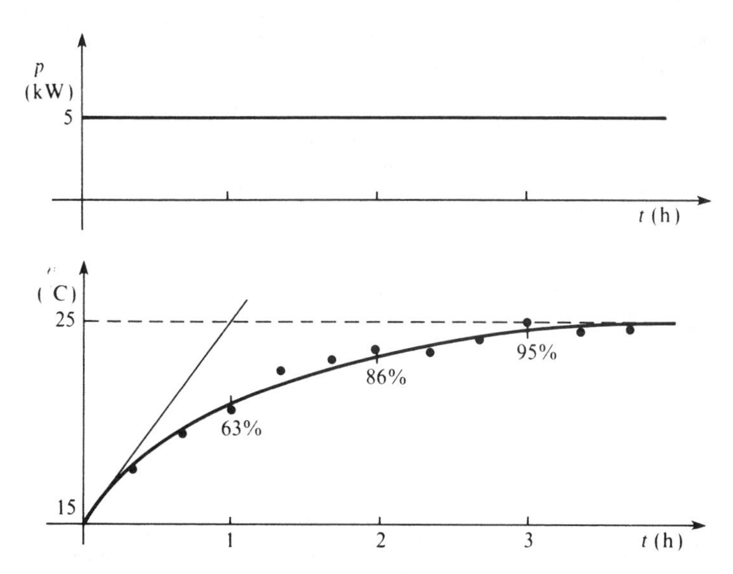


Figure 1.3 Warming of the room when heating device is switched on. Time courses of power of heating ($p$) and temperature in the room ($v$), where dots are measured values and the curve is the response of the developed model.

Without analysing the possibilities for our process modelling, let us try to obtain the model in the simplest possible way. Though not complex, the model must of course satisfy all the requirements. The experiment proceeds as follows:

- With the unheated room at relatively constant temperature (for example 15 °C), switch the heating device on ($t = 0, p = 5$ kW).
- The time course of the temperature must then be measured (Fig. 1.3).

As can be seen in Fig. 1.3, the response of the process is of the proportional type expected due to the fact that an increase of temperature also causes an increase of heat loss (through the doors, windows, walls, etc.). The measurements show that the temperature reaches the steady state at 25 °C, which means that at this temperature the heat produced by the heating device equals the heat lost to the environment. If we suppose that the temperature derivative is proportional to the difference between the heating power and the heating lost, the behaviour described can be approximated by the following first order differential equation:

$$\frac{\mathrm{d}}{\mathrm{d}t}(v - v_{\mathrm{e}}) + \frac{1}{T}(v - v_{\mathrm{e}}) = \frac{k}{T}p \tag{1.11}$$

where $v$ = temperature in the room (°C); $v_{\mathrm{e}}$ = temperature of the environment (°C) which is constant and in our case equals 15 °C; $p$ = power of the heating device (kW), in our case 5 kW; $k$ = gain of the first order system; $T$ = time constant of the first order system.

From the measurements shown in Fig. 1.3 the gain $k$ and the time constant $T$ must therefore be determined. The latter can be estimated with the aid of the tangent at time $t = 0$ (as shown in the diagram) or by a percentage of the final value of the response at one, two and three time constants. Such estimation is of course possible only for the first order system. In the way described the time constant was estimated to be

$$T = 1\ \mathrm{h} \tag{1.12}$$

The gain of our proportional system is defined as the relation between the change of temperature in the steady state and the change of input signal (in our case heating). Thus the gain is calculated to be

$$k = \frac{\Delta\vartheta}{\Delta p} = \frac{10\,^\circ\mathrm{C}}{5\ \mathrm{kW}} = 2\,^\circ\mathrm{C/kW} \tag{1.13}$$

Thus the linear model obtained, which only describes the system behaviour for a certain temperature interval ($15\,^\circ\mathrm{C} \leqslant \vartheta \leqslant 25\,^\circ\mathrm{C}$), for constant environmental temperature is given by the following differential equation:

$$\dot{\vartheta} + (\vartheta - \vartheta_e) = 2p \tag{1.14}$$

Usually, a model which describes the conditions relative to a working point (in our case the environmental temperature) is used. So the corresponding temperature is

$$\vartheta_w = \vartheta - \vartheta_e \tag{1.15}$$

The final model is given in the form

$$\dot{\vartheta}_w + \frac{1}{T}\vartheta_w = \frac{k}{T}p \tag{1.16}$$

or, in our case,

$$\dot{\vartheta}_w + \vartheta_w = 2p$$

At this point, one should be satisfied by the model in Eqn (1.16), which can also be given in the transfer function form

$$\frac{\Theta_w(s)}{P(s)} = \frac{k}{sT + 1} = \frac{2}{s + 1} \tag{1.17}$$

In practice, however, it turns out that the linear model can only be used in the case where the thermostat controller which detects the actual temperature is not too far from the heating device. If this is not the case, the problem is better described by a differential equation of higher order which can again be approximated by the first order model combined with the so-called dead time or time delay. We thus obtain for that case

$$\dot{\vartheta}_w + \vartheta_w = 2p_d \tag{1.18}$$

and

$$p_d(t) = p(t - T_d) \tag{1.19}$$

where $T_d$ is dead time. If system is described with the aid of the transfer function it can be written as

$$\frac{\Theta_w(s)}{P(s)} = \frac{2}{s+1}\exp(-T_d s) \tag{1.20}$$

Description of the system in transfer function form is usually connected with representation of the system structure by block diagrams in the theoretical modelling procedure. In studying control problems (as in our case) the block diagram approach is convenient due to the fact that different signals in the loop can be obtained by simulation (in the case of overall transfer function it is impossible), and this is important in some cases. In this approach the input–output relationship can be given explicitly by transfer functions, or either the static characteristic of a particular element of the system or its typical time response (usually on the step input) may be drawn.

Let us try to compose the block diagram of our heating control problem.

The desired temperature in the room ($\vartheta_r$) can be set on the thermostat controller. The input to the latter is the difference ($e$) between the desired and measured temperature in the room:

$$e = \vartheta_r - \vartheta \tag{1.21}$$

Taking into account the hysteresis characteristic (Fig. 1.4), which is given through the relationships

$$u(t) = \begin{cases} 0 & e < -\dfrac{\Delta y}{2} \\ 1 & e > \dfrac{\Delta y}{2} \\ u(t - \Delta t) & -\dfrac{\Delta y}{2} \leqslant e \leqslant \dfrac{\Delta y}{2} \end{cases} \tag{1.22}$$

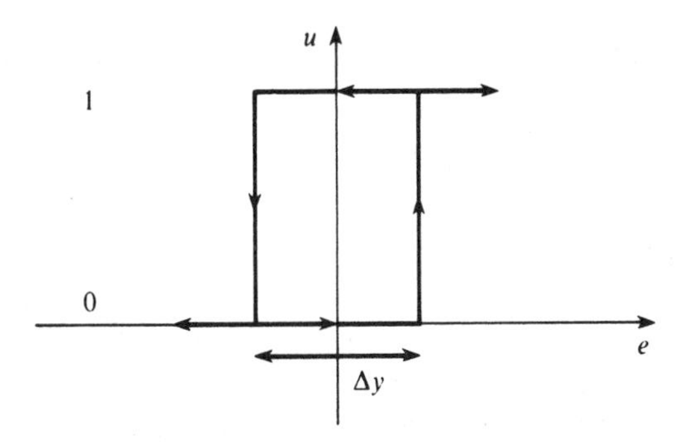


Figure 1.4 Hysteresis characteristics.

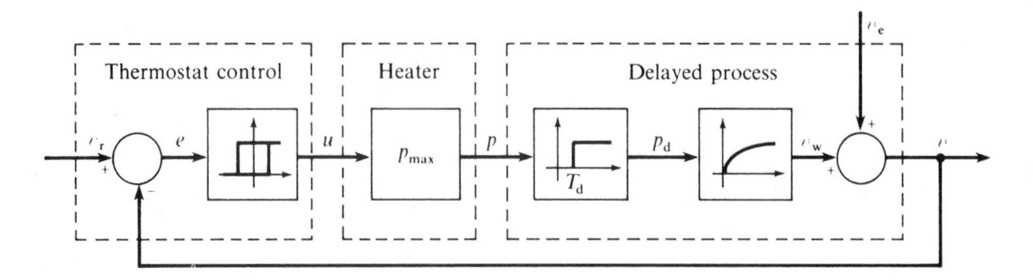


Figure 1.5 Block diagram of the heating control problem.

where $u(t - \Delta t)$ is the previous value of $u$, the output ($u$) of the thermostat controller is either 0 or 1 (switch off–switch on). The latter, together with the switching system can therefore be modelled as the gain $p_{max}$. The input–output relation is thus

$$p = p_{max} u \tag{1.23}$$

The signal obtained, $p$, represents the input for the process modelled with Eqn (1.16) for $T_d = 0$ and with Eqn (1.18) for $T_d \neq 0$ respectively. The absolute temperature in the room is calculated using the relationship

$$v = v_w + v_e \tag{1.24}$$

which is obtained from Eqn (1.15). The block diagram of our problem is represented in Fig. 1.5, from which the structure of the system is evident. The input–output relationships for particular signals and blocks are given in graphic form. This completes the preparation for the simulation problem and efficient study of the following aspects:

- determination of the hysteresis width which results in nonexcessive switching of the heater and in low oscillations of the temperature;
- determination of the minimum heating power which assures satisfactory control (relatively quick adjustment of temperature to the new desired value, disturbance rejection, etc.);
- study of the disturbance influence on the controlled temperature (several kinds of input and output disturbances can be simulated);
- determination of the optimal time course of reference (desired temperature) signal which, together with a suitable temperature profile in the room, ensures optimal energy saving;
- study of the influence of thermostat location on the temperature in the room (especially in the case of poor air mixing);
- study of the influence of different time constants for warming and cooling on the temperature in the room (in the case of more precise modelling).

**Problem 1.3 Example of the prey and predator problem**

The so-called *population dynamics models* or exponential models which are used in the study of growth and decay characteristics of different types of populations are well known in the literature (Neelamkavil, 1987; Cook, 1986). They can also be extended to the example of two species of populations. One possibility here is the *Lotka–Volterra ecosystem model* or, as it is also called, the *prey and predator problem*.

If the populations are treated as continuous variables, the logarithmic growth of one population depends only on the other population. Consequently, it can be seen that the logarithmic growth of predators is directly proportional to the population of prey, while the logarithmic growth of prey is inversely proportional to the population of predators. The population of prey is denoted by $x_1$ and population of predators by $x_2$. Assuming linear relations, the following mathematical model can be developed:

$$\frac{\dot{x}_1}{x_1} = (a_{11} - a_{12}x_2) \tag{1.25}$$

$$\frac{\dot{x}_2}{x_2} = (a_{21}x_1 - a_{22}) \tag{1.26}$$

or

$$\dot{x}_1 = (a_{11} - a_{12}x_2)x_1 \tag{1.27}$$

$$\dot{x}_2 = (a_{21}x_1 - a_{22})x_2 \tag{1.28}$$

It is also assumed here that self-inhibition is absent and that in the absence of prey the predator population declines, while in the absence of predators the prey population increases. Here, positive constants $a_{11}$ and $a_{22}$ depend on corresponding growth rates, whereas the positive constants $a_{12}$ and $a_{21}$ represent mutual inhibition factors and are proportional to the size of the other population.

Let us observe the populations of rabbits (prey) and foxes (predator). The constants in Eqns (1.27) and (1.28) can be estimated with the aid of the following four assumptions:

1. Every pair of rabbits produces an average of ten young in one year.
2. Every fox captures an average of twenty-five rabbits in one year
3. The average age of foxes is five years, which means that 20% of foxes die yearly.
4. On average, the number of young foxes which survive depends on the available food (this means that the number of young foxes surviving is equal to the number of rabbits divided by 25).

For a time unit of one year the constants $a_{11}$ and $a_{22}$ can be calculated from assumptions 1 and 3 in the following way:

$$a_{11} = \left.\frac{\dot{x}'_1}{x_1}\right|_{a_{12}=0} \approx \frac{\Delta x'_1/\Delta t}{x_1} = \frac{10}{2} = 5 \tag{1.29}$$

$$a_{22} = -\left.\frac{\dot{x}'_2}{x_2}\right|_{a_{21}=0} \approx -\frac{\Delta x'_2/\Delta t}{x_2} = -\frac{-1}{5} = 0.2 \tag{1.30}$$

As mentioned above, constants $a_{12}$ and $a_{21}$ are also dependent on the area in which both populations are observed. In other words, they depend on the average number of rabbits and foxes. If the observed area is, for instance, 50 $km^2$ where the average number of rabbits is $\bar{x}_1 = 500$ and foxes $\bar{x}_2 = 100$, assumption 2 gives

$$a_{12} = -\left.\frac{\dot{x}''_1}{\bar{x}_1\bar{x}_2}\right|_{a_{11}=0} \approx -\frac{\Delta x''_1/\Delta t}{\bar{x}_1\bar{x}_2} = -\frac{-25\bar{x}_2}{\bar{x}_1\bar{x}_2} = \frac{25}{500} = 0.05 \tag{1.31}$$

and assumption 4 gives

$$a_{21} = -\left.\frac{\dot{x}''_2}{\bar{x}_1\bar{x}_2}\right|_{a_{22}=0} \approx -\frac{\Delta x''_2/\Delta t}{\bar{x}_1\bar{x}_2} = -\frac{\bar{x}_1/25}{\bar{x}_1\bar{x}_2} = \frac{1}{25\bar{x}_2} = \frac{1}{2500} = 0.0004 \tag{1.32}$$

For an area of observation one hundred times larger with one hundred times larger populations, then both constants in Eqns (1.31) and (1.32) are one hundred times smaller.

As has been shown, the mathematical model obtained in Eqns (1.27) and (1.28) is the set of first order nonlinear differential equations. Such a model for the estimated values of constants prepares us for further studies.

## BIBLIOGRAPHY


ABURDENE, M.F. (1988), *Digital Continuous System Simulation*, Wm. C. Brown Publishers, Dubuque, Iowa.

CELLIER, F.E. (1982), *Progress in Modelling and Simulation*, Academic Press, London.

CELLIER, F.E. (1991), *Continuous System Modeling*, Springer-Verlag, Heidelberg.

COOK, P.A. (1986), *Nonlinear Dynamical Systems*, Prentice Hall, Englewood Cliffs, NJ.

EYKHOFF, P. (1974), *System Identification, Parameter and State Estimation*, John Wiley & Sons, London.

EYKHOFF, P. (1985), 'Biomedical identification: overview, problems and prospects', *7th IFAC/IFORS Symposium on Identification and System Parameter Estimation*, in H.A. Barker and P.C. Young (eds.), York, UK, Pergamon Press, Vol. 1, Oxford, pp. 37–44.

FASOL, K.H. and H.P. JÖRGL (1979), 'Modelling and identification', in *Tutorials on System Identification at the Fifth IFAC – Symposium on Identification and System Parameter Estimation* (R. Isermann, ed.), Darmstadt, pp. I-1–I-13.

GILOI, W.K. (1975), *Principles of Continuous System Simulation*, B.G. Teubner, Stuttgart.

GODFREY, K.R. and J.J. DISTEFANO III (1985), 'Identifiability of model parameters', *7th IFAC/IFORS Symposium on Identification and System Parameter Estimation*, in H.A. Barker and P.C. Young (eds), York, UK, Pergamon Press, Vol. 1, Oxford, pp. 89–114.

JACKSON, A.S. (1960), *Analog Computation*, McGraw-Hill, London.

KORN, G.A. and J.V. WAIT (1978), *Digital Continuous System Simulation*, Prentice Hall, Englewood Cliffs, NJ.

KOSKOSSIDIS, D.A. and C.J. BRENNAN (1984), 'A review of simulation techniques and modelling', *Proceedings of the 1984 Summer Computer Simulation Conference* (W.D. Wade, ed.), Vol. 1, Boston, pp. 55–8.

LJUNG, L. (1987), *System Identification: Theory for the User*, Prentice Hall, Englewood Cliffs, NJ.

LUNZE, J. (1988), *Robust Multivariable Feedback Control*, Prentice Hall, London.

MEYER, W.J. (1985), *Concepts of Mathematical Modelling*, McGraw-Hill, Singapore.

NEELAMKAVIL, F. (1987), *Computer Simulation and Modelling*, John Wiley, NY.

NORTON, J.P. (1986), *An Introduction to Identification*, Academic Press, London.

ÖREN, T.I. (1974), 'A bibliography of bibliographies on modelling, simulation and gaming', *Simulation*, **23** (3), pp. 90–5 and **23** (4), pp. 115–16.

ROGERS, A.E. and CONNOLLY, T.W. (1960), *Analog Computation in Engineering Design*, McGraw-Hill, NY.

SHEARER, J.L. and B.T. KULAKOWSKI (1990), *Dynamic Modelling and Control of Engineering Systems*, Macmillan Publishing Company, NY.

SÖDERSTRÖM, T. and P. STOICA (1989), *System Identification*, Prentice Hall, NY.

SPRIET, J.A. and G.C. VANSTEENKISTE (1982), *Computer Aided Modelling and Simulation*, Academic Press, London.

WEBER, T.W. (1973), *An Introduction to Process Dynamics and Control*, John Wiley, London.

WELLSTEAD, P.E. (1979), *Introduction to Physical System Modelling*, Academic Press, London.

# 2

# Basic Methodologies in Solving Problems with Simulation

This chapter discusses some fundamental simulation methodologies. The first part introduces the indirect approach to solving differential equations and represents the essence of continuous systems simulation. Together with direct and implicit methods, it is supported by the simulation-tool-independent simulation scheme. Two methods of transfer function simulation, which is very important in solving many problems, are discussed in detail. The second part of the chapter discusses the two types of discrete systems which deal with the nature of events appearance. The first type, which is described by difference equations, uses a similar approach to that of continuous systems, while discrete event simulation is not within the scope of this book and is therefore introduced only for distinguishing purposes.

## 2.1 CONTINUOUS SYSTEMS SIMULATION

Though the 'differential analyser scheme' for the solution of differential equations dates back to Lord Kelvin, the general approach and attendant ideas concerning simulation have their origins in analog computation. It is not the intention of this chapter to investigate the formal descriptions or the kinds of unifying concepts for general continuous system simulation, such as concepts of continuous recursion or 'continuous automaton' (Giloi, 1975). But, on the other hand, we wish to present a general approach in the sense that the latter are described in a manner which is, as far as possible, independent of the simulation tool used. For this reason all the examples will be presesented, together with the mathematical form, by the corresponding *simulation scheme*. This scheme is not an analog scheme because no sign changes due to the electrical properties of the elements are supposed at the moment. However, it is similar to some schemes used for block oriented digital simulation languages programming. For our purposes only four kinds of blocks are introduced for such a scheme and these are shown in Fig. 2.1.

Signs for the blocks in Fig. 2.1 can be defined through the signs of the corresponding parameters (sign of the constant $k$ in the gain block, which assures multiplication with constant) or they can be placed at the block input (absence of

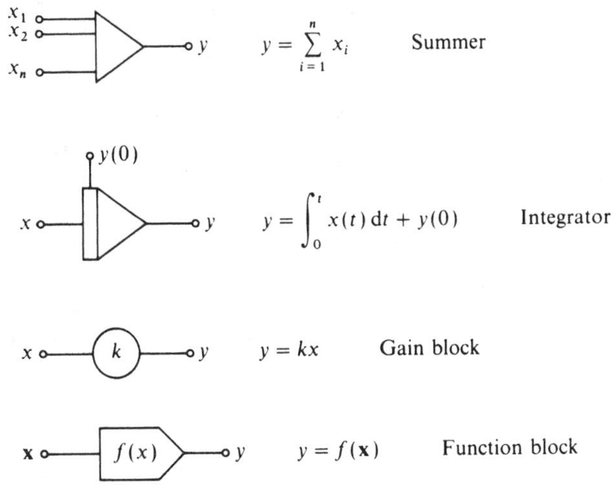

$y = \sum_{i=1}^{n} x_i$ Summer

$y = \int_0^t x(t)\,dt + y(0)$ Integrator

$y = kx$ Gain block

$y = f(\mathbf{x})$ Function block

Figure 2.1 Possible blocks in the simulation scheme.

sign means positive sign). Note that our summer has an arbitrary number of inputs while the integrator has only one. The purpose of the function block where $\mathbf{x}$ can be a vector (multiplier, divider, signal source, function generator, etc.) can be defined by the corresponding mathematical symbol, by suitable abbreviation or by some simple graphic representation of its function.

As mathematical models of dynamic systems are most often described by sets of different kinds of ordinary or partial differential equations, the main approach to continuous simulation is the so-called *indirect approach*, which originated in the differential analyser concept. It integrates the highest derivative in the differential equation as many times as its order. Note here that this principle also demonstrates the essence of simulation. Obtaining the solution of a differential equation from its analytical solution and calculation of the values of the independent variable is not simulation. Though the results calculated in such a way may sometimes be more accurate and need less time to obtain, it is well known that the percentage of differential equations (especially of those which describe real processes) that are analytically solvable is negligible.

So the general scheme for solving differential equations with simulation (especially different kinds of ordinary differential equations) can be summarized in three steps as follows:

Step 1. The differential equation must be rearranged so that the highest derivative represents the left-hand side of the equation, while all remaining terms are concentrated on the right-hand side. This is already the case if the system is given in the state space description form.

Step 2. Then the scheme, consisting of a cascade of as many integrators as the order of the differential equation, must be drawn. It is assumed that the input of the first integrator of the cascade represents the highest order derivative and so the following multiple integration yields all lower order derivatives as well as the solution of the differential equation itself. The assumption of the known highest derivative is based on the fact that it exists only in implicit form as given by the equation in Step 1, which leads to the next step.

Step 3. Using the (so far virtually) given lower derivatives and solution of the differential equation, all the right-hand side terms and their sum can be generated and are equal to the highest order derivative. Depending on the form of these terms, all kinds of operations (blocks in the scheme), excepting integration, may be used to obtain them. The entire right-hand side of the equation from Step 1 must therefore be fed into the first integrator of the cascade through the summer, forming a corresponding number of loops; this is very important from a structural point of view and it completes the simulation scheme. Such a connection of elements is often called a canonical structure.

Such rearrangement of differential equations enables their direct use in the application of some simulation tools, while the application of others is based on special simulation schemes which may, depending on the particular simulation tool characteristics, be slightly modified versions of those mentioned above. This problem will be discussed in detail in Sections 4.3 and 4.6.

The simplicity of this procedure is best illustrated through some examples which will show that it can be successfully applied to linear or nonlinear ordinary differential equations with constant or time variable coefficients.

### Example 2.1 Heating control problem

As the first example, let us take the heating control problem introduced in Problem 1.2. Here we shall use only the mathematical model of the process in the form

$$T\dot{\vartheta}_w + \vartheta_w = kp \tag{2.1}$$

Let us apply the indirect approach:

Step 1 involves rearrangement of Eqn (2.1) into the form

$$\dot{\vartheta}_w = -\frac{1}{T}\vartheta_w + \frac{k}{T}p \tag{2.2}$$

In Step 2 only one integration must be performed for the linear first order differential equation, as shown in Fig. 2.2.

Step 3 generates the right-hand side of Eqn (2.2), which completes the simulation scheme depicted in Fig. 2.3.

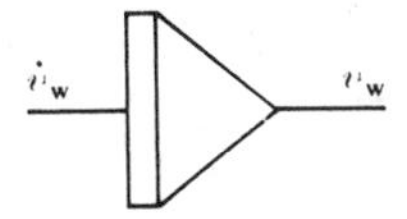

Figure 2.2 Simulation scheme for Step 2.

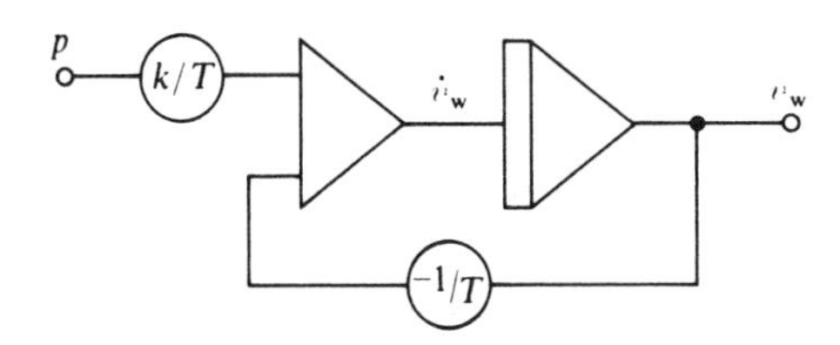


Figure 2.3 Complete simulation scheme for the first example.

As we can see from Fig. 2.3, the summer and two gain blocks are used to obtain the right-hand side of Eqn (2.2). The signs of both coefficients are included in the gain blocks. □

**Example 2.2 Car suspension system**

The second problem is the example of the car suspension system (introduced in Problem 1.1) simulation described with the linear third order differential equation

$$\dddot{y}_1 + \frac{k_1 + k_2}{f}\ddot{y}_1 + \frac{k_2}{M}\dot{y}_1 + \frac{k_1 k_2}{Mf}y_1 = 0 \qquad y_1(0) = -y_{10} \tag{2.3}$$

or, in simplified notation,

$$\dddot{y} + a\ddot{y} + b\dot{y} + cy = 0 \qquad y(0) = -d \tag{2.4}$$

Step 1. Rearrangement of Eqn (2.4) yields

$$\dddot{y} = -a\ddot{y} - b\dot{y} - cy \qquad y(0) = -d \tag{2.5}$$

Step 2. Multiple integration is shown in Fig. 2.4.
Step 3. Completion of the simulation scheme is shown in Fig. 2.5.

Note that the system in this example is excited by the initial condition and that in this case we put the signs of the coefficients in the summer (Fig. 2.5). All the loops go from the outputs of integrators to the input of the first integrator through the summer. □

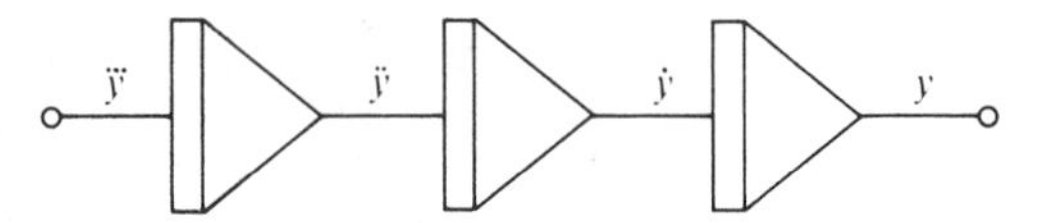


Figure 2.4 Simulation scheme for Step 2.

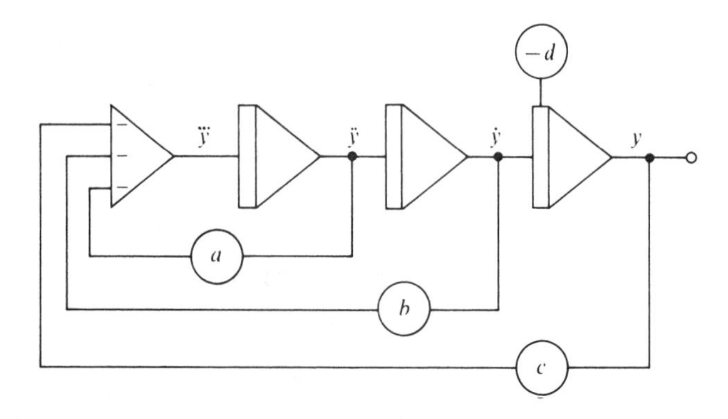


Figure 2.5 Complete simulation for the second example.

**Example 2.3 Prey and predator problem**

The example of the prey and predator problem, which was introduced in Problem 1.3, shows how the system of differential equations which also have nonlinear characteristics can be solved using the indirect approach. The mathematical model is given by

$$\left.\begin{aligned} \dot{x}_1 &= a_{11}x_1 - a_{12}x_1x_2 \qquad & x_1(0) &= x_{10} \\ \dot{x}_2 &= a_{21}x_1x_2 - a_{22}x_2 \qquad & x_2(0) &= x_{20} \end{aligned}\right\} \tag{2.6}$$

Step 1. Not needed since the equations are already given in the corresponding, i.e. state space description, form.
Step 2. Integration for both equations in the system (2.6) is shown in Fig. 2.6.
Step 3. Completion of the simulation scheme is shown in Fig. 2.7.

As can be seen from Figs. 2.6 and 2.7, the procedure for the simulation of a nonlinear system of differential equations is the same as that for a simple linear one. First, the corresponding integrations are made, then the loops for particular equations are completed and, last, the interconnections between the equations are realized. The nonlinearity of equations is reflected in this case in the use of two multiplication blocks which are introduced for systematic reasons. It is obvious that only one multiplication generating $x_1x_2$ could be used. □

The *direct approach* is the opposite of the indirect approach discussed above and

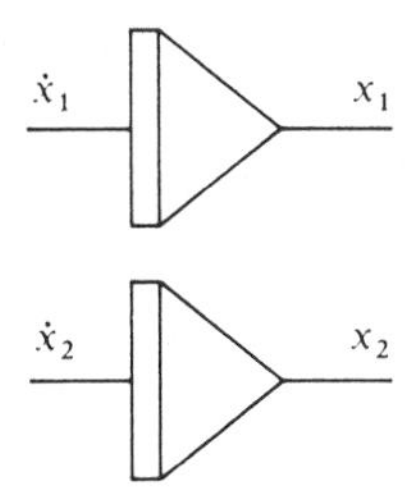


Figure 2.6 Simulation scheme for Step 2.

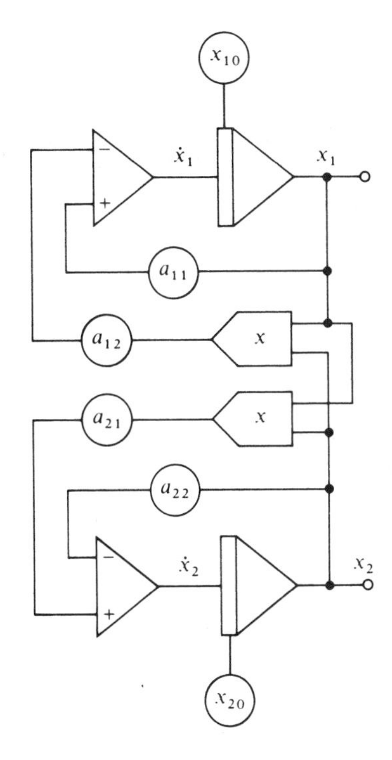


Figure 2.7 Complete simulation scheme for the prey and predator problem.

is used only for the realization of some expressions in the form of formulas of algebraic relations. If required, the simulation scheme which contains the correspondingly connected blocks for particular operations can also be developed.

Also worth mention is the idea of *implicit function generation*, which involves the solution of an implied rather than a given equation (Jackson, 1960). If we assume that the indirect approach makes the solving of differential equations easy, the implicit approach is justified. We are looking for the differential equation with a solution which is equal to the analytic function to be generated. Such a differential equation can be obtained by repeated differentiations of the given relation with respect to the independent variable. The differentiations are complete only when the higher order derivatives and the function itself are in the relation. The question, however, is whether this is possible. The simplest example is the generation of the exponential function

$$y = \exp(-at) \tag{2.7}$$

where the first derivative gives the required relationship

$$\dot{y} = -a\exp(-at) = -ay \tag{2.8}$$

Note that in every step we must try to express as many terms as possible using the previous derivatives or the function itself. Also, the initial conditions of the differential equation obtained must be determined, In our case $y(0) = 1$. The differential equation obtained (in our case Eqn (2.8) with the corresponding initial condition) is then solved using the indirect approach. In some cases, the logarithms of both sides of the given relation can be taken prior to differentiation.

Continuous systems simulation plays an important role in the analysis and synthesis of complex processes. It also enables study of the behaviour of particular parts of the system. Thus, the concept of transfer function is often used in systems analysis and methods of simulation are closely related to the indirect approach to the solving of differential equations. A transfer function is, in fact, a linear ordinary differential equation written in short notation. If the Laplace transform operation is performed on the equation, a function $G(s)$ in the complex frequency variable $s$ is obtained as the ratio of output transform $Y(s)$ and input transform $U(s)$:

$$G(s) = \frac{Y(s)}{U(s)} \tag{2.9}$$

The function $G(s)$ is termed the transfer function and completely characterizes the performance of the system. In applying Laplace transformation zero initial conditions are always assumed. The importance of transfer function description of the system lies in the fact that transfer functions can be manipulated algebraically in a much easier way than can the original differential equations. If it is convenient we can combine transfer functions describing particular parts of the system to form an overall transfer function using the rules of so-called block diagram algebra. If $j\omega$ is substituted for the complex variable $s$, the frequency response of the system can be obtained directly from the transfer function.

Many approaches exist for the *simulation of transfer functions* (Jackson, 1960). We shall only discuss two methods which in our opinion are the most straightforward and the most usable for computer simulation. Both will be presented by the aid of the corresponding example.

**Example 2.4 Transfer function**

Two approaches are used to simulate the following transfer function:

$$\frac{Y(s)}{U(s)} = \frac{as^3 + bs^2 + cs + d}{s^3 + es^2 + fs + g} \tag{2.10}$$

The first approach is sometimes called the *nested form* method and is very useful for transfer functions given in unfactored form. It is performed in several steps as follows:

- Cross multiplying in Eqn (2.10):

$$(s^3 + es^2 + fs + g)Y = (as^3 + bs^2 + cs + d)U \tag{2.11}$$

- Arranging all terms with the same power of $s$:

$$s^3(Y - aU) + s^2(eY - bU) + s(fY - cU) + (gY - dU) = 0 \tag{2.12}$$

- Solving for the highest derivative (the highest power of $s$) of the output:

$$s^3 Y = s^3 aU - s^2(eY - bU) - s(fY - cU) - (gY - dU) \tag{2.13}$$

- Dividing by the highest power of $s$:

$$Y = aU - \frac{1}{s}(eY - bU) - \frac{1}{s^2}(fY - cU) - \frac{1}{s^3}(gY - dU) \tag{2.14}$$

- Rearranging in the 'nested form' gives the final form of the equation in Laplace transform notation:

$$Y = aU + \frac{1}{s}\left\{(bU - eY) + \frac{1}{s}\left[(cU - fY) + \frac{1}{s}(dU - gY)\right]\right\} \tag{2.15}$$

It is important to note the structure where the subsidiary variables are assigned:

$$\left.\begin{aligned} U_0 &= \frac{1}{s}(dU - gY) \\ U_1 &= \frac{1}{s}(cU - fY + U_0) \\ U_2 &= \frac{1}{s}(bU - eY + U_1) \end{aligned}\right\} \tag{2.16}$$

Eqn (2.15) therefore becomes

$$Y = aU + U_2 \tag{2.17}$$

The corresponding simulation scheme, generated from the term, which has to be integrated three times (variable $U_0$), is shown in Fig. 2.8, where the scheme shown represents the canonical structure which in control theory is called the *observable canonical form* (Ogata, 1990). The coefficients of the transfer function appear as separate gain blocks. The introduction of subsidiary variables simplifies both

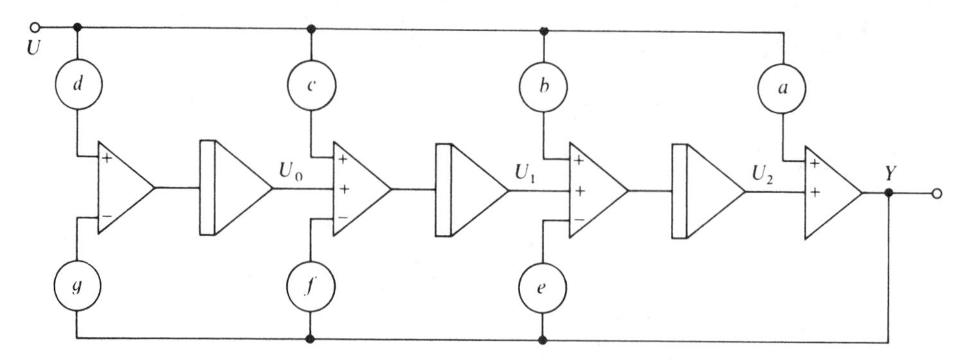


Figure 2.8 Simulation scheme for the 'nested form' method.

generation of the simulation scheme and direct programming of equation oriented digital simulation languages.

The second approach is called the *partitioned form* method and gives the *controllable canonical structure* of the simulation scheme. Like the first approach, it also uses a number of integrators equal to the order of the transfer function. Using the transfer function (2.10) the second approach can be divided into the following steps:

- Partitioning of the transfer function into

$$\frac{W(s)}{U(s)} = \frac{1}{s^3 + es^2 + fs + g} \tag{2.18}$$

and

$$\frac{Y(s)}{W(s)} = as^3 + bs^2 + cs + d \tag{2.19}$$

- Rearranging Eqn (2.18) in the indirect approach form:

$$s^3W = U - es^2W - fsW - gW \tag{2.20}$$

and generating according to Eqn (2.19):

$$Y = as^3W + bs^2W + csW + dW \tag{2.21}$$

The corresponding simulation scheme is shown in Fig. 2.9, where it can be seen that the structure consists of two parts. One obtains the denominator using the indirect approach and the other generates the numerator using the direct approach.

An advantage of such a scheme is that the derivatives of $W$ can also be used in the realization of some other transfer function with the same denominator. □

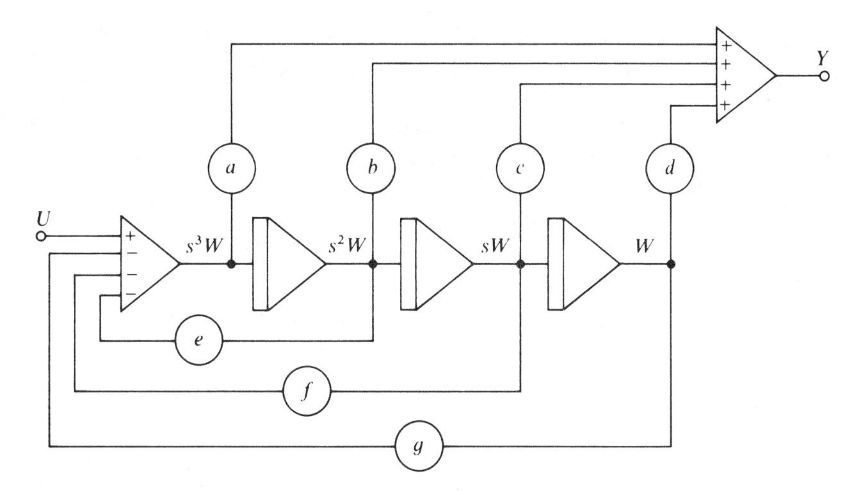


Figure 2.9 Simulation scheme for the 'partitioned form' method.

## 2.2 DISCRETE SYSTEMS SIMULATION

Different types of systems can be observed all around us and can be divided into two distinct groups: continuous and discrete systems. The simulation of *continuous systems*, which are described by differential equations, is the main subject of this book. The basic methodologies for the simulation of such systems were described in Section 2.1.

In contrast to continuous systems, where all the changes occur smoothly, changes in *discrete systems* take place instantaneously, i.e. in discrete steps or in a series of discrete events. Between events, the state of the system remains constant. These events may occur periodically (e.g. changes in the control signal of a discrete controller) or randomly (e.g. request for a connection in a telephone system). The first approach to discrete system simulation is very similar to continuous systems simulation, which is dynamic and, in general (with exception of disturbance signals which are stochastic), deterministic. It actually represents the simulation of difference equations. The second approach is, in general, dynamic and stochastic by nature and is called discrete event simulation. There are, however, systems where all types are mixed or combined and are simulated by a combination of all the approaches. In the next two sections the simulation of difference equations and discrete event simulation will be reviewed briefly.

### 2.2.1 Simulation of Difference Equations

Difference equations describe systems where all changes occur in discrete steps which are periodical. Typical examples of such systems are *sampled systems*, which arise from sampling continuous time analog signals at discrete values of time. With sampled systems the discretization applies to time, so the terms involved in difference equations are the values of variables at moments $kT, kT + T, kT + 2T, \ldots, kT + nT$, where $T$ is the sampling period. Difference equations can, however, also be obtained if systems having periodic structure are treated. In such systems the independent variable may be length or position.

In this book the most straightforward notation will be used and only the location of the sample in the sequence of samples will be used. For sampled systems, $y(kT)$ will be denoted as $y(k)$ and the sampling time $T$ will be involved implicitly in the difference equation. A linear difference equation of order $n$ has the form

$$y(k+n) + a_1 y(k+n-1) + \cdots + a_n y(k)$$
$$= b_0 u(k+n) + b_1 u(k+n-1) + \cdots + b_n u(k) \tag{2.22}$$

where $u(k)$ is the forcing series. Analogous to the use of the Laplace transformation with differential equations, the $\mathscr{Z}$ transformation is used with difference equations. If it is performed on Eqn (2.22) the transfer function of the system is obtained as the ratio of output transform $Y(z)$ and input transform $U(z)$:

$$G(z) = \frac{Y(z)}{U(z)} = \frac{b_0 z^n + b_1 z^{n-1} + \cdots + b_n}{z^n + a_1 z^{n-1} + \cdots + a_n} \tag{2.23}$$

where

$$\mathscr{Z}\{y(k+i)\} = z^i Y(z) \tag{2.24}$$

Eqn (2.23) represents the transfer function of the system described by the difference equation (2.22).

Discrete systems described by difference equations can be realized by the corresponding hardware. In this case the physical layout of a combination of arithmetic and storage operations that produce the given transfer function or difference equation is built. This layout is called the *realization scheme* and can also be used to simulate the discrete system on a digital computer. Simulation of a discrete system can also be called 'realization in software'. In this case the realization scheme is used to make the approach, as far as possible, independent of the programming language used. The blocks of the realization scheme for discrete systems are the same as those of the simulation scheme shown in Fig. 2.1 for continuous systems, with the exception of the integrator, which is replaced by the unit delay block shown in Fig. 2.10. Some authors represent the summer in realization schemes by a circle rather than by a triangle but for reasons of clarity the latter will be used in this book.

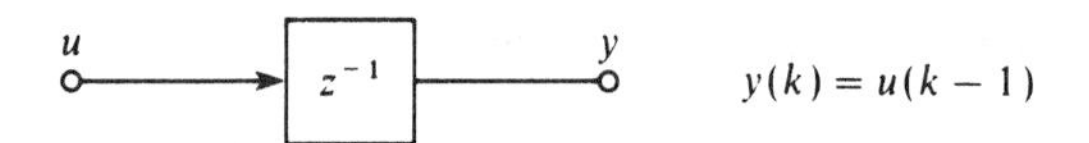


Figure 2.10 The unit delay block.

Observing that the unit time delay is the discrete counterpart of the integrator (and prediction the discrete counterpart of differentiation), the general scheme for solving difference equations with simulation can be summarized in a similar way as was the scheme for solving differential equations given in Section 2.1:

Step 1. The difference equation must be rearranged so that the highest prediction represents the left-hand side of the equation, while all remaining terms are concentrated on the right-hand side.

Step 2. Then the scheme, consisting of a cascade of as many unit time delays as the order of the difference equation, must be drawn. As for continuous systems, it is assumed that the input of the first unit delay block in the cascade is the highest prediction and so the following unit delay blocks yield all lower predictions along with the solution of the difference equation itself. The assumption of the known highest prediction is based on the fact that it exists only in implicit form as given by the equation in the first step, which leads to the next step.

Step 3. Using the (so far virtually) given lower predictions and the solution of the difference equation, all the right-hand side terms and their sum can be generated and are equal to the highest prediction. As for the continuous case, all kinds of operations except unit time delays may be used to obtain the highest prediction, which completes the realization scheme.

All arithmetic operations in the realization scheme run in parallel and the time delay operations (realized by storage elements) introduce the dynamics into the system. A difference equation could be simulated by a digital simulation language, but due to the very simple scheduling algorithm the program can be easily derived from the realization scheme in any general language. For simplicity, the BASIC (Beginners' All-purpose Symbolic Instruction Code) will be used here. The reason for this choice is that it is available on every home computer and that it can easily be translated by the reader into his favourite program language.

The following examples will illustrate the realization scheme approach to solving difference equations by simulation.

### Example 2.5 Fibonacci equation

The Fibonacci equation representing the growth of a species has the following form:

$$y(k+2) = y(k+1) + y(k) \qquad y(1) = y(2) = 1 \tag{2.25}$$

For a difference equation of order $n$ the initial conditions are $n$ successive values of its solution. In our case ($n = 2$) these are the values $y(1)$ and $y(2)$, which are both equal to 1.

Fibonacci Eqn (2.25) can easily be evaluated by the program shown in Fig. 2.11, where the Fibonacci numbers are first evaluated in the first `FOR–NEXT` loop and

```
10 DIM Y(100)
20 LET Y(1)=1
30 LET Y(2)=1
40 FOR K=3 TO 100
50 LET Y(K)=Y(K-1)+Y(K-2)
60 NEXT K
70 FOR K=1 TO 100
80 PRINT K,Y(K)
90 NEXT K
100 END
```

Figure 2.11 Program for the evaluation of the Fibonacci numbers.

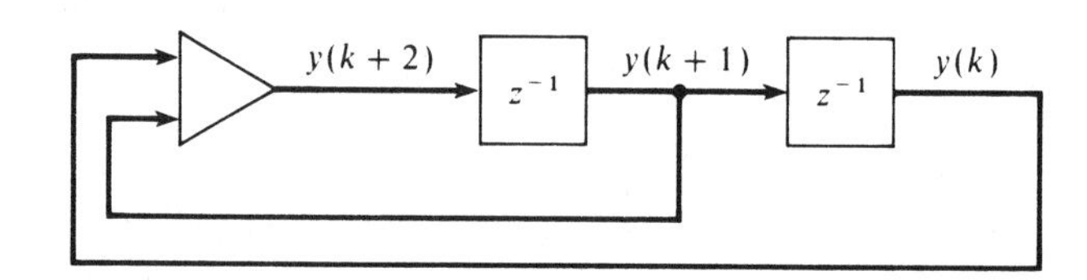


Figure 2.12 Realization scheme for the Fibonacci equation.

then written on the screen in the second one. The main drawback of this off-line approach is the memory required. Even if the `PRINT` statement could be included in the first `FOR–NEXT` loop the `DIM` statement would still be required. If, for example, a digital filter or digital controller were used in real time applications, they would need to run for an indefinite period of time. In this case every computer would run out of memory and the approach given in Fig. 2.11 would be useless.

With the simulation approach the realization scheme is first drawn according to the given rules. Since the highest (second) prediction in Eqn (2.25) is already on the left-hand side, no rearrangement is necessary. In the second step the cascade of two time delay units is drawn, and in the third step $y(k+1)$ and the solution $y(k)$ are added, giving $y(k+2)$ according to Eqn (2.25). This completes the realization scheme shown in Fig. 2.12.

A program will now be designed to simulate the Fibonacci equation. With the simulation approach the program for the simulation of the difference equation consists of two parts:

1. All arithmetic operations which are shown in the realization scheme are coded appropriately. Here suitable names must be given to variables which appear explicitly in the program. In our case the solution $y(k)$ is denoted by `Y`, its first prediction $y(k+1)$ by `Y1`, and the second prediction $y(k+2)$ by `Y2`. There is only one arithmetic operation (addition) which in BASIC is represented by the following statement:

```
LET Y2=Y1+Y
```

It should be pointed out that according to the realization scheme all operations are performed in parallel, but on a digital computer operations can only be performed in the sequence. Therefore, the order of operations performed in the sequence must be carefully determined in order to avoid the use of a variable prior to its evaluation. The starting points of the sorting procedure are the outputs of the delay units which remain constant during evaluation of the highest prediction and are changed only in periodic discrete moments. This sorting procedure is the same as the sorting procedure for digital continuous simulation systems described in Chapter 3.

2. The outputs of the delay units are changed only in periodic discrete moments ($k=0, 1, 2, \ldots$). This periodicity can be realized by a FOR–NEXT loop, which also determines the duration of the simulation. Every transition of the loop represents a discrete moment during which delay operations are performed. The delay operation is actually a shift operation, where the inputs of the delays at the moment $k$ become the valid outputs for the moment $k + 1$. According to the realization scheme, all shifts are performed simultaneously but, due to the sequential operation of digital computers, they are performed in sequence. As they are placed in cascade, the order of their execution is important (the shift operations must be performed backwards). In our case the shift operations are coded in BASIC in the following form:

```
LET Y=Y1
LET Y1=Y2
```

The first part of the program defines static relations defined by the realization scheme, while the second part introduces the dynamics. Since the static relations must be fulfilled at every step, the first part is included at the beginning of the loop. Prior to the loop, the initial conditions must be set and after evaluation of static relations the solution of the difference equation is written on the output device. This completes the BASIC program, which is shown in Fig. 2.13.

It must be pointed out that no DIM statement is required and that the memory requirements do not depend on the duration of the simulation. In our case the simulation run is terminated after the evaluation of 100 Fibonacci numbers.

```
10 LET Y=1
20 LET Y1=1
30 FOR K=1 TO 100
40 LET Y2=Y1+Y
50 PRINT K,Y
60 LET Y=Y1
70 LET Y1=Y2
80 NEXT K
90 END
```

Figure 2.13 Simulation program for the Fibonacci equation.

Other simulation termination conditions can be programmed by, for example, `IF . . . GO TO . . .` statements, or an infinite loop can even be obtained, as shown in the next example. □

**Example 2.6 Butterworth filter**

This example demonstrates the simulation of a discrete transfer function. The second order low pass Butterworth filter with cut off frequency 50 Hz has the following transfer function at a sampling frequency of 500 Hz:

$$G(z) = \frac{Y(z)}{U(z)} = \frac{0.063\,96z^2 + 0.127\,92z + 0.063\,96}{z^2 - 1.1683z + 0.4241} \tag{2.26}$$

The procedure for the simulation of continuous transfer functions described in Section 2.1 can be carried over to discrete systems. Both approaches, the nested and the partitioned forms, can be used but only the second one will be used here.

According to this approach the transfer function (2.26) is first partitioned into

$$\frac{W(z)}{U(z)} = \frac{1}{z^2 - 1.1683z + 0.4241} \tag{2.27}$$

and

$$\frac{Y(z)}{W(z)} = 0.063\,96z^2 + 0.127\,92z + 0.063\,96 \tag{2.28}$$

In the second step, Eqn (2.27) is first rearranged into

$$z^2W = U + 1.1683zW - 0.4241W \tag{2.29}$$

and then the output $Y$ is generated according to Eqn (2.28):

$$Y = 0.063\,96z^2W + 0.127\,92zW + 0.063\,96W \tag{2.30}$$

The corresponding realization scheme is shown in Fig. 2.14. The program, in BASIC, is now obtained immediately in the same way as in the preceding example and is shown in Fig. 2.15. There are three peculiarites in this program which deserve detailed explanation. The first one is the loop obtained by an unconditional GO TO statement, which causes the program, once started, to run forever (in practice, of course, until it is terminated by an interrupt). The absence of a termination condition can be an advantage if the filter is used in an industrial project as, for example, part of the control loop. For that same reason the input of the filter is supposed to be a sampled signal obtained by an analog to digital (A/D) converter, and the output of the filter is supposed to be fed to a digital to analog (D/A) converter. The corresponding part of the program is the second peculiarity. The implementation of the A/D and D/A converters mainly depends on the hardware used and is beyond the scope of this book. So these procedures are only indicated in the remark statements given in the

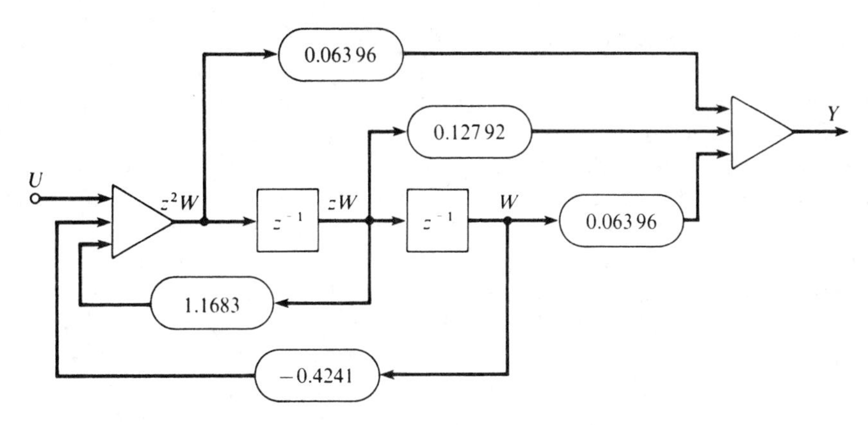


Figure 2.14 Realization scheme for the second order Butterworth filter.

```
10 LET W=0
20 LET ZW=0
30 REM SYNCHRONIZATION OF THE LOOP (HARDWARE DEPENDENT –
31 REM NOT GIVEN HERE)
40 REM SAMPLING OF THE INPUT SIGNAL BY AN A/D CONVERTER –
41 REM HARDWARE DEPENDENT –
42 REM HERE THE INPUT IS GIVEN AS A RANDOM FUNCTION
50 LET U=RND
60 REM REALIZATION OF ALGEBRAIC RELATIONS
70 LET Z2W=U+1.1683*ZW–0.4241*W
80 LET Y=0.06396*Z2W+0.12792*ZW+0.06396*W
90 REM OUTPUT Y IS PUT ON A D/A CONVERTER –
91 REM HARDWARE DEPENDENT –
92 REM HERE IT IS WRITTEN ON THE SCREEN
100 PRINT U,Y
110 REM REALIZATION OF TIME DELAYS
120 LET W=ZW
130 LET ZW=Z2W
140 REM END OF THE LOOP
150 GO TO 30
```

Figure 2.15 BASIC program for the second order Butterworth filter.

program shown in Fig. 2.15; and a random signal produced by an internal random generator is used as the input of the filter, the output being written on the screen. The third peculiarity of the program shown in Fig. 2.15 is the synchronization of the loop. With the coefficients used the transfer function (2.26) represents a low pass filter with cut off frequency 50 Hz if the sampling frequency is 500 Hz. This means that the loop has to be executed every 2 ms exactly, and the synchronization must be implemented to ensure this. The synchronization procedure is based on a real time clock in the computer and is also hardware dependent; thus it is only indicated in a remark statement in the program shown in Fig. 2.15. □

### 2.2.2 Discrete Event Simulation

A typical example of discrete event simulation is the classical single-server–single-queue system. A server, denoted as a small post office, provides a service for customers, one at a time on a first come–first served basis, as shown in Fig. 2.16. Customers arrive at random and wait in the queue if the server is busy. The state of the system consists of the number of customers in the queue and the state (busy or idle) of the server. The events, changing the state of the system, are the arrivals and departures (after completion of service) of customers. The server becomes idle if the number of customers in the queue is zero (i.e. if it is empty) after completion of servicing a customer. It becomes busy again after arrival of the next customer. If the queue is not empty, the number of customers in the queue is increased by one after the arrival of a customer and is decreased by one after completion of the service. The parameters of the system are the probability distributions of the interarrival time and the service time. The standard results of such a single-server–single-queue system are the probability distributions of the number of customers in the queue (queue length), number of customers in the system, waiting time in the queue, time spent in the system and utilization of the server.

Similarly, as simple differential or difference equations can be solved analytically, so also can simple discrete event problems. However, the problems in real life are too complex to be solved analytically and thus simulation is the appropriate tool. The simulation scheme for discrete event simulation can be represented in the form of flowcharts, as shown in Fig. 2.17, for the simple single-server–single-queue system.

The essential difference between continuous and discrete systems is the representation of the time. In continuous systems time runs smoothly and there is an infinitesimal change of the system's state in every infinitesimal portion of time. In discrete systems, changes only occur within a particular set of values of time. There is, however, also a difference in the representation of time between the systems, described by difference equations and discrete event systems. The changes in difference equations occur periodically and such systems are suitable for *interval-oriented*

Figure 2.16 Single-server-single-queue system.

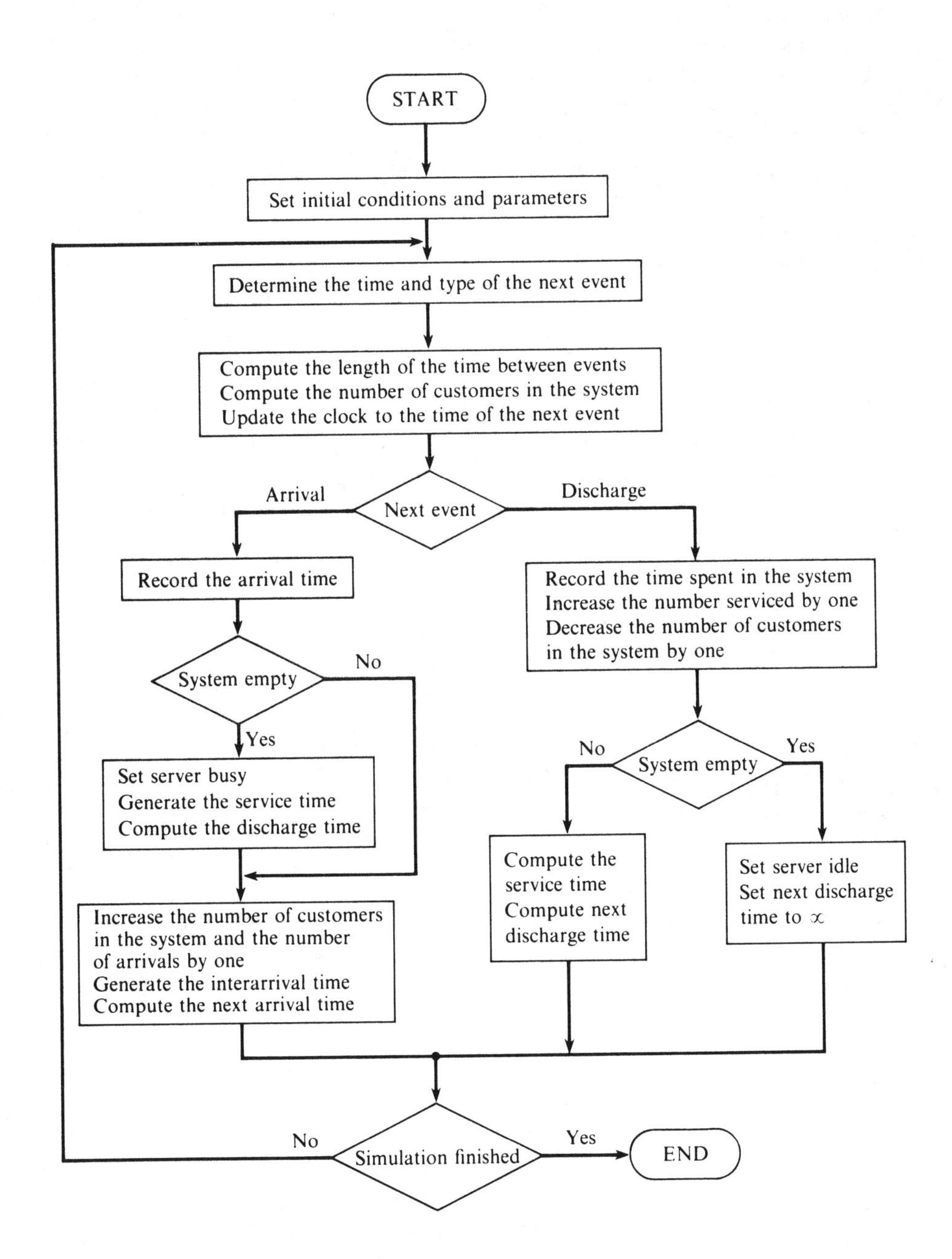


Figure 2.17 Flow chart of the single-server-single-queue system.

*simulation*, while changes in discrete event systems occur randomly and thus such systems are suitable for *event oriented simulation*.

There are two events in our single-server–single-queue system, i.e. the arrival and discharge of the customer. As shown in Fig. 2.17 after selecting the next event the length of time and the number of customers in the system between events are computed and the clock is updated to the value determined by the next event. In the case of arrival of a new customer his arrival time is first recorded; then if the system is empty then the server becomes busy and the service time is generated from a random generator with appropriate distribution and the next discharge time is computed. Then the number of customers in the system and the number of arrivals are increased by one. As the last task the interarrival time is generated from a random generator with appropriate distribution and the next arrival time is computed.

On discharge of the customer, his time spent in the system is first recorded, the number of serviced customers is increased by one and the number of customers in the system is decreased by one. If the system is not empty, then the service time is generated from a random generator with appropriate distribution and the next discharge time is computed; otherwise the server becomes idle and the next discharge time is set to infinity.

The initial conditions are set at the beginning of the simulation and the termination condition is tested at each event. The occurrence of events in the system is simulated by generating the next event time, which is the minimum time of next arrival (computed on the basis of random interarrival time) and next discharge (computed on the basis of random service time). This strategy of event scheduling is one of many possible simulation strategies. Other strategies are described in Section 3.3 where discrete simulation languages are treated. As the main subject of this book is the simulation of continuous systems, we shall not go into the details of discrete event simulation. The reader interested in such simulation is referred to the appropriate literature, e.g. Fishman (1973), Neelamkavil (1987), and many others.

## 2.3 PROBLEMS

### Solved problems

#### Problem 2.1 Modified car suspension system

As the first problem let us take a slightly modified model of the car suspension system shown in Fig. 2.18. In this case the half mass of axle with mass of wheel is not neglected as it was in Example 2.2. Also, the system is now excited by the reaction of the ground (when, for instance, the car is driven over an obstacle, a curb say). The motion of the car body is given by $y_1$ and the motion of the wheel by $y_2$, while the ground reaction is given by $z$, which in this case is a constant (height of the

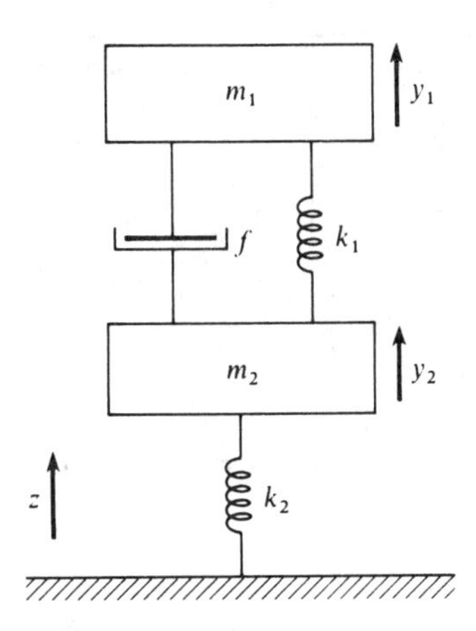


Figure 2.18 Mechanical model of the modified car suspension system.

curb). The mathematical model is given in the form

$$m_1\ddot{y}_1 + f(\dot{y}_1 - \dot{y}_2) + k_1(y_1 - y_2) = 0 \tag{2.31}$$

$$m_2\ddot{y}_2 + f(\dot{y}_2 - \dot{y}_1) + k_1(y_2 - y_1) + k_2(y_2 - z) = 0 \tag{2.32}$$

Using the indirect approach, such a system of linear second order differential equations is solved in the following way:

Step 1. Rearrangement:

$$\ddot{y}_1 = -\frac{f}{m_1}(\dot{y}_1 - \dot{y}_2) - \frac{k_1}{m_1}(y_1 - y_2) \tag{2.33}$$

$$\ddot{y}_2 = +\frac{f}{m_2}(\dot{y}_1 - \dot{y}_2) + \frac{k_1}{m_2}(y_1 - y_2) + \frac{k_2}{m_2}(z - y_2) \tag{2.34}$$

Steps 2 and 3. The simulation scheme generation is shown in Fig. 2.19.

### Problem 2.2 Applications of the implicit method

This problem presents the function generation using the implicit method. Consider

$$y = At^2 \exp(-bt) \tag{2.35}$$

Sequential differentiation gives

$$\dot{y} = 2At \exp(-bt) - At^2 b \exp(-bt) = 2At \exp(-bt) - by \tag{2.36}$$

Note that the second term on the right-hand side of Eqn (2.36) is immediately

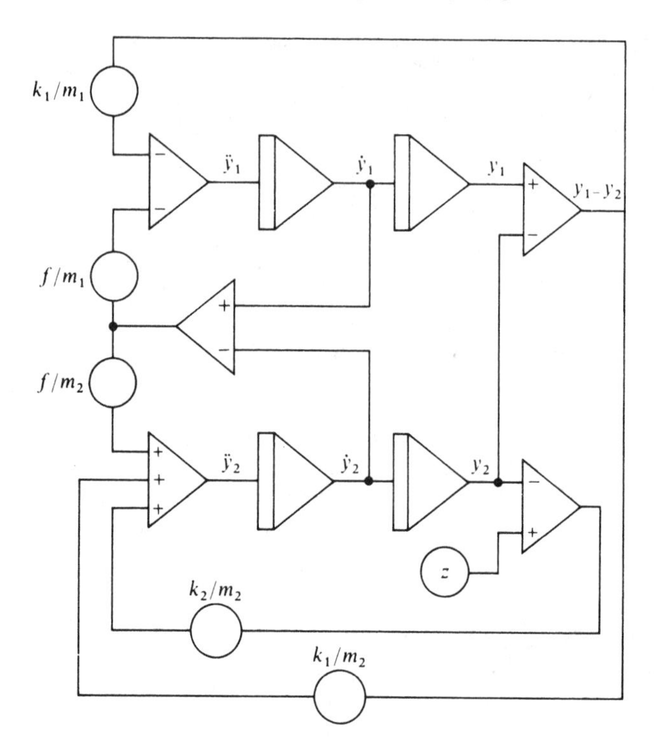


Figure 2.19 Simulation scheme for the car suspension system.

substituted using Eqn (2.35):

$$\ddot{y} = 2A\exp(-bt) - 2Abt\exp(-bt) - b\dot{y} \tag{2.37}$$

The second term on the right-hand side of Eqn (2.37) can be expressed with the aid of Eqn (2.36) as

$$2At\exp(-bt) = \dot{y} + by \tag{2.38}$$

and again immediately substituted in Eqn (2.37):

$$\ddot{y} = 2A\exp(-bt) - b\dot{y} - b^2y - b\dot{y} \tag{2.39}$$

At this point one could finish with the differentiation generating the term $2A\exp(-bt)$ in Eqn (2.39) with the separate implicit function generation procedure. However, a more compact solution is obtained by proceeding with the differentiations, if this is profitable of course:

$$\dddot{y} = -2Ab\exp(-bt) - 2b\ddot{y} - b^2\dot{y} \tag{2.40}$$

Again, first term on the right-hand side of Eqn (2.40) can be expressed using

Eqn (2.39) as

$$2A\exp(-bt) = \ddot{y} + 2b\dot{y} + b^2y \tag{2.41}$$

and substituted in Eqn (2.42) to obtain, finally,

$$\dddot{y} = -3b\ddot{y} - 3b^2\dot{y} - b^3y \tag{2.42}$$

with the following initial conditions:

$$y(0) = 0 \qquad \dot{y}(0) = 0 \qquad \ddot{y}(0) = 2A$$

which must not be forgotten. Note that in Eqn (2.42) only $y$ and its derivatives are included and that it enables simulation using the indirect approach.

### Problem 2.3 Mathieu's differential equation

The third problem shows that the indirect approach remains the same when solving linear ordinary differential equations with time variable coefficients. The only additional task is the generation of the coefficients which are functions of the independent variable (in most cases time). Consider, for instance, Mathieu's equation

$$\ddot{y} + [a - 2\nu\cos(\omega t)]y = 0 \qquad y(0) = y_0 \qquad \dot{y}(0) = \dot{y}_0 \tag{2.43}$$

which is used in the study of frequency modulated oscillations. Rearranging the equation for the indirect approach yields:

$$\ddot{y} = -ay + 2\nu\cos(\omega t)y \qquad y(0) = y_0 \qquad \dot{y}(0) = \dot{y}_0 \tag{2.44}$$

The corresponding simulation scheme is shown in Fig. 2.20. Due to the fact that the term $2\nu\cos(\omega t)$ is time dependent, it is sometimes convenient for some simulation tools to generate it with the aid of the implicit approach:

$$x = 2\nu\cos(\omega t) \tag{2.45}$$

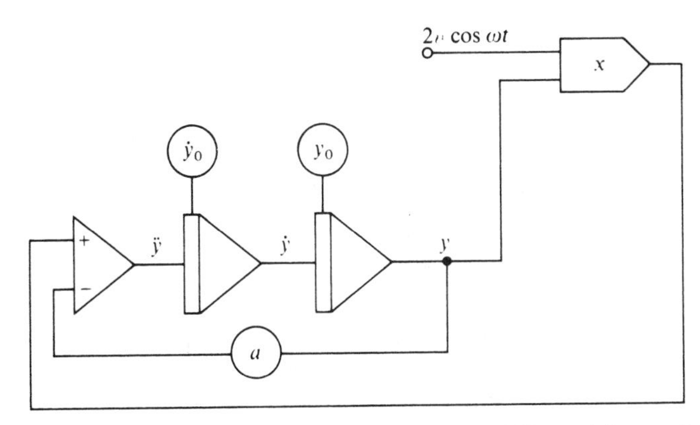


Figure 2.20 Simulation scheme for Mathieu's differential equation.

$$\dot{x} = -2\upsilon\omega \sin(\omega t) \tag{2.46}$$

$$\ddot{x} = -2\upsilon\omega^2 \cos(\omega t) = -\omega^2 x \tag{2.47}$$

with initial conditions $x(0) = 2\upsilon$, $\dot{x}(0) = 0$. So the rearranged differential equation for the realization of $2\upsilon \cos(\omega t)$ using the indirect approach has the form

$$\ddot{x} = -\omega^2 x \qquad x(0) = 2\upsilon \qquad \dot{x}(0) = 0 \tag{2.48}$$

while the simulation scheme is given in Fig. 2.21.

### Problem 2.4 Transfer functions in the factored form

A system is sometimes given as a transfer function in factored form, presenting particular parts of the system. Due to the fact that such a formulation enables more profound study of the system's behaviour, the simulation is also performed for corresponding parts of the transfer function which are then connected to the complete scheme:
Consider

$$G_p(s) = \frac{Y(s)}{U(s)} = \frac{1 - 4s}{(1 + 4s)(1 + 10s)} \tag{2.49}$$

which can be written as

$$G_p(s) = G_1(s)G_2(s) \tag{2.50}$$

where

$$G_1(s) = \frac{1 - 4s}{1 + 4s} \tag{2.51}$$

$$G_2(s) = \frac{1}{1 + 10s} \tag{2.52}$$

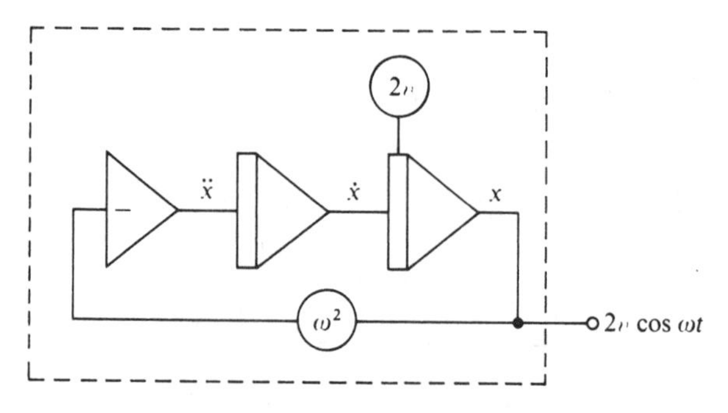


Figure 2.21 Simulation scheme for $2\upsilon \cos(\omega t)$ generation with the implicit approach.

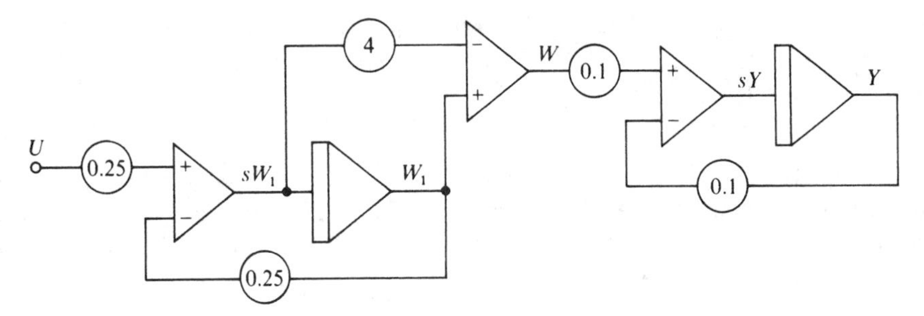


Figure 2.22 Simulation scheme for the transfer functions in factored form.

Such a choice is based on the physical background of the problem.

With the 'partitioned form' method we obtain

$$G_1(s) = \frac{W}{U} = \frac{1 - 4s}{1 + 4s} \tag{2.53}$$

$$\frac{W_1}{U} = \frac{1}{1 + 4s} \tag{2.54}$$

$$sW_1 = 0.25U - 0.25W_1 \tag{2.55}$$

$$\frac{W}{W_1} = 1 - 4s \tag{2.56}$$

$$W = W_1 - 4sW_1 \tag{2.57}$$

and

$$G_2(s) = \frac{Y}{W} = \frac{1}{1 + 10s} \tag{2.58}$$

$$sY = 0.1W - 0.1Y \tag{2.59}$$

which gives the simulation scheme shown in Fig. 2.22, where the product in Eqn (2.50) is depicted as the cascade connection of simulation schemes for $G_1$ and $G_2$.

### Problem 2.5 Discrete PID controller

In this problem the discrete proportional-integral-derivative (PID) controller is programmed in PASCAL. It is supposed that the function AD and the procedure DA, which establish the connection of the analog to digital and digital to analog converters are available. It is also supposed that the AD function also synchronizes the required sampling time, i.e. that the program waits (or lies dormant) until the time for the next passage of the loop. It is beyond the scope of this book to provide these subprograms, which are of course hardware-dependent.

The discrete PID controller has the following transfer function:

$$G(z) = \frac{q_0 + q_1 z^{-1} + q_2 z^{-2}}{1 - z^{-1}} \tag{2.60}$$

given in negative powers of the complex variable $z$, which is a common way of representing the discrete transfer functions. The coefficients $q_0$, $q_1$ and $q_2$ are determined by the gain $K$, the integration time $T_I$ and the derivative time $T_D$ of the PID controller as follows (Isermann, 1989):

$$q_0 = K\left(1 + \frac{T_D}{T_s}\right) \tag{2.61}$$

$$q_1 = -K\left(1 + 2\frac{T_D}{T_s} - \frac{T_s}{T_I}\right) \tag{2.62}$$

$$q_2 = K\frac{T_D}{T_s} \tag{2.63}$$

where $T_s$ is the sampling time. The nested form approach will be used here and, according to this, Eqn (2.60) is first written in positive powers of $z$:

$$G(z) = \frac{Y(z)}{U(z)} = \frac{q_0 z^2 + q_1 z + q_2}{z^2 - z} \tag{2.64}$$

The next steps in designing the realization scheme are the same as those for the simulation scheme of continuous systems:

- Cross multiplying in Eqn (2.64):

$$z^2 Y - zY = q_0 z^2 U + q_1 zU + q_2 U \tag{2.65}$$

- Arranging all terms with the same power of $z$:

$$z^2(Y - q_0 U) - z(Y + q_1 U) - q_2 U = 0 \tag{2.66}$$

- Solving for the highest prediction (the highest power of $z$):

$$z^2 Y = z^2 q_0 U + z(Y + q_1 U) + q_2 U \tag{2.67}$$

- Dividing by the highest power of $z$:

$$Y = q_0 U + z^{-1}(Y + q_1 U) + z^{-2} q_2 U \tag{2.68}$$

- Rearranging in 'nested form'

$$Y = q_0 U + z^{-1}[(Y + q_1 U) + z^{-1}(q_2 U)] \tag{2.69}$$

The realization scheme which represents the observable canonical structure is shown in Fig. 2.23.

The PASCAL program shown in Fig. 2.24 is written in the same way as the BASIC programs given in Examples 2.5 and 2.6. It includes many comments, so the

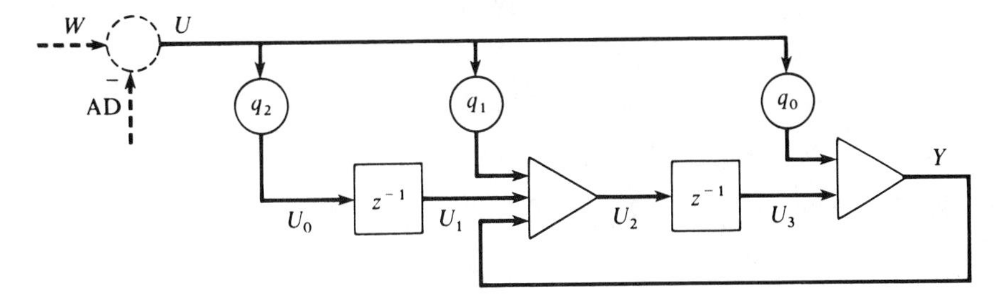


Figure 2.23 Realization scheme for the discrete PID controller.

```
Program PID_controller;
const q0=2;q1=-2;q2=1;  {definition of discrete PID controller parameters
                         for gain=1, integration time=1 and derivative time=1}
var U,U0,U1,U2,U3,Y:real;
function AD:real;extern;
procedure DA(Y:real);extern;
begin
 U1:=0;U3:=0;
 repeat
 begin
  U:=W-AD;          {evaluation of the PID controller input}
  {PART 1: algebraic relation due to the realization scheme:}
  Y:=q0*U+U3;       {evaluation of the PID controller output}
  U0:=q2*U;         {evaluation of the first time delay unit input}
  U2:=q1*U+U1+Y;    {evaluation of the second time delay unit input}
  DA(Y);            {PID controller output is put on the D/A converter}
  {PART 2: realization of the unit delay operations:}
  U1:=U0;
  U3:=U2;
 end
 until 1=0;         {an infinity loop is realized by this condition}
end.
```

Figure 2.24 Realization of the PID controller in PASCAL.

only statement, which needs an explanation is the evaluation of the PID controller input `U` which is obtained as a difference between the reference signal `W` and the process output supplied by the analog to digital converter (hatched connections in Fig. 2.23).

## Exercises

### Problem 2.6

Use the implicit function generation technique for the following examples:

(a) $y = t \sin t$

(b) $y = \sin[\exp(-at)]$
(c) $y = At\exp(-at)\sin t$
(d) $y = it\exp(it),\ i = \sqrt{-1}$
(e) $\dot{y} = a\sqrt{1 + y^2} \qquad y(0) = 1$

Also, generate simulation schemes using the indirect approach.

**Problem 2.7**

Use the indirect approach in solving the following differential equations:

(a) Bessel's equation:

$$t^2\ddot{y} + t\dot{y} + (t^2 - n^2)y = 0$$

$$n = 1 \qquad y(0) = 0.5 \qquad \dot{y}(0) = 0$$

(b) Legendre's equation:

$$(1 - t^2)\ddot{y} - 2t\dot{y} + n(n + 1)y = 0$$

$$n = 1 \qquad y(0) = 1 \qquad \dot{y}(0) = 0$$

(c) Weber's equation:

$$\ddot{y} + \left(1 + \frac{t^2}{4}\right)y = 0 \qquad y(0) = 20 \qquad \dot{y}(0) = 5$$

(d) Rayleigh's equation:

$$\ddot{z} + r\left[\frac{\dot{z}^3}{3}\right] - \dot{z} + z = 0$$

$$z(0) = 0.3 \qquad r = 0.5 \qquad \dot{z}(0) = 0$$

(e) Van der Pol's equation:

$$\ddot{y} + r(y^2 - 1)\dot{y} + y = 0$$

$$r = 0.7 \qquad y(0) = 4 \qquad \dot{y}(0) = 0.4$$

(f) $\ddot{x} + 40\dot{x} + 400(x + ry) = 0$

$$\ddot{y} + 40\dot{y} + 400(y - rx) = 0$$

$$r = 0.5 \qquad \dot{x}(0) = \dot{y}(0) = 10 \qquad x(0) = y(0) = 0$$

(g)

$$\ddddot{y} + 6\dddot{y} + 12.5\ddot{y} + 60\dot{y} + 2500y = 0$$

$$y(0) = 9 \qquad \dot{y}(0) = 0 \qquad \ddot{y}(0) = -600 \qquad \dddot{y}(0) = 600$$

(h) $\dot{x}_1 = x_3$

$\dot{x}_2 = x_4$

$\dot{x}_3 + 50x_1 - 10\cos(x_2) + 0.1x_3 - 175 = 0$

$\dot{x}_4 + 10\dfrac{\sin(x_2)}{x_1} + 0.1x_4 = 0$

$x_1(0) = 5 \qquad x_2(0) = 45 \qquad x_3(0) = x_4(0) = 0$

**Problem 2.8**

Use the two approaches for the simulation of the following transfer functions:

(a) $G(s) = \dfrac{s - 1}{s^2 + 3s + 2}$

(b) $G(s) = \dfrac{s + 2}{s^3 + s^2 + 7s + 1}$

(c) $G(s) = \dfrac{s^2 + 5s + 1}{s^2 + s + 1}$

(d) $G(s) = \dfrac{1 - (T/2)}{1 + (T/2)} \qquad T = 4$

**Problem 2.9**

Use the two approaches discussed for the simulation of analog filters with the following transfer functions:

(a) Second order Butterworth lowpass with cut off frequency 50 Hz:

$$G(s) = \frac{98\,696}{s^2 + 444s + 98\,696}$$

(b) First order Butterworth bandstop with centre frequency 50 Hz and bandwidth 10 Hz:

$$G(s) = \frac{s^2 + 98\,696}{s^2 + 63s + 98\,696}$$

(c) Second order Chebyshev bandpass with 1 dB ripple in the passband, centre frequency 10 Hz and bandwidth 5 Hz:

$$G(s) = \frac{1088s^2}{s^4 + 34.49s^3 + 8983s^2 + 136\,146s + 15\,585\,454}$$

(d) Second order Chebyshev lowpass with stopband ripple 50 dB down from the peak of the passband and cut off frequency 10 Hz:

$$G(s) = \frac{0.003\,16s^2 + 24.97}{s^2 + 7.055s + 24.97}$$

(e) Second order elliptic lowpass with 1 dB ripple in the passband, 30 dB stopband and cut off frequency 50 Hz:

$$G(s) = \frac{0.035\,56s^2 + 110\,530}{s^2 + 338.5s + 110\,530}$$

(f) Second order elliptic lowpass with 3 dB ripple in the bandpass, stopband 50 dB down from the peak value in the passband and cut off frequency 10 Hz:

$$G(s) = \frac{0.004\,453s^2 + 2804}{s^2 + 40.43s + 2804}$$

**Problem 2.10**

Use the two approaches discussed for the simulation of digital filters with the following transfer function which are discrete versions of the analog filters of Problem 2.9 for a sampling rate of 500 Hz!

(a) $$G(z) = \frac{0.1454z^{-1} + 0.1079z^{-2}}{1 - 1.1582z^{-1} + 0.4115z^{-2}}$$

(b) $$G(z) = \frac{1 - 1.6335z^{-1} + 0.9924z^{-2}}{1 - 1.5227z^{-1} + 0.8816z^{-2}}$$

(c) $$G(z) = \frac{0.0021z^{-1} - 0.0022z^{-2} - 0.0020z^{-3} + 0.0021z^{-4}}{1 - 3.8982z^{-1} + 5.7310z^{-2} - 3.7659z^{-3} + 0.9333z^{-4}}$$

(d) $$G(z) = \frac{0.0032 - 0.0063z^{-1} + 0.0032z^{-2}}{1 - 1.9859z^{-1} + 0.9860z^{-2}}$$

(e) $$G(z) = \frac{0.0356 + 0.1074z^{-1} + 0.1667z^{-2}}{1 - 1.1985z^{-1} + 0.5081z^{-2}}$$

(f) $$G(z) = \frac{0.0045 - 0.0034z^{-1} + 0.0097z^{-2}}{1 - 1.9116z^{-1} + 0.9223z^{-2}}$$

## BIBLIOGRAPHY

FISHMAN, G.S. (1973), *Concepts and Methods in Discrete Event Digital Simulation*, John Wiley, NY.

GILOI, W.K. (1975), *Principles of Continuous System Simulation*, B.G. Teubner, Stuttgart.

ISERMANN, R. (1989), *Digital Control Systems*, 2nd edn, Springer, Berlin.

JACKSON, A.S. (1960), *Analog Computation*, McGraw-Hill, London.

NEELAMKAVIL, F. (1987), *Computer Simulation and Modelling*, John Wiley, Chichester.

OGATA, K. (1990), *Modern Control Engineering*, Prentice Hall, Englewood Cliffs, NJ.

STANLEY, W.D. (1975), *Digital Signal Processing*, Reston Publishing, Reston, VA.

# 3

# Basic Concepts in Continuous System Simulation

Modern simulation methodologies can only be used with the aid of modern simulation systems. So in this chapter we will describe the basic features and classification of simulation systems. The concept of digital continuous simulation systems will be described briefly, enabling the implementation of simulation concepts by simulation as well as general purpose programming languages. The chapter will conclude with a description of the software structure of modern simulation languages.

## 3.1 BASIC FEATURES AND CLASSIFICATION OF SIMULATION SYSTEMS

Modern simulation tools enable simulation of dynamical systems in such a way that the user can concentrate mainly on the problems of modelling and simulation and, to a lesser extent, on the programming problems and techniques. From the user's point of view the following are important properties when analysing different simulation tools:

- the time needed to become familiar with a simulation tool;
- user friendliness;
- the time needed to develop the simulation model;
- the possibilities in the phase of model changing;
- the capability during run time (level of interactiveness, accuracy, speed, possibilities for real time simulations with hardware-in-the-loop, ...);
- the possibilities for model documentation, results documentation, etc.

With regard to the software and hardware equipment used in modern digital simulation, simulation tools can be divided into three main groups. *Simulation systems on general purpose digital computers* can be included in the first. These range from microcomputers to supercomputers that work on the so-called von Neumann principle. The second group is represented by complex *digital computer systems which are based on parallel execution of operations*. Such simulation systems demand special hardware and software equipment and are used in particular for the simulation of complex problems and for special real time simulations. The *analog–hybrid systems* form the third group. Although at several times in the past it seemed that analog–hybrid simulation was dying, these systems were once again revived in the 1980s, including the most powerful analog–hybrid multiprocessor system SIMSTAR and some other smaller and cheaper systems. Analog–hybrid systems are used for special problems (real time simulations with hardware-in-the-loop, optimization, simulation of discontinuous systems, algebraic loops, etc.) but, also, their pedagogical effectiveness cannot be neglected.

In this book consideration is given to digital simulation systems on general purpose computers. The essential features of hybrid systems will be presented in Section 4.6. Special purpose simulation systems will be discussed in Section 6.2.

Digital simulation systems on general purpose computers significantly influenced the development of simulation as a modern technique in the 1980s. The installation of an appropriate simulation system on one or several computers of the same type, or even on different computers, represents the cheapest simulation possibility for a wide range of users. Thus simulation can be used successfully in educational as well as in research processes. Such systems have perhaps only one important limitation: they cannot usually be used for real time simulations.

Schmidt (Schmidt, 1986) proposed the following classification of simulation systems:

- simulation packages; and
- simulation languages.

*Simulation packages* represent the older types of simulation systems. They are composed of a main program and a library of subroutines. The user must describe the simulation model in the language of the package (e.g. FORTRAN, PASCAL) by introducing the subroutine calls which carry out the desired functions of the simulation program. The user can also include his own modules in the library. The advantage of such a system is that the source code is accessible and the user is usually very familiar with it. This gives him a high degree of flexibility. On the other hand, the disadvantage is that the user must be familiar with programming in the language in which the package is written. This fact may lead indirectly to a number of errors being made during the model development.

In comparison with packages, *simulation languages* realize their functions by the assistance of program statements which must be appropriately processed before simulation. The advantage of a simulation language is that it is much easier to use, which is extremely important for unskilled users. But the distinct disadvantage of the languages arises from the fact that the language elements cannot be modified or expanded easily.

The following can be derived from a comparison of languages and packages:

- The unexperienced user who would like to solve standard problems quickly and easily should use a language.
- Some advanced and more sophisticated simulation studies which require nonstandard approaches should be handled by packages because of the frequent requirement for expandability and flexibility.

## 3.2 THE CONCEPT OF DIGITAL CONTINUOUS SIMULATION SYSTEMS

In order to simulate continuous systems on a digital computer, the independent variable must be discretized so that differential equations become difference equations. As all variables are functions of the independent variable $t$, they are defined only in discrete values (moments) of the independent variable. If simulation operates with constant independent-variable increments $\Delta t$ (calculation interval), these points can be expressed as $t_i = t_0 + i\,\Delta t$, $i = 0, 1, 2, \ldots, i_{max}$. Thus the simulation run begins at time $t = t_0$ and terminates at time $t = t_{max} = t_0 + i_{max}\,\Delta t$.

The basic equation describing the simulated system originates in the indirect approach introduced in Section 2.1. The first equation corresponds to the first step and the second equation to the second step of the indirect approach (see p. 27):

$$\dot{\mathbf{x}}(t) = \mathbf{f}(t, \mathbf{x}(t)) \tag{3.1}$$

$$\mathbf{x}(t) = \int \dot{\mathbf{x}}(t)\,\mathrm{d}t \tag{3.2}$$

Eqns (3.1) and (3.2) can be simulated using the simulation scheme shown in Fig. 3.1 where the vector $\mathbf{x}$ represents the state variables and $\dot{\mathbf{x}}$ their derivatives.

To simulate the scheme given in Fig. 3.1 on a digital computer, every simulation system must possess the following two important features:

1. Eqn (3.2) must be realized with a numeric integration algorithm. Integration is the heart of each simulation system. It can be realized in a very simple manner but it can also be very complicated and sophisticated, which is the case for modern numerically powerful simulation tools. Euler's algorithm is very suitable for elucidating the numerical integration procedure:

$$\mathbf{x}(t + \Delta t) = \mathbf{x}(t) + \mathbf{f}(t, \mathbf{x}(t))\Delta t \tag{3.3}$$

   It can be seen that the integration procedure uses current values of states $\mathbf{x}(t)$ and derivatives $\mathbf{f}(t, \mathbf{x}(t))$ to evaluate the states for the next calculation interval $\mathbf{x}(t + \Delta t)$.
2. Due to the fact that, unlike analog computers, general purpose digital computers do not perform parallel computations, all the loops shown in Fig. 3.1 must be broken in some way so that all the variables in each loop can be calculated sequentially. The loops are broken at the outputs of

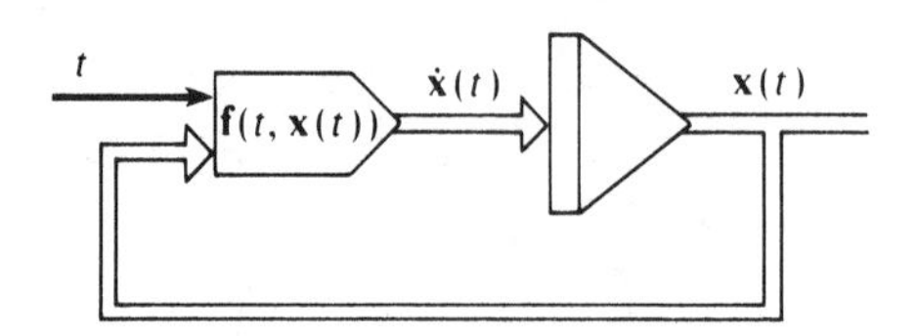


Figure 3.1 The basic scheme for simulation.

so-called memory blocks (sometimes also referred to as blocks with delay attributes which define states). In continuous simulation, such blocks are usually integrators but in the general case they can also be blocks having the property that their current output value is not influenced by their current input value (e.g. delay, zero order hold, lag function, transfer function with numerator degree less than denominator degree, etc.).

If the loop of the system, which is represented by the scheme shown in Fig. 3.1, is broken at the output of the integrator and the continuous integration is replaced by the numeric one, the basic digital simulation concept is evident and is shown in Fig. 3.2. At the beginning of a simulation run the initialization procedure must be accomplished, which means that, along with other procedures, the initial values of states $\mathbf{x}(t)$ must be defined.

The scheduling algorithm of continuous simulation languages is very simple in comparison with that of discrete event simulation languages and is given by a two-step iteration: the first step consists of the evaluation of all derivatives (evaluation of all model definition equations) and the second includes the integration procedure, which evaluates the state variables for the next calculation interval. This two-step iteration is usually implemented in simulation systems with two subprograms (DERIV, INTEG).

The basic concept of digital simulation systems is shown in Fig. 3.3. The simulation run is executed by the integration procedure (INTEG subprogram) call after presimulation operations. During simulation the integration procedure requires many evaluations of state derivatives (depending on the integration algorithm). In the prescribed time instants the control is given to the OUTPUT subprogram which

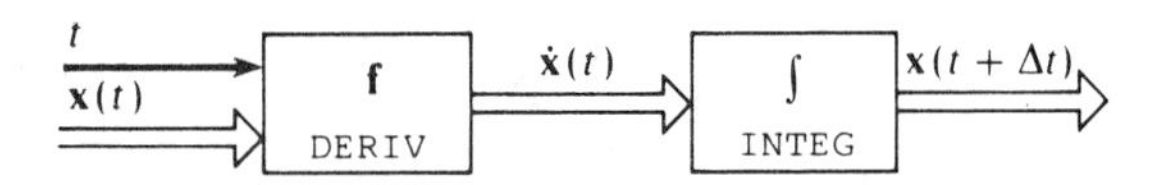


Figure 3.2 The basic digital simulation concept.

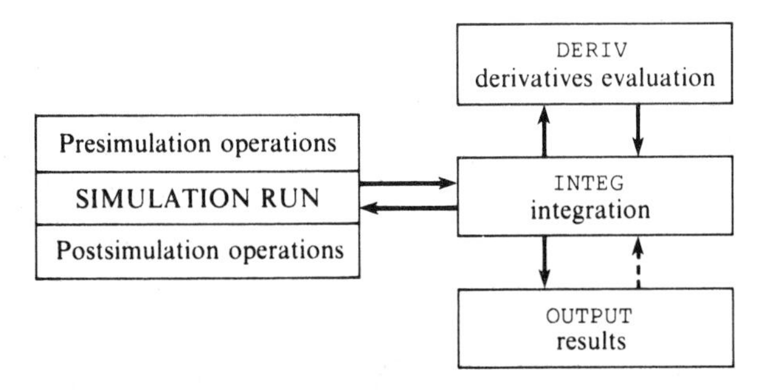


Figure 3.3 The basic concept of digital simulation systems.

supplies the user with simulation results. After the simulation run some postsimulation operations are executed (e.g. the processing of simulation results).

The concept described is common to simulation languages and packages. As simulation languages represent a more convenient tool for a wider spectrum of problems, the following chapters will focus on this type of simulation system.

## 3.3 BASIC FEATURES AND CLASSIFICATION OF DIGITAL SIMULATION LANGUAGES

Simulation languages are modern simulation tools. They are widely used because they are relatively cheap. Languages are rather independent of the hardware and system software equipment, which represents their most important advantage. Unfortunately, they also have disadvantages, namely, the very limited simulation speed sometimes makes the languages unsuitable for real time applications.

Regarding the universality in the phase of model definition and the executable system properties simulation languages can be divided into the following

- general purpose simulation languages; and
- special purpose or problem oriented simulation languages.

*General purpose simulation languages* (e.g. ACSL, CSSL IV, TUTSIM, CSMP) use a model definition which is quite independent of the model type. *Special purpose or problem oriented simulation languages*, on the other hand, are used for efficient modelling and simulation in special applications (e.g. simulation languages for the simulation of electric and electronic circuits – SPICE, PSPICE, ECAP, etc.).

In the development and use of simulation languages the following questions frequently appear:

- Is it reasonable to develop a simulation language of high generality?
- Is it possible to estimate the degree of simulation language generality precisely?
- Is the use of a general purpose simulation language justified from a practical point of view?

There is no doubt that the use of general purpose simulation languages is advantageous because models can be defined in a very direct way using physical laws and mathematical relations. Thus these languages can be used in practically all fields (technical and nontechnical, e.g. economics, sociology) but their users must have a basic knowledge of modelling and simulation methods. It is also important that users are aware of the fact that these languages are general because they can deal with general mathematical and physical relations and not because they would be convenient for solving general (all) problems. Thus general purpose simulation languages are suitable for solving problems which are not very specific and not very complex. On the other hand, problem oriented languages can be used for routine solution of specific and complex problems in the absence of a deep understanding of modelling and simulation problems. But they are inconvenient for the solution of general purpose problems.

One of the significant characteristics of modern simulation languages is the possibility of user defined submodels for solving specific problems (e.g. macro language). In this way a completely new and much more convenient language can be generated for specific problems. However, the efficiency of mathematical algorithms and some other functions of simulation languages to which the user has no access is questionable. Such languages are often called *high level general purpose simulation languages* due to the fact that complex operators can be used for model descriptions. *Low level general purpose simulation languages*, however, possess only basic operators, the elements for program control, the appropriate numerical algorithms and possibilities for the presentation of results.

Because this book is mainly devoted to general purpose problem solving, emphasis will be placed on general purpose languages which attain such a level that they can also be used for the modelling and simulation of relatively complicated problems in all domains of science and technology. Their use is especially important in the field of education.

With regard to the type of model simulated, simulation languages can be divided into the following:

- continuous languages;
- discrete languages; and
- combined languages.

The latter classification is important in the modelling sense as well as in the sense of simulation tool development because different types of language demand very different concepts in simulation language development.

Although the accent of this book is given to continuous simulation systems, the important features of all three approaches will be given briefly because any simulation of a real dynamic system usually demands elements of combined simulation, although the problem is treated as continuous.

### 3.3.1 Continuous Simulation Languages

Continuous simulation languages are used for the simulation of continuous dynamic systems. They also possess some possibilities for combined simulation, but only in the sense of the realization of discontinuities and not in the sense of correct numerical treatment of such phenomena.

With respect to the types of differential equations, which can be ordinary or partial, simulation languages can be classified as follows:

- languages for simulation problems using ODE (Ordinary Differential Equations); and

- languages for simulation problems described using PDE (Partial Differential Equations).

*The ODE simulation languages* attained a high level of perfection. On the other hand, it has been difficult up until now to find a general purpose *simulation language for PDE*, although the first of these products came on to the market around 1970. Numerical problems in PDE systems are much more complicated, as can be seen from the definition of integration algorithm efficiency for solving specific problems (Cellier, 1979):

$$\eta(a, p) = \frac{t_{CPU}(a^*, p)}{t_{CPU}(a, p)} \tag{3.4}$$

where $\eta$ = the efficiency of integration algorithm; $a$ = the notation of the used integration algorithm; $a^*$ = the notation of the optimal algorithm for a problem; $p$ = the notation of the simulation problem; $t_{CPU}$ = the CPU time needed to solve a problem.

In solving ODE problems, the efficiency $\eta$ rarely achieves a value lower than 0.01 for any problem and any algorithm except in the case of very stiff and oscillating problems. In solving PDE problems the efficiency $\eta$ is much lower. Taking into account the most frequently used methods for elliptic problems, it can be shown that the value of $\eta$ is between $10^{-3}$ and $10^{-6}$, which means that it is much more important to use an appropriate integration algorithm correctly for PDE than for ODE problems. Methods for solving PDE problems are so heterogeneous that it is questionable if it is possible to design a general purpose PDE simulation language at all.

Another classification is important. A language can be one of the following two types:

- interpreter oriented; or
- compiler oriented.

In an *interpreter oriented language*, the simulation program is processed into tables that describe the structure and parameters of the model. Using these tables the interpreter performs simulation, executing those blocks that describe the model in appropriately sorted order. The user can change the structure and parameters of the model interactively without reprocessing. This interpretative way represents some advantages in the phase of model development where the structure must be changed several times. One disadvantage of the approach is a decrease of simulation speed, which is unsuitable for the simulation of tested and complex models. Interpreter oriented languages are particularly unsuitable when the simulation run represents only part of a complex experiment (e.g. optimization, robustness analysis, etc.). Another disadvantage is the fact that the user is not able to include his own blocks because the interpreter language consists of an executable program. The user thus cannot link with a simulation language library, which makes such languages rather unflexible for complex modelling and simulation. Yet another disadvantage of interpreter oriented languages is that they are almost always block oriented, which means that the model can be described only by relatively simple, elementary blocks. For model description the user must define the types of blocks, their parameters and connections. Thus such programs are huge, difficult to survey, nonmodular and represent poor model documentation. Some of the problems stated above are solved using an enormous number of blocks which are included in a language. Such languages are called compact languages, which means that the user rarely needs to include a new block. On the other hand, such huge numbers of blocks represent difficulties when learning the simulation language.

*Compiler oriented languages* have fewer disadvantages. Some of them process the source program directly into machine code. Because such processing is rather fast, they enable a high level of interactiveness, even in the case of model structure changes. However, such languages are nonexpandable and are therefore limited in solving extremely complex problems. Thus many modern simulation languages translate the source model into a high level general purpose programming language (e.g. FORTRAN, PASCAL). Using this concept the language is easily expanded with new translator target language (e.g. in FORTRAN or PASCAL) written simulation modules. Compiler languages usually offer more and better integration methods, the possibility for solving algebraic loops, optimization possibilities, etc. They also offer greater portability than other computers and operating systems, faster simulation and great flexibility in solving complex problems. The executable program generated by the processing procedure is independent of the simulation language itself, so these languages are very suitable for building application programs. However, the processing procedure, which requires two compilations and linking, takes a great deal of time, so that this approach is inferior, especially in cases where the modelling demands frequent changes in model structure. Therefore, compiler oriented languages are not very convenient in the phase of model development and testing.

Whereas interpreter oriented simulation languages are almost always block oriented, compiler languages can be divided into *block oriented* (explicit block structure) and *equation oriented* languages (implicit block structure). In equation oriented languages the model can be described with arbitrary mathematical expressions, which means that differential equations can be included directly in a program. So the modelling is much easier and faster, the simulation programs are shorter, much more understandable and modular. Some modern simulation languages possess interpreter and compiler oriented modes of operation.

Fig. 3.4 shows the complete classification of simulation systems for general purpose computers. In this book only the systems that are illustrated in this diagram will be discussed in more detail.

To demonstrate the applicability of equation oriented continuous systems simulation languages we shall represent this type of language by two examples already introduced in Section 1.5. The prey and predator species ecosystem model and the heating control problem will be simulated, analysed and discussed using a CSSL'67 standard simulation language (Strauss, 1967). Examples of simulation languages will be described in Sections 4.2 and 4.3 and the CSSL'67 standard will be briefly discussed in Section 4.1.

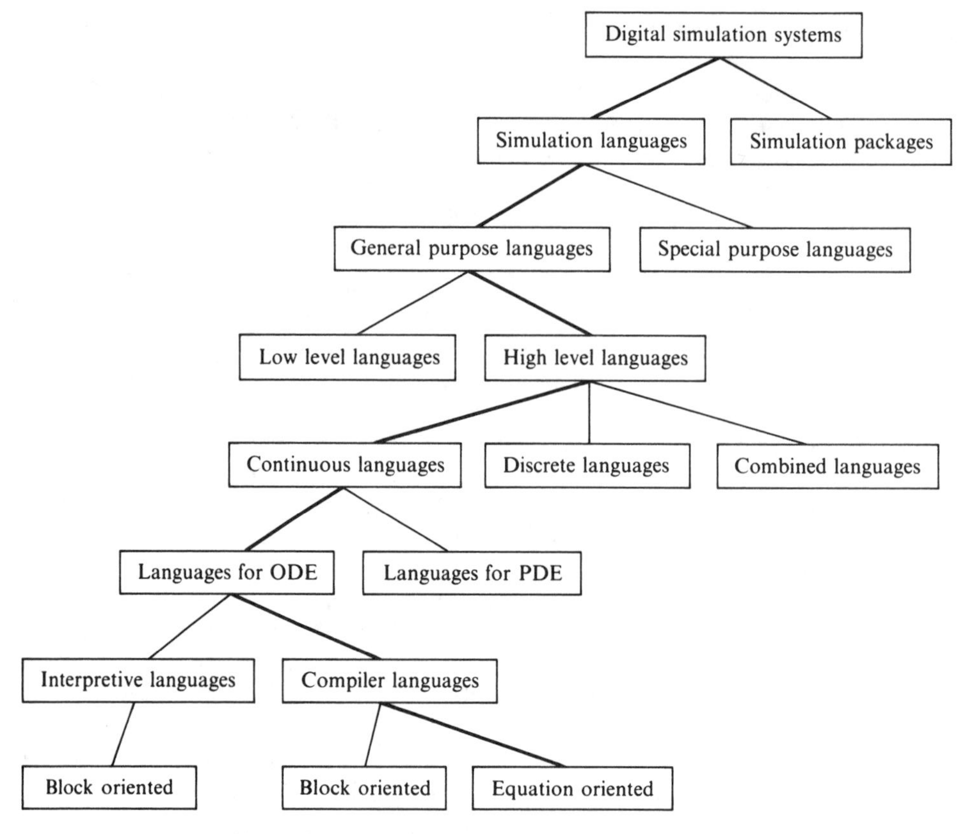


Figure 3.4 Types of digital simulation systems.

**Example 3.1 Prey and predator problem**

The purpose of this example, introduced in Section 1.5, is to acquaint the reader with digital simulation languages. The problem with model constants $a_{11} = 5$, $a_{12} = 0.05$, $a_{21} = 0.0004$, $a_{22} = 0.2$ and with the initial number of rabbits and foxes $x_{10} = 520$, $x_{20} = 85$ will be simulated with the aid of the simulation scheme shown in Fig. 2.7 or by using Eqn (2.6) directly. Before writing the simulation program the simulation program variables must be defined. It is often convenient to define them in such a way that there is obvious correspondence with the physical variables. The following correspondence with the simulation scheme shown in Fig. 2.7 is chosen:

| | | | |
|---|---|---|---|
| $x_1$ | RAB | $\dot{x}_1$ | RABDOT |
| $x_2$ | FOX | $\dot{x}_2$ | FOXDOT |
| $x_{10}$ | RAB0 | $x_{20}$ | FOX0 |
| $a_{11}$ | A11 | $a_{12}$ | A12 |
| $a_{21}$ | A21 | $a_{22}$ | A22 |

The simulation program consists of statements. The order of statements in CSSL standard languages is arbitrary but for good documentation purposes the following order is recommended:

1. parameter definition statements;
2. structure definition statements;
3. simulation run control statements.

The first statement in the simulation program can be the PROGRAM statement, giving a name to the program:

```
PROGRAM PREY_AND_PREDATOR
```

The constants of the simulation model can then be defined as follows:

```
"Model's constants
  CONSTANT A11=5,A12=0.05,A21=0.0004,A22=0.2
  CONSTANT RAB0=520,FOX0=85
```

As we can see, the comment introduced with a " sign can be used. Such comments enable good documentation of the simulation model.

After parameter definition the order of the statements is arbitrary. These statements can be written using the simulation scheme shown in Fig. 2.7. But due to the fact that CSSL standard languages are equation oriented, Eqn (2.6), which describe the corresponding derivatives, can be used directly:

```
RABDOT=A11*RAB-A12*RAB*FOX
FOXDOT=A21*RAB*FOX-A22*FOX
```

The derivatives must be integrated with INTEG statements:

```
RAB=INTEG(RABDOT,RAB0)
FOX=INTEG(FOXDOT,FOX0)
```

The first parameters of INTEG statements represent the input variables and the second parameters represent the initial conditions (initial number of rabbits and foxes).

After model structure definition the simulation run control parameters must be defined. It is very important to choose an appropriate duration of simulation run, i.e. the correct interval of the independent variable (time). Since years are included in the model constants ($a_{11} = 5$ is the number of young per rabbit in a year), the independent variable unit is one year. Because we want to obtain information for at least one ecological period, which is in this case approximately seven years, a simulation run duration of ten years is chosen. The termination condition is specified with the TERMT statement:

```
TERMT T.GE.10
```

which means that the termination condition is fulfilled when the simulation time is equal to, or exceeds, ten years. However, if it is intended to change the duration of the simulation interactively during particular simulation runs, the logical termination condition must be defined as a function of a variable with a value which is initialized in the CONSTANT statement:

```
CONSTANT TFIN=10
TERMT T.GE.TFIN
```

The communication interval, which defines the time intervals when the user can obtain results is defined by the statement

```
CINTERVAL CI=0.01
```

which means that the user obtains results every 0.01 year (3.65 days). Finally, we must define which variables are to be presented on the screen during simulation (OUTPUT statement) and which variables are to be stored on the data file (PREPAR statement) for further processing:

```
OUTPUT 100,RAB,FOX
PREPAR 2,RAB,FOX
```

```
''-----------------------------------------------
PROGRAM PREY AND PREDATOR
''-----------------------------------------------
''Model's constants
CONSTANT A11=5,A12=0.05,A21=0.0004,A22=0.2
CONSTANT RABO=520,FOXO=85
''-----------------------------------------------

''The structure of the model
RABDOT=A11*RAB-A12*RAB*FOX
FOXDOT=A21*RAB*FOX-A22*FOX
RAB=INTEG(RABDOT,RABO)
FOX=INTEG(FOXDOT,FOXO)
''-----------------------------------------------

''The length of the simulation run
CONSTANT TFIN=10
TERMT(T.GE.TFIN)
''-----------------------------------------------
''The definition of the communication interval
CINTERVAL CI=0.01
''-----------------------------------------------
''Output specification
OUTPUT 100, RAB,FOX
PREPAR 2,RAB,FOX
''-----------------------------------------------
END
```

Figure 3.5 The CSSL standard program for the prey and predator problem.

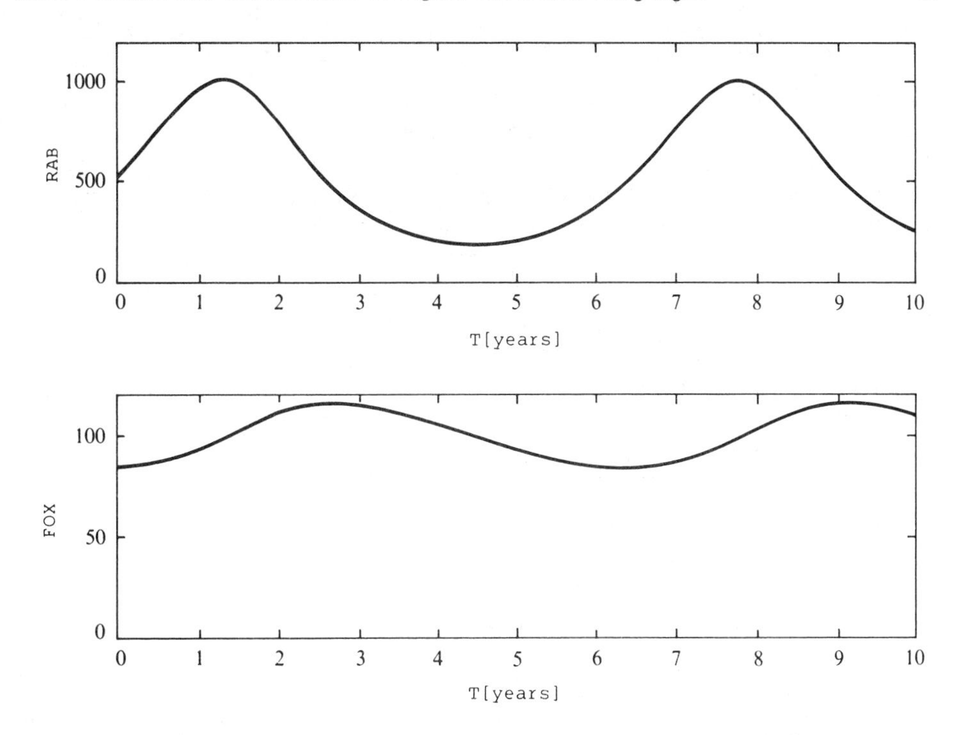


Figure 3.6 The time histories of the rabbits and foxes population.

The number `100` in the first statement means that each one-hundredth simulation result is presented on the screen (meaning one presentation every year) and the number `2` in the second statement means that every second result (every 7.3 days) is stored on the data file. The simulation program must be terminated by the `END` statement

```
END
```

CSSL-type languages are compiler oriented. So the model described is first translated, usually into high level general purpose language modules (e.g. FORTRAN) and data base files, and then into an execution program using the standard compiler and linker. After that, the simulation run commences and the results (appropriate value of time, the number of rabbits and foxes) are presented on the screen. They can also be presented in graphic form after termination of the simulation run.

The complete simulation program for the prey and predator problem is shown in Fig. 3.5, while Fig. 3.6 depicts the time histories of the rabbit and fox populations. It can be seen that the ecological period is approximately seven years and that the maximum fox population increases and decreases corresponding to the maximum and minimum of the rabbit population. □

### Example 3.2 Heating control problem

The purpose of this example is to improve the reader's knowledge of the CSSL-type simulation language and to show the sequence of steps which are usually performed in a simulation study.

The heating control problem was introduced in Section 1.5. The block diagram of the problem without the additional time delay is shown in Fig. 3.7.

Eqn (2.1) describes the working point model of the process. The width of the hysteresis function is $\Delta y = 1\ ^\circ\text{C}$ and the power of the heater is $p = p_{max} = 5$ kW. The working point temperature (the environment temperature) is $\vartheta_e = 15\ ^\circ\text{C}$, the gain of the process model is $k = 2\ ^\circ\text{C/kW}$ and its time constant is $T = 1$ h. The reference temperature is defined by Table 3.1 as a function of time. Again, the names of the simulation program variables, corresponding to the physical variables, must be defined as follows:

| | | | |
|---|---|---|---|
| $\vartheta_r$ | THR | $\vartheta_e$ | THE |
| $\vartheta$ | TH | $\vartheta_w(0)$ | THWO |
| $\dot{\vartheta}_w$ | THWD | $\vartheta_w$ | THW |
| $e$ | E | $p$ | P |
| $p_{max}$ | PMAX | $\Delta y$ | DELTAY |
| $u$ | U | $T$ | TIMCON |
| $k$ | GAIN | | |

The model parameters (constants, breakpoints of function generators) are usually defined at the beginning of the simulation program:

```
"Model's constants
   CONSTANT DELTAY=1,PMAX=5,GAIN=2,TIMCON=1
   CONSTANT THE=15,THWO=1
"Function generator's breakpoints
   TABLE REF,1,9,...
    0.,  5.99,  6.,  8.99,  9., 14.99, 15., 20.99, 21.,...
   15., 15.,   20., 20.,   18., 18.,   20., 20.,   15.
```

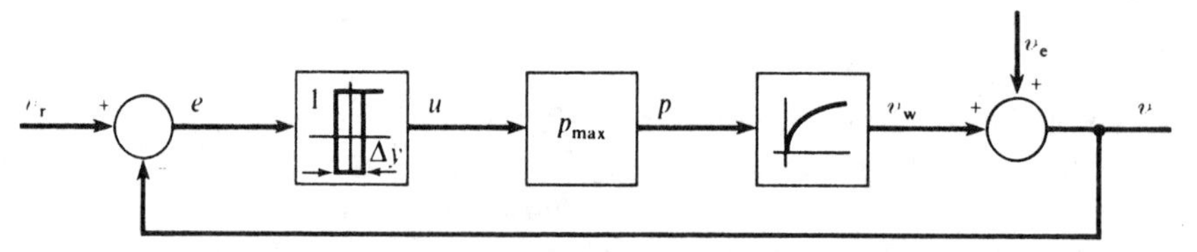


Figure 3.7 Block diagram of the heating control problem.

Table 3.1 The room reference temperature

| $t(h)$ | 0 | 5.99 | 6 | 8.99 | 9 | 14.99 | 15 | 20.99 | 21 |
|---|---|---|---|---|---|---|---|---|---|
| $\vartheta_r[^\circ\text{C}]$ | 15 | 15 | 20 | 20 | 18 | 18 | 20 | 20 | 15 |

The function generator is specified by its name, by the number of independent variables, by the number of breakpoints and by the set of values of independent and dependent variables in the breakpoints. The model structure must then be described in an arbitrary statements order. The reference signal is realized by function generator call

```
THR=REF(T)
```

where `T` is the independent variable (time) and `REF` is the name of the function generator, defined in the `TABLE` statement.

The error defined by the statement

```
E=THR-TH
```

is fed into the bang-bang controller thermostat, characteristics of which can be modelled using the hysteresis function with the width equal to $\Delta y$, low value equal to zero and high value equal to one:

```
U=HSTRSS(E,-DELTAY/2.,DELTAY/2.,0.,1.,0.)
```

The last parameter in this function is the initial condition, which is used for output value if the initial control error $e$ lies within the range $-\Delta y/2 \leqslant e \leqslant \Delta y/2$.

The heater can be simulated as a gain:

```
P=PMAX*U
```

The process is simulated with the aid of the simulation scheme shown in Fig. 2.3 or by using Eqn (2.2):

```
THWD=-1./TIMCON*THW+GAIN/TIMCON*P
THW=INTEG(THWD,THW0)
```

The second parameter of the `INTEG` function is the initial condition of the working point model. To obtain the absolute value of the temperature, the working point temperature must be added to the working point model temperature:

```
TH=THW+THE
```

After model structure definition, the simulation run control parameters must be defined. Since observation of relations over twenty-four hours is desired, the simulation run termination condition can be specified by the statements

```
CONSTANT TFIN=24
TERMT T.GT.TFIN
```

The integration algorithm is at the heart of each simulation system. If it is not specified explicitly, the default algorithm (usually Runge–Kutta with adaptive calculation interval) is chosen. But in models that include the hysteresis function, an algorithm with fixed calculation interval is recommended and can be specified by the statement

```
ALGORITHM IALGOR=1,JALGOR=5
```

The first parameter (IALGOR=1) is used for the selection of the integration method initialization procedure and the second parameter (JALGOR=5) for the selection of the integration method.

The communication interval during which the user can obtain results is defined by the statement

```
CINTERVAL CI=0.02
```

which means that the results can be obtained every 0.02 h (1.2 min). It must be emphasized again that all units in the program must be consistent, which means that all parameters that include time must be expressed in the same units (e.g. model constants, the duration of simulation run, the communication interval). If the statement CINTERVAL is not used, the default value (1) of the communication interval is used. Finally, the variables to be presented on screen during simulation and those to be stored on data file are defined by

```
OUTPUT 10,THR,P,TH
PREPAR THR,P,TH
```

The simulation program must be terminated by the statement END:

```
END
```

After the processing procedure the simulation run commences and the results (the room temperature $\vartheta$ and the power of the heater $p$), which can be obtained in graphic form after the simulation run, are shown in Fig. 3.8. The heater is switched on approximately every 30 min and the room temperature oscillates within a 1 °C boundary. If we double the hysteresis width to 2 °C, the results of the simulation are shown in Fig. 3.9. The oscillations in this case are too high for comfort in the room and the switching period increases to approximately 50 min.

The study is more realistic if we take into account the dead time which appears if the thermostat sensor is placed some distance from the heater. The problem is illustrated by the block diagram in Fig. 1.5. Two additional simulation program variables in obvious correspondence with the physical variables must be defined as follows:

$p_d$ PD
$T_d$ TDELAY

This equation is

```
CONSTANT TDELAY=0.1,WORK=50*0
ARRAY WORK(50)
  PD=DELAY(P,TDELAY,WORK,CI)
```

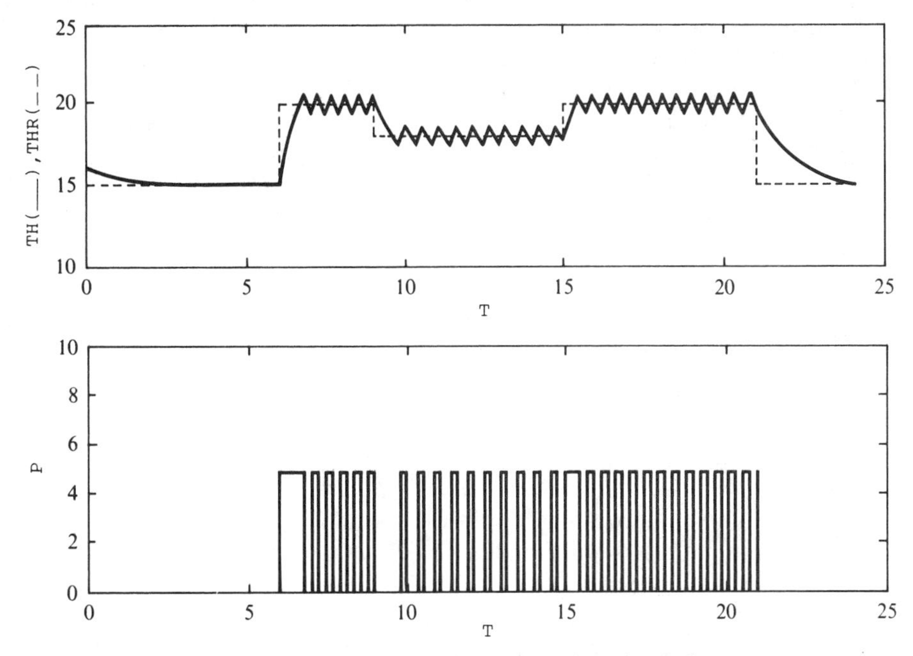


Figure 3.8 The results of room heating simulation.

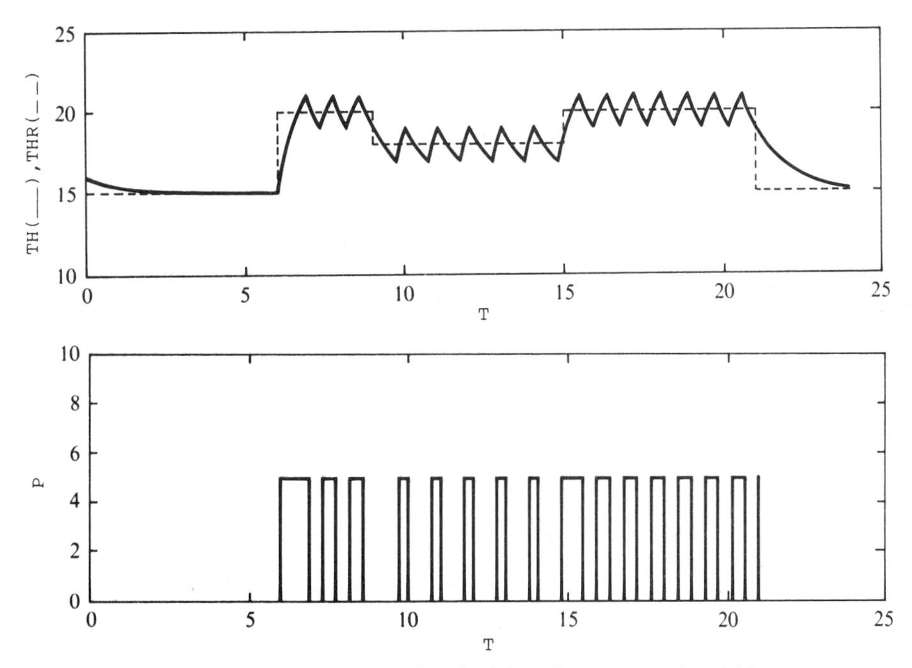


Figure 3.9 The results with double value hysteresis width.

The first parameter of the DELAY operator is the variable to be delayed. The second parameter is the dead time, the third parameter is the working array which is used for dead time realization and the fourth parameter is the sampling (working array shift) interval. The initial values of the working array and its dimension must be specified with CONSTANT and ARRAY statements. To obtain the best possible approximation of the dead time the minimal sampling interval equal to the communication interval CI is chosen. Because the variable $p_d$, which is the output of the DELAY operator, now represents the input of the process, the model process equation

```
THWD=-1./TIMCON*THW+GAIN/TIMCON*P
```

must be changed into equation

```
THWD=-1./TIMCON*THW+GAIN/TIMCON*PD
```

Fig. 3.10 represents the complete simulation program, while Fig. 3.11 shows the simulation results when dead time $T_d = 0.1$ h is introduced.

Comparing these results with those in Fig. 3.8 we can see that the oscillating boundary is wider (approximately 2 °C) and the results are similar to those shown in Fig. 3.9. We can conclude that the temperature oscillating width is influenced by the width of hysteresis and by the dead time. It is evident, that the thermostat is too far from the heater for control of comfort.

Now several simulation runs can be performed without reprocessing the source simulation program. All variables initialized with CONSTANT statements can be changed interactively. It is very simple to simulate an increase in power of the heater: only the constant PMAX must be changed interactively to the new value. Fig. 3.12 represents the results of the simulation if a 10 kW heater is used. Note that the temperature increases faster but the oscillating boundary is wider than in Fig. 3.11 (approx. 3 °C).

The influence of the environmental temperature is examined by changing the model constant THE. Fig. 3.13 shows the results when this temperature is 17 °C. It is obvious that it is impossible to reach a reference temperature which is lower than the environmental temperature without additional cooling.

The next step in our simulation example is the replacement of the linear process with the nonlinear one. The assumption that the model has the same time constant in the phase of heating and cooling is only an approximation. We can also study the situation simply and efficiently when the time constant of the cooling phase differs from that of the heating phase. The PROCEDURAL block represents an efficient method of introducing nonlinearities (in our case the time constant switching mechanism). In our problem the cooling and heating phases can be identified by the model temperature derivative. To obtain the time constant $T = 1$ in the heating phase ($\dot{\vartheta}_w \geq 0$) and the time constant $T = 4$ in the cooling phase ($\dot{\vartheta}_w < 0$), the definition of the time constant

```
CONSTANT TIMCON=1
```

```
''---------------------------------------------------------
''HEATING CONTROL PROBLEM
''---------------------------------------------------------
''Model's constants
    CONSTANT DELTAY=1,PMAX=5,GAIN=2,TIMCON=1
    CONSTANT THE=15,TDELAY=0.1,THW0=1
    CONSTANT WORK=50*0
''Function generator's breakpoints
    TABLE REF,1,9,...
      0.,  5.99,  6.,  8.99,  9., 14.99, 15., 20.99, 21.,...
      15., 15.,   20., 20.,   18., 18.,   20., 20.,   15.
''---------------------------------------------------------
''Working array for delay realization
    ARRAY WORK(50)
''---------------------------------------------------------

''The structure of the model
       THR=REF(T)
       E=THR-TH
       U=HSTRSS(E,-DELTAY/2.,DELTAY/2.,0.,1.,0.)
       P=PMAX*U
       PD=DELAY(P,TDELAY,WORK,CI)
       THWD=-1./TIMCON*THW+GAIN/TIMCON*PD
       THW=INTEG(THWD,THW0)
       TH=THW+THE
''---------------------------------------------------------

''The duration of the simulation run
    CONSTANT TFIN=24
    TERMT T.GT.TFIN
''---------------------------------------------------------
''The definition of the integration algorithm
''and the communication interval
    ALGORITHM IALGOR=1,JALGOR=5
    CINTERVAL CI=0.02
''---------------------------------------------------------
''Output specification
    HDR HEATING CONTROL SYSTEM
    OUTPUT 10,THR,P,TH
    PREPAR THR,P,TH
''---------------------------------------------------------
    END
```

Figure 3.10 Simulation program of the heating control system.

must be replaced by the following part of the program:

```
CONSTANT T1=1, T2=4
PROCEDURAL (TIMCON=THWD)
        TIMCON=T1
        IF (THWD.LT.0) TIMCON=T2
END
```

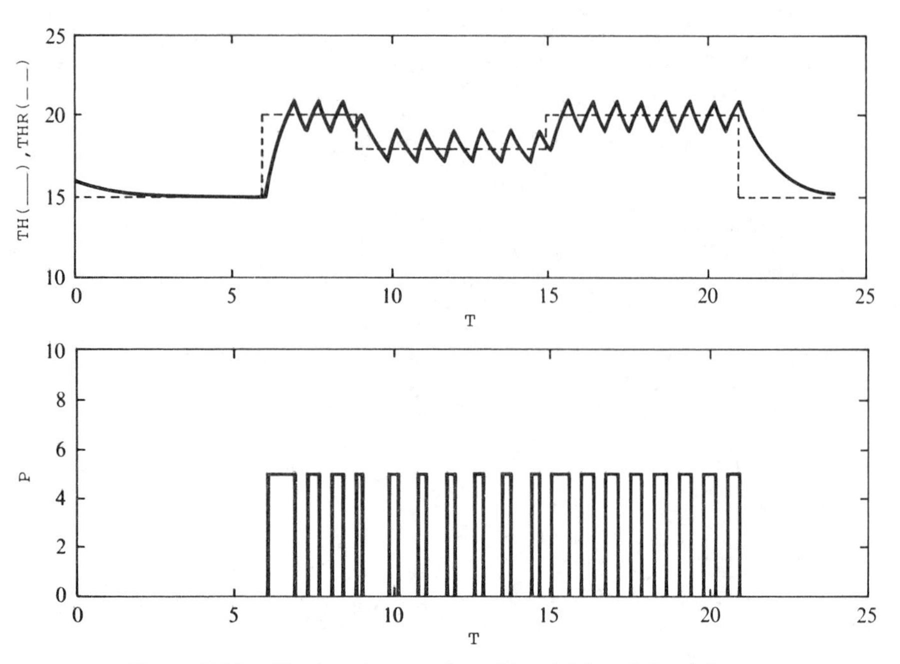


Figure 3.11 The heating results with additional dead time.

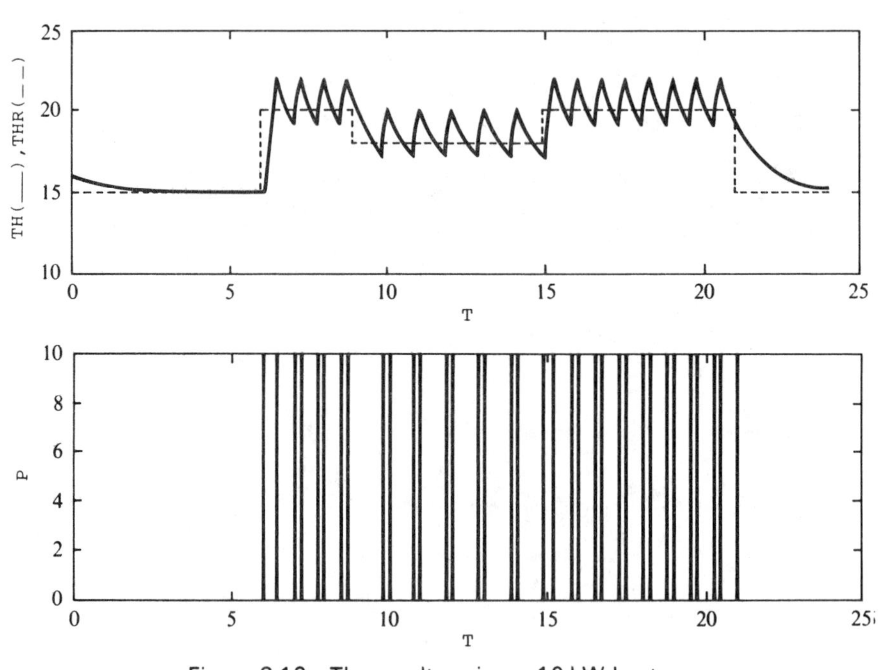


Figure 3.12 The results using a 10 kW heater.

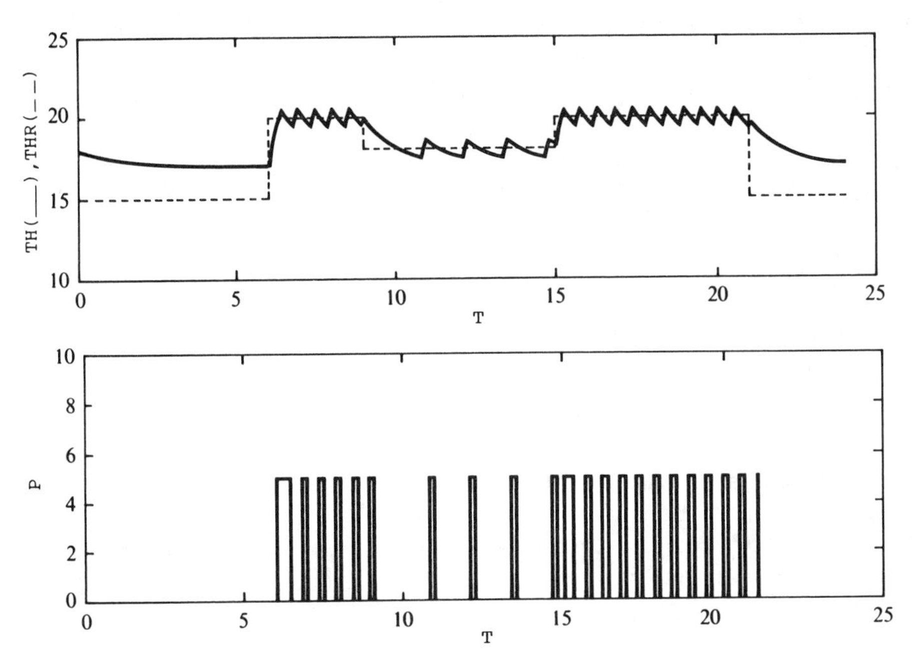


Figure 3.13 The results at higher environmental temperature.

The PROCEDURAL block is introduced with the PROCEDURAL statement, which contains the list of output block signals (to the left of the equals sign) and the list of input block signals (to the right of the equals sign). However, in this case the so-called algebraic loop appears because in order to evaluate the temperature derivative $\dot{\vartheta}_w$ the time constant $T$ is needed and to evaluate $T$ the temperature derivative $\dot{\vartheta}$ is needed. Such modelling is inefficient because solution of the algebraic loop solving is time consuming. If we imagine the problem without the additional time delay, the condition for switching the time constants can be realized directly by testing the control signal $u$ because a positive control signal always obtains a positive temperature derivative:

```
CONSTANT T1=1, T2=4
PROCEDURAL (TIMCON=U)
        TIMCON=T1
        IF (U.LE.0) TIMCON=T2
END
```

□

The prey and predator species ecosystem model and heating control problem were simulated and analysed using simulation language SIMCOS (Zupančič, 1989). However, note that any language that is based on a continuous systems simulation language standard (e.g. CSSL IV, ACSL) can be used in a very similar way.

### 3.3.2 Discrete Simulation Languages

Discrete simulation languages are used for the simulation of discrete systems, which are described by difference equations and discrete events. The differences between difference equation and discrete event simulation have already been given in Section 2.2. However, when talking about discrete simulation languages, the discrete event simulation languages are often meant. So in this section the basic principles of discrete event simulation languages will be given briefly. A discrete event simulation language can be one of the following types:

- interval oriented; or
- event oriented

In *interval oriented* simulation languages the clock (independent variable) is advanced from $t$ to $t + \Delta t$ synchronously with uniform, fixed time increment. One disadvantage of this method is that $\Delta t$ must be small enough not to introduce significant errors in the simulation. Another drawback appears in the case where the interval between two events is large compared to the fixed time increment because the simulation goes through several unproductive phases which do not influence the system states. The realization of interval oriented simulation languages is relatively simple because the activation time for each event is maintained in a list used for events scheduling.

Modern discrete event simulation languages are *event oriented*. Here, the clock is incremented asynchronously from time $t$ to the next event time (variable time increment method). So the periods between events are inactive and therefore consume no time, even though the activities between events do consume time in the real world. There are several ways in which to implement the *next event time increment method*. The method of implementation is called *the strategy* of discrete event simulation language. The three major approaches concerning these strategies are as follows (Hooper, 1987; Neelamkavil, 1987):

- event scheduling;
- activity scanning;
- process interaction.

Modern simulation languages can contain two or three different strategies.

Common to all strategies is the execution of an appropriate model block (model subprogram) that changes the states of the system. All strategies also contain the concept of conditional and unconditional events. Unconditional events are executed at precisely defined moments, while the execution of conditional events depends upon the system states. The general structure of discrete event simulation languages is represented in Fig. 3.14.

The strategy of *event scheduling* demands that unconditional events are executed sequentially as they are programmed. The program chooses from the list of events the one that appears earliest and the time is automatically increased to that time. All conditions except those dealing with time must be included in the blocks that describe the events. The simulation is terminated when all events are executed.

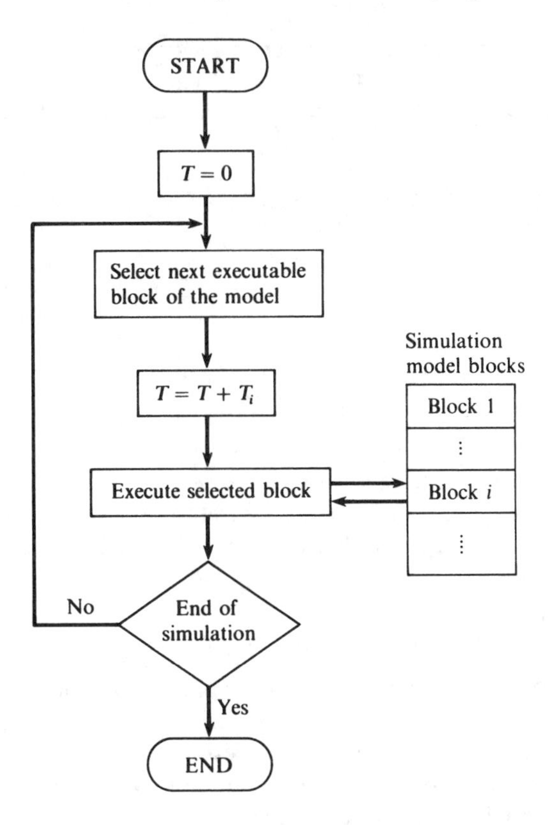


Figure 3.14 The general structure of discrete event simulation languages.

The activity is defined as a state of the system and is bounded by two successive events. The simulation proceeds from event to event by activity scanning if the *activity scanning* strategy is used. Every activity possesses a logical condition which depends on the simulation time and on the states of the system. At each time the status of all the activities in the model is scanned. The activity with appropriate boolean condition is executed and the events are scheduled implicitly. The majority of modern languages do not use this strategy.

The objects of the discrete model and the corresponding events are formed into processes by using the *process interaction strategy*. In this case the behaviour of the system is described by a sequence of mutually exclusive activities, only one activity being commenced at any particular instant. Processes may overlap and their interactions are used for model description. So the emphasis here is given to the scheduling of processes, and events are executed indirectly activating a process (scheduling and executing blocks that describe a model) which is at the top of the list at an appropriate time instant. Processes can be interrupted and reactivated. Conflicts due to process overlap are solved by using the wait and delay procedures.

Table 3.2 The classification of discrete event simulation languages with regard to the strategies

| *Event scheduling* | *Activity scanning* | *Process interaction* |
|---|---|---|
| GASP (II, IV) | AS | GPSS (/360, V, /H) |
| SIMPSCRIPT (I.5, II, II.5) | CSL | Q-GERT |
| SLAM, SLAM II | ECSL | SIMSCRIPT II.5 |
| SIMAN | ESP | SLAM, SLAM II |
| | SIMON | SIMAN |
| | | SIMULA |

Modelling and simulation with the process interaction method is simpler because the program structure is very similar to the model structure but it offers less flexibility compared to the event scheduling method.

Table 3.2 shows the classification of discrete event simulation languages with regard to the strategies mentioned. We can see that some languages involve different strategies.

### 3.3.3 Combined Simulation Languages

Combined simulation languages have the capacity for continuous and discrete simulation (integration, discrete event simulation). For many years the general opinion was that such languages were unnecessary because continuous simulation languages with minimum discrete (or better discontinuity) elements were sufficient for simulations of predominantly continuous real problems, the discrete languages being sufficient for discrete event problems. Most of the combined languages and packages were developed from pure discrete event languages (e.g. GASP IV from GASP II). The reason for this is that discrete event simulation languages have much more complex structural concepts but, on the other hand, continuous simulation languages have more complex numeric algorithms. So for combined simulation it is necessary to expand the discrete event language structure with an integration algorithm. However, most practical situations that demand the combined simulation elements arise from problems that are simulated with continuous simulation languages, while problems simulated with discrete event tools rarely demand continuous elements (integration). Thus the developers of some continuous simulation languages expanded their products to include efficient combined simulation elements (e.g. ACSL).

The major feature of combined simulation languages is that along with the elements for continuous and discrete simulation they also include mechanisms for transition from continuous to discrete simulation and vice versa. Because discontinuities appear frequently during simulation, such languages must have the appropriate numerical algorithms for numerically accurate treatment of discontinuity. Only a few simulation languages have solved this problem in an efficient manner (GASP V, COSY, SYSMOD). The problems of combined simulation systems are described by Cellier (Cellier, 1979).

## 3.4 THE SOFTWARE STRUCTURE OF CONTINUOUS SIMULATION LANGUAGES

A simulation language usually consists of:

- the part for experiment (model) description;
- the processor;
- the run time simulation system, sometimes called the simulator;
- the postprocessor for the presentation of simulation results;
- the supervisor program.

The basic structure of a simulation language is represented in Fig. 3.15.

### 3.4.1 Experiment Description

In most simulation systems the method and the model description are not strictly separate. The reason for this is that not long ago the simulation run was the only experiment to be performed with simulation systems. In classical simulation languages (CSSL IV, ACSL, etc.) that are still commercially attractive, the simulation program

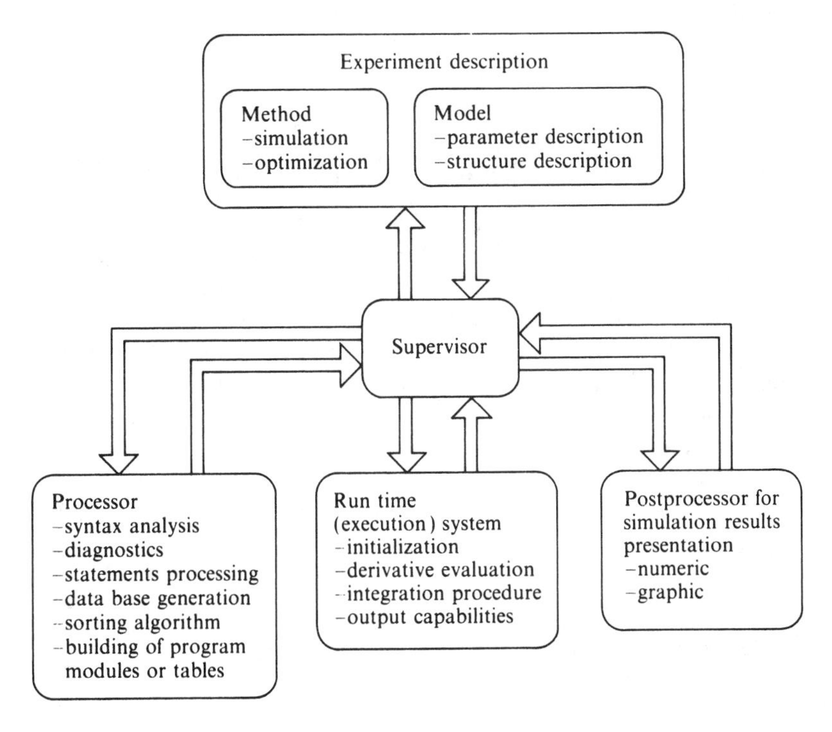


Figure 3.15 Software structure of a simulation language.

consists of one module. The following items must be defined in this program module:

- the structure description of the model;
- the model parameters;
- the simulation run control parameters;
- the output parameters.

**Structure description of the model**

In respect of model structure definition, the simulation languages differ considerably. Some of them are block oriented with very elementary functions. More sophisticated modern simulation languages allow very powerful modelling by direct input of differential equations, nesting expressions, etc.

**Model parameters**

Most languages distinguish between the following two types of parameters:

- model constants;
- function generator data.

**Simulation run control parameters**

These parameters can usually control the following features:

- The simulation (integration) algorithm.
- The communication interval, which defines the points of the independent variable at which simulation results can be presented to the user.
- The calculation interval (step size), which is used as a step for the integration algorithm. The choice of this interval depends on the model time constants and the permissible relative and absolute errors during simulation.
- The initial and final values of the independent variable. The simulation run termination condition can usually also be specified as an arithmetic or logical expression.
- The relative and absolute error tolerances if integration methods with adaptive calculation interval are used.

**Output parameters**

The output parameters define which variables are requested as outputs (the results that can be observed by the user), which output medium is used, and what are the formats of numeric data and plots. The results can generally be presented during the simulation run, and after it in numeric or graphic form.

Different possibilities exist for experimenting with the simulation model. Most compiler oriented languages have so-called sections (e.g. `INITIAL`, `TERMINAL`) which are formally part of the model. The operations before and after the simulation run are specified by these sections. In this way, typical functions as, for example, the definitions of input signals for the simulation or criterion functions for the optimization procedures, must be included in the model. Such structure is linear and unnatural. Some recent languages have achieved the true hierarchical concept, the so-called model–method–experiment concept (Breitenecker and Solar, 1986). Using this concept, models and methods are independent (the user can use or create the library of models and methods) and an experiment represents an execution, i.e. the use of a method on a model. These new concepts are influenced by the new software engineering proposals and will be mentioned in Section 6.1.

### 3.4.2 The Processor

The processor of a simulation language consists of one or several program modules which translate the model (and experiment) description into modules with an appropriate data base in the case of compiler oriented languages, or into data files in the case of interpreter oriented languages. The basic features of interpreter and compiler oriented languages have already been described in Section 3.2. Thus the processor generates the vital parts of the simulation run time (execution) system.

The processor first performs the appropriate syntax analysis. Program statements are then processed in different ways. The processing of statements or blocks which realize the states of the model (the memory blocks or blocks with a delay attribute) are processed specifically, so that the sorting algorithm can sort the statements in an order which enables simulation of a virtually parallel system. Loops without statements or blocks with delay attribute are treated as algebraic loops and must be detected by the sorting algorithm. In many simulation languages processing is terminated with appropriate diagnostics, while in some languages an algebraic loop implicit solver can be used. This procedure will be discussed as a specific feature of digital simulation in Section 5.2.

During statements processing the model data base is first generated. Then either an internal table which has to be interpreted during the simulation run or machine code execution program, or program modules in an intermediate language (usually high level languages such as FORTRAN, C, PASCAL) are built. The last possibility demands appropriate compiling and linking in the intermediate languages. Good processors include efficient diagnostics in all processing phases.

### 3.4.3 Run Time (Execution) System

The run time system enables simulation execution. The framework can be either a fixed program (interpreter oriented languages) or it can be generated by the simulation processor (compiler oriented languages). Sometimes this part of the simulation language is also called the simulator and represents the most important part of the language. The main properties of the execution system are those of accuracy, numeric stability and the calculation efficiency of the simulation (integration) algorithm. These properties will be described as specific features of simulation languages in Section 5.1. We will now describe the general software concept of the run time system. The flow chart of this concept is represented in Fig. 3.16.

The definition of parameters is performed at the beginning of the simulation run. The program must include the model constants, the function generator data, the simulation run control parameters and the output specifications. Usually all these

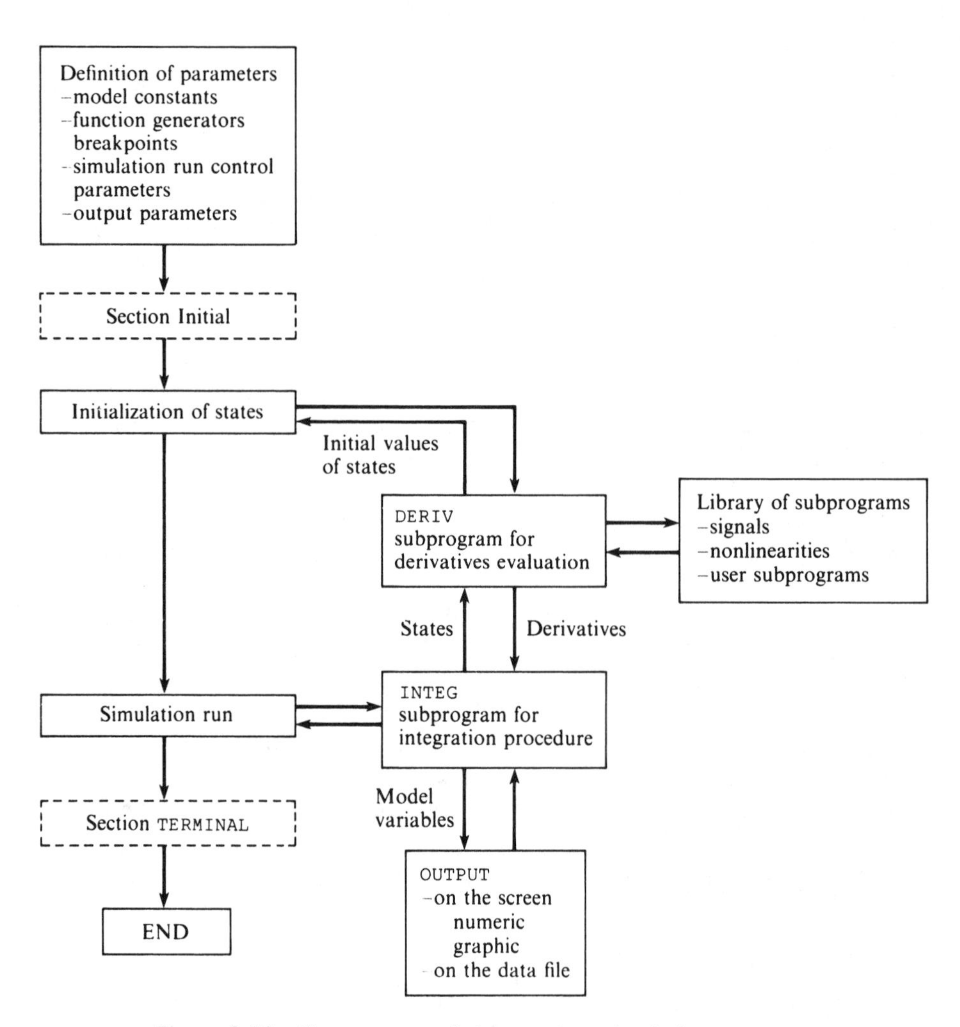


Figure 3.16 The concept of the run time simulation system.

data are stored in one or several files which are read in. Some simulation systems include the so-called `INITIAL` section which performs the operations to be executed before the simulation run. If this is the case the `INITIAL` section is executed next. After that, the initial values of the model states are evaluated. If expressions are allowed for the definitions of the initial states, the evaluation is realized by calling the `DERIV` subprogram. This subprogram consists of all the model equations. The actual simulation run is executed by calling the `INTEG` subprogram, which requires many evaluations of the state derivatives (i.e. once in one calculation interval in the Euler integration procedure, four times in the Runge–Kutta fourth order procedure, etc.) during the complete simulation run. In the so-called communication intervals, control is given to the output routine which performs the required output specifications. After the simulation run the `TERMINAL` section is executed in some simulation systems. This section specifies the operations to be performed after the simulation run.

### 3.4.4 Postprocessor for Simulation Results Presentation

Results can be obtained on-line, i.e. during the simulation run, but a complex analysis of the simulation results must be performed after the simulation. The system for the presentation of the simulation results after simulation must enable the numeric and graphic representation of results. It is often convenient for the user to be able to combine the results of one or several simulation runs, or of one or several experiments, i.e. the results from different data files. The following features of graphic postprocessors are very important:

- drawing several graphs on one screen;
- automatic scaling;
- changing the ranges of variables;
- zooming;
- tracing (obtaining numeric results from moving curves by arrow keys);
- different interpolations between two points (linear, zero order hold);
- linear or logarithmic scales on both axes, with or without grid.

All these functions must be realized in such a manner that the user can work with the postprocessor system in a highly interactive and user friendly way.

### 3.4.5 The Supervisor

The supervisor is a central part of a simulation language because it interfaces all the parts of the simulation language presented in Fig. 3.15. In some parts it is closely linked with the user interface through which the user can work with the simulation language in an efficient and user friendly way. The most important functions of the supervisor (together with the user interface) are as follows:

- Selection of the simulation model. In compiler oriented languages models can be in either source or in compiled form.
- Selection of the method of experiment (e.g. simulation, optimization, etc.).
- The capacity for new model or method definitions or for editing existing ones.
- Starting the simulation language processor (also, in the case of compiler oriented languages, the intermediate language compiler and linker with simulation and intermediate language libraries).
- Execution of the simulation runs or other experiments.
- User friendly editing of model constants, function generator data, output requirements, simulation run control parameters (without having to edit the source program and recompile it again in the case of compiler oriented languages).

Efficient supervisors include all the above-mentioned functions, together with the user interface, so that highly interactive and user friendly work is possible with a simulation language. The user interface is of major importance in new generation simulation languages and will be presented briefly in Section 6.1.

## BIBLIOGRAPHY

BOSCH, P.P.J. VAN DEN (1987), *Simulation Program PSI* (manual), Delft University of Technology.

BREITENECKER, F. and D. SOLAR (1986), 'Models, methods and experiments – modern aspects of simulation languages', *Proceedings of the 2nd European Simulation Congress*, Antwerp, Belgium, pp. 195–9.

CELLIER, F.E. (1979), *Combined Continuous/Discrete System Simulation by Use of Digital Computers: Techniques and Tools*, PhD, Swiss Federal Institute of Technology, Zurich.

GAUTHIER, J.E. (1987), 'ACSL and simulators', *Proceedings of the 1987 Summer Computer Simulation Conference*, Orlando, USA, pp. 73–7.

HOOPER, J.W. (1987), 'Strategy-related characteristics of discrete event languages and models', *Simulation*, **46** (4), pp. 153–8.

HUBER, R.M. and A. GUASCH (1985), 'Towards a specification of a structure for continuous system simulation languages', *Proceedings of the 11th IMACS World Congress*, Oslo, Norway, pp. 109–13.

KORN, G.A. and J.V. WAIT (1978), *Digital Continuous System Simulation*, Prentice Hall, NJ.

MITCHEL & GAUTHIER ASSOC. (1981), *ACSL: Advanced Continuous Simulation Language* (user guide/reference manual).

MOLNAR, I. (1983), 'Some problems in research of general simulation systems', *Proceedings of the 1st European Simulation Congress*, Aachen, W. Germany, pp. 88–92.

NEELAMKAVIL, F. (1987), *Computer Simulation and Modelling*, John Wiley & Sons, Chichester.

NILSEN, N.R. (1984), *The CSSL IV Simulation Language* (reference manual). Simulation Service, Chatsworth, California, USA.

NILSEN, N.R. (1985), 'Recent advances in CSSL IV', *Proceedings of the 11th IMACS World Congress*, Oslo, Norway, Vol. 3, pp. 101–3.

RIMVALL, M. and F.E. CELLIER (1986), 'Evaluation and perspectives of simulation languages following the CSSL standard', *Modeling, Identification and Control* (Norwegian Research Bulletin), **6** (4), pp. 181–99.

SAUCEDO, R. and E.E. SHIRRING (1986), *Introduction to Continuous and Digital Control Systems*, Macmillan Applied System Science, NY.

SCHMIDT, G. (1980), *Simulationstechnik*, R. Oldenbourg, Munich.

SCHMIDT, B. (1986), 'Classification of simulation software', *Syst. Anal. Model. Simul.* (Benelux Journal), **3** (2), pp. 133–40.

STRAUSS, J.C. (1967), 'The SCi continuous system simulation language', *Simulation* No. 9, pp. 281–303.

ZUPANČIČ, B. (1989), *SIMCOS – The Language for Continuous and Discrete Systems Simulation* (users' manual), Faculty of Electrical and Computer Engineering, University of Ljubljana, Slovenia.

# 4

# Tools for the Simulation of Systems with Predominantly Continuous Characteristics

Simulation is an effective, powerful and universal approach to solving problems in different areas but it is useless without powerful simulation tools. In this chapter we will give an historical overview of the development of computer simulation tools, with an emphasis on digital simulation languages. This overview will be followed by a description of some conceptually different simulation languages. The use of general purpose programming languages, CAD packages and analog–hybrid systems will also be presented.

## 4.1 HISTORICAL OVERVIEW OF COMPUTER SIMULATION TOOL DEVELOPMENT

### 4.1.1 Development of Simulation Systems

Different models to represent (simulate) real objects (e.g. models of the universe) were understood centuries ago. However, our interest will focus only on modern simulation, which began with the development of computers.

In 1930 Howard Aiken from Harvard and George Stibitz from the Bell Telephone Laboratory developed the first electrical relay computer. In 1946 John Mauckly and Prisper Eckert from the University of Pennsylvania developed ENIAC (Electronic Numerical Integrator and Calculator), which possessed electronic tubes instead of relays. This was the first electronic digital computer.

But these first computers were not suitable for simulation applications due to their poor calculation capabilities, small memory and very poor input–output devices. Fortunately, George A. Philbrick from the Foxboro company developed 'an automatic control analyser' in 1937–38 which was in fact the first special purpose analog computer. In 1943 John R. Regazzini, Robert H. Randal and Frederick A. Russel from the University of Columbia developed an analog computer which was described as 'an electronic system for obtaining an engineering solution for integro-differential equations of physical systems'.

In those days, analog computers were used for the simulation of continuous systems and digital computers were used for the simulation of discrete event systems. The first were extremely fast while the advantage of the second was their greater accuracy. The analog computer, TRIDAC, which was is use in Great Britain by around 1955, represents one of the first efficient simulation tools used in Europe. It had electronic, mechanical and hydraulic components.

The first ideas for linking the analog and the digital computer originated in the 1950s as advantages and disadvantages of both types of computer became evident. The first digital simulation languages came on to the market in the mid-1950s.

The 1960s saw great progress in the field of analog and digital computer development, and complex simulation studies became a reality. In 1963 Electronic Associates (EAI) developed HYDAC, the first analog computer with some digitally controlled functions and logic. True hybrid systems of general purpose digital and general purpose analog computers came on to the market in the mid-1960s. Digital simulation languages were also becoming popular and in 1967 (Strauss, 1967) the standard for continuous system simulation languages was accepted.

Through the 1970s digital computers became much more popular than analog computers for most simulation applications due to their low price, high interactiveness and efficient calculation capabilities. Analog computers were used only for special (real time simulation, in aeronautics, for complicated control tasks, etc.) as well as for pedagogic purposes. The hybrid computers of Electronic Associates (EAI 580, EAI 680, EAI 2000, EAI 1000) were well known at that time. In the mid-1970s, scientists at Applied Dynamics International (ADI) decided that analog–hybrid systems were insufficient for increasingly more complex modelling and simulation applications, and they began to develop a new digital simulation oriented multiprocessing system. So the processor AD-10 came on to the market at the end of the 1970s. Digital simulation languages have progressed further, influenced by the development of numerical methods, mathematical libraries and, later, complex CAD packages. Thus the languages of that time were already numerically robust, very flexible and highly interactive.

The expansion of micro and personal computers influenced the development of simulation tools dramatically in the 1980s. Very powerful simulation languages were transferred to PCs and simulation as a modern methodology became easily accessible to almost everybody. The development of complex CAD packages, particularly of those from the computer aided control system design area (MATRIXx, CTRL/C, MATLAB with control toolboxes, etc.) was also very important as all these packages included the capacity for complex simulation. Supercomputers and parallel processors (e.g. transputers) were used for special simulation applications. Highly interactive workstations with powerful graphics and array processors (e.g. XANALOG) represented ideal simulation tools for time critical (e.g. real time) simulations as their calculation speed became comparable with analog–hybrid computers. However, the manufacturers of analog–hybrid systems progressed still further. Electronic Associates developed a completely new type of hybrid system–multiprocessor system SIMSTAR (1983). It incorporated only one disadvantage of the older hybrid systems: high price. Using the simulation language STARTRAN, simulation using SIMSTAR became practically the same as that using a digital simulation language.

### 4.1.2 Development of Simulation Languages

The first steps towards digital simulation were made by the users of analog computers, who initially solved problems in a way that was very similar to problem solving on an analog computer. Thus the first simulation languages were close to the concepts of analog computation.

The year 1955 is regarded as the first appearance of digital simulation languages. In this year Selfridge proposed the idea of the block oriented digital simulation language. Users had to sign variables by numbers, the sorting algorithm was not yet known, and the inconvenient Simpson integration method was in use.

Because of the limited capabilities of hardware and software equipment and numerical algorithms, the common disadvantages of subsequent simulation languages (MIDAS, COBLOC, PACTOLUS, CSMP, DYNAMO, etc.) were as follows:

- Languages were block oriented, so the concept of programming was very similar to that on an analog computer. Due to huge programs many possibilities for errors existed.
- The programs were ambiguous so they were not convenient for model documentation.
- Languages were nonexpandable.
- Simulation was numerically unreliable.

In 1965 IBM developed the language DSL 90 (Dynamic Simulation Language), which was very important because it represented the first example of an equation oriented compiler simulation language. It contained the so-called PROCEDURE block by which a block could be described with FORTRAN statements. From DSL 90 the first powerful simulation languages, running on mainframe computers (CSMP 360 and later CSMP III), were developed. These compiler oriented and FORTRAN based languages possessed function generators, macro language, algebraic loop solvers, seven integration methods, approximately sixty operators for model description and enabled the simulation of very complex systems.

Because of the increasing number of simulation languages attempts to standardize continuous simulation languages were made in the mid-1960s. The main disadvantage was that models were not compatible and transportable between different languages. So the committee for continuous systems simulation languages was established in 1965 by the Simulation Council (SCi), and the very popular standard CSSL'67 (Continuous Systems Simulation Language) was issued in 1967 (Strauss, 1967). The concept of this standard was strongly influenced by successful languages developed up to that time, especially by MIDAS and DSL 90.

The following three major design goals were taken into account by the CSSL standardization committee:

- CSSL standard language must provide a simple and obvious programming tool suitable for unskilled users. Thus a comprehensive syntax for differential equation description and for block definitions, powerful diagnostics, sorting algorithms, user hidden integration procedures and interactive run time commands are necessary.
- The language must provide the skilled user with a very flexible tool which enables modelling and simulation of large and complex systems. This goal is met by translating the source model into a high level procedural language. Thus users can add new operators (functions) programmed in the target language. This feature made a CSSL'67 standard simulation language functionally open-ended.
- In anticipation of future technological advances (e.g. graphics capabilities, interactiveness of computer systems) the CSSL standard language had to provide the possibilities for flexible expansion. The last goal made the standard open, enabled later inclusion of many new features into this type of language and significantly prolonged its use. But, on the other hand, this was also the reason why many CSSL dialects arose.

Taking into account the goals mentioned above, the CSSL'67 standard proposed structural and functional elements. *Structural elements* were defined long before such terms as 'structured programming' or 'data structures' appeared. These elements are as follows:

- Elements that enable sorting: this feature makes simulation 'quasi parallel'.
- Macro language which can be used to avoid repetitive code in model description.
- Model (derivative) sections that can contain different integration algorithms.
- Structural program with sections `INITIAL` (procedural program before simulation run), `DYNAMIC` (nonprocedural part for model description) and `TERMINAL` (procedural program after simulation run).
- Flexibility of execution: execution can be controlled by a program, by a run time command language or by both approaches.

Besides structural elements, only basic *functional elements* were defined by the CSSL'67 standard. Integration and derivative operators, implicit algebraic loop solver, delay operator, operators for standard signals and nonlinearities and function generators were considered. Constants, variables and arrays were provided and statements for the definition of integration method, calculation interval, absolute and relative errors, duration of communication interval and simulation run, as well as statements for output specification, were forseen.

The standard CSSL'67 influenced further development of compiler oriented continuous simulation languages considerably. The languages which included these standard elements have been the most successful commercially up until now. The

Table 4.1 Survey of continuous systems simulation languages

| | | |
|---|---|---|
| 1955 | SELFRIDGE | first simulation language |
| 1956 | | compilation concept (Stein, Rose, Parker) |
| 1963 | MIDAS | first really usable simulation language; sorting algorithm; integration method with step size control |
| 1964 | COBLOC | several integration algorithms; free format of simulation model coding |
| 1965 | DYNAMO | diagnostics; macro language; presentation of results with graphs; detection of algebraic loop |
| 1965 | DSL 90 | compiler equation orientation, PROCEDURE block |
| 1967 | CSMP 360, CSMP III | first powerful simulation languages; function generators; 60 operators; solving algebraic loop |
| 1967 | DSL 1130/1180 | first interactive features; use of digital plotter and storage oscilloscope |
| 1967 | | standard CSSL'67 |
| 1968 | MIMIC | first language based on CSSL'67 |
| 1969 | CSSL III | first powerful CSSL language |
| 1970 | SL-1 | real time facilities |
| 1972 | CSSL IV | up to now one of the most powerful simulation languages |
| 1972 | CSMP III | efficient graphical presentation of results |
| 1973 | HL-1 | programming of hybrid system |
| 1975 | SIMNON | good interactive properties; derivatives need not be integrated by INTEG statement due to the states and its derivatives declarations |
| 1975 | ACSL | very powerful language, commercially very successful |
| 1975 | DARE P | implemented on minicomputer, portable; the attempt of model and experiment separation |
| 1976 | DARE/ELEVEN | the combination of equation and block oriented model segments for fast real time simulations |
| 1980 | MICRODARE, DESIRE, DESKTOP | ultra fast compiler |
| 1981 | | new recommendations CSSL'81 (SCS, IMACS) |
| 1983 | ISIM | implemented on eight- and sixteen-bit microcomputers; modular programming |
| 1983 | HYBSIS, STARTRAN | simulation languages for hybrid computers; simulation in real time |
| 1984 | ADSIM, PARSIM | simulation languages for special purpose computers; simulation in real time |
| 1984 | ESL, SYSMOD, COSMOS | new generation simulation languages |

most important languages that followed the CSSL'67 standard strictly were CSSL III and, later, CSSL IV (Nilsen, 1984) and ACSL (Advanced Continuous Simulation Language, Mitchel & Gauthier, 1981).

Many other languages were either less influenced or not at all influenced by the CSSL'67 standard, among these being SIMNON (Elmqvist, 1975; Åström, 1985a), the family of DARE languages (Korn and Wait, 1978), TUTSIM (TUTSIM, 1983) and PSI (van den Bosch, 1987).

The first usable simulation languages for modern hybrid systems were developed in the 1980s, e.g. HYBSIS (Kleinert *et al.*, 1983), and STARTRAN (Landauer, 1988), as well as some other simulation languages which run on special purpose simulation computers, e.g. ADSIM (Grierson, 1986) and PARSIM (Bruijn and Soppers, 1986).

We have mentioned those languages that contributed considerably to the development of digital simulation up until the mid-1980s. But the need for a new CSSL standard that would also take into account the modern concepts of software engineering intensified. Due to very different interests, two working groups (IMACS and SCS groups) could only make recommendations for the new simulation languages, but, based on their recommendations, some new simulation languages (ESL, SYSMOD, COSMOS) began to develop in the mid-1980s. The features of these languages will be discussed in Section 6.1.

The important characteristics of continuous systems simulation languages are represented briefly in Table 4.1. Emphasis is given to those properties that can be treated as novelties appearing in particular simulation languages but the table does not include all existing languages. For detailed information the reader is referred to the literature (Divjak, 1975; Cellier, 1983; Rimvall and Cellier, 1986; Bausch-Gall, 1987; and Zupančič, 1989a).

## 4.2 EXAMPLES OF EQUATION ORIENTED SIMULATION LANGUAGES

As mentioned in Section 3.3, compiler languages are usually equation oriented, which means that differential equations can be included directly in a program. Thus the computer implementation of a model is much easier and faster, the simulation programs are shorter, more modular and can be used efficiently for model documentation purposes.

Due to the strong influence of the CSSL'67 standard on the development of equation oriented languages, emphasis must be placed here on a language of this type. ACSL (Mitchel & Gauthier, 1981) and CSSL IV (Nilsen, 1984) are two of the most popular CSSL languages, but we will describe our own language, SIMCOS, here as the models which use only basic properties are completely compatible with other CSSL languages and thus the presentation of SIMCOS will provide a good basis for CSSL programming.

Besides the CSSL-type language SIMCOS, a successful language which does not obey CSSL'67 standard (SIMNON) will be presented as an alternative language.

### 4.2.1 CSSL Standard Based Language SIMCOS

The simulation language SIMCOS (Zupančič *et al.*, 1986; Zupančič, 1989a, b; and Zupančič *et al.*, 1991) is a CSSL-type equation oriented language which was developed at the Faculty of Electrical and Computer Engineering in Ljubljana. It works as a compiler. The model, which is coded in CSSL syntax, is processed by the compiler into FORTRAN modules and a model data base. The FORTRAN modules are further processed by a FORTRAN compiler and subsequently linked with appropriate libraries into an executable simulation program.

The supervisor program automatically handles all the above procedures and is able, together with a highly interactive user interface, not only to simulate the model but also to perform simple experiments (e.g. change of model constants, output specifications, function generators breakpoints) and complex ones (e.g. parameter studies and optimization).

Using the CSSL input language feature the simulation model (program) must be written in a file using the constructs for model description.

Statements as basic constructs are divided into the following:

- basic statements;
- simulation run control statements;
- model representation statements;
- output statements.

In the following review, only the most important statements will be mentioned.

#### Basic statements

These statements provide definitions of some basic properties of the simulation model (declaration of variables, constant definitions, etc.). The first statement of the program is usually the statement `PROGRAM`, which is optional and defines the name of a model. The last statement must be the `END` statement. Comments can be introduced by the `COMMENT` statement or with the character " in the first column.

The `ARRAY` statement declares the names of arrays, their dimension and size (maximum number of elements in each dimension). It has the form

```
ARRAY var(dim[,dim][,dim]) [,var(dim[,dim][,dim])]...
```

`var` ... array variable name; `dim` ... array variable size. In this and in the subsequent definitions the square brackets ([ ]) represent optional entities.

The `CONSTANT` statement can be used for definition/initialization of variables and one-dimensional arrays and has the form

```
CONSTANT var=const [,var=const]...
```

`var` ... variable or one-dimensional array name; `const` ... variable or one-dimensional array values (separated by commas).

The `TABLE` statement defines the function generator for a function with one or two independent variables and has the form

```
TABLE name,n,dim,data
```

`name` ... function generator name; `n` ... number of independent variables (integer constant 1 or 2); `dim` ... number of independent variable(s) points (integer constants dim1, dim2) (in the case of one independent variable, only one constant is specified); `data` ... values of independent variable(s) in breakpoints, followed by the corresponding values of dependent variables (real or integer constants).

The `TERMT` statement defines the condition for simulation run termination. It is written in the form

```
TERMT logic expression
```

If the logic expression becomes `TRUE` the simulation run is terminated.

### Simulation run control statements

Simulation run control statements are used for the definition of names and values of fixed variables. They define the parameters for the independent variable, the simulation algorithm and the output activity. Most of these statements have the form

```
STATEMENT var=const
```

`var` ... variable name; `const` ... the value of the variable (real or integer constant).

The `VARIABLE` statement defines the name of the independent variable and its initial value. The `CINTERVAL` statement defines the communication interval in units of the independent variable (usually time). The `NSTEPS` statement defines the number of calculation intervals in the communication interval when using fixed calculation interval methods and the initial number of calculation intervals in the communication interval when using adaptive calculation interval methods. The `MERROR` and `XERROR` statements determine the relative and absolute integration errors allowed. Both statements are meaningful only for integration methods with adaptive calculation interval and prescribe the maximum allowed error (tolerance). The `ALGORITHM` statement determines the method of initialization of the simulation algorithm (`var1=const1`) and the simulation (usually integration) algorithm (`var2=const2`). The user can choose between the following methods: discrete simulation, the Euler method, the Runge–Kutta–Gill method, the Runge–Kutta–Merson method, the Rosenbrock semiimplicit method, the extrapolation method with linear implicit midpoint rule, and the Gear-stiff and Adams– Moulton predictor corrector methods. Some algorithms are also available for real time simulations.

All these simulation control statements are not usually used, and in this case the compiler assigns default names and default values to these variables.

## Model representation statements

The simulated model is defined by the corresponding representation statements. The statement order is not important due to the fact that a parallel system is simulated. The compiler has a built-in sorting algorithm, which sorts the statements in such a way that all input variables are defined prior to execution of a particular statement.

Representative statements define the value of one or several output variables as the result of certain operations on the set of input variables. Simulation oriented operators (e.g. integrators and procedural blocks), standard FORTRAN-like arithmetic assignment statements (where all system and user functions can be used), statements for signal generation, statements for the realization of nonlinearities and statements for the realization of discrete transfer functions can be used as representative statements.

### *Simulation oriented statements*

The `INTEG` operator represents the basic statement of the simulation language and realizes the simulation algorithm (integration or delay procedure). The operator is usually in the form

```
var=INTEG(expr1,expr2)
```

`var` ... the integrator output variable name (state); `expr1` ... the integrator input (derivative); `expr2` ... the initial condition.

The `PROCEDURAL` block introduces a group of FORTRAN statements for the realization of a certain structure (block) describing the desired input–output relations. The first statement (head) of the `PROCEDURAL` block has the form:

```
PROCEDURAL (list1=list2)
```

`list1` ... output variables list; `list2` ... input variables list.

Every procedural block must begin with the `PROCEDURAL` statement and terminate with the statement `END`. The sorting algorithm treats the procedural block as a representation statement and sorts it according to the input–output variables list. The statement order inside the procedural block is not changed.

### *Statements for the realization of signals and nonlinearities*

Fig. 4.1 represents the signals and nonlinearities, which are built-in in the simulation language SIMCOS. These functions are included in the library and are called in the

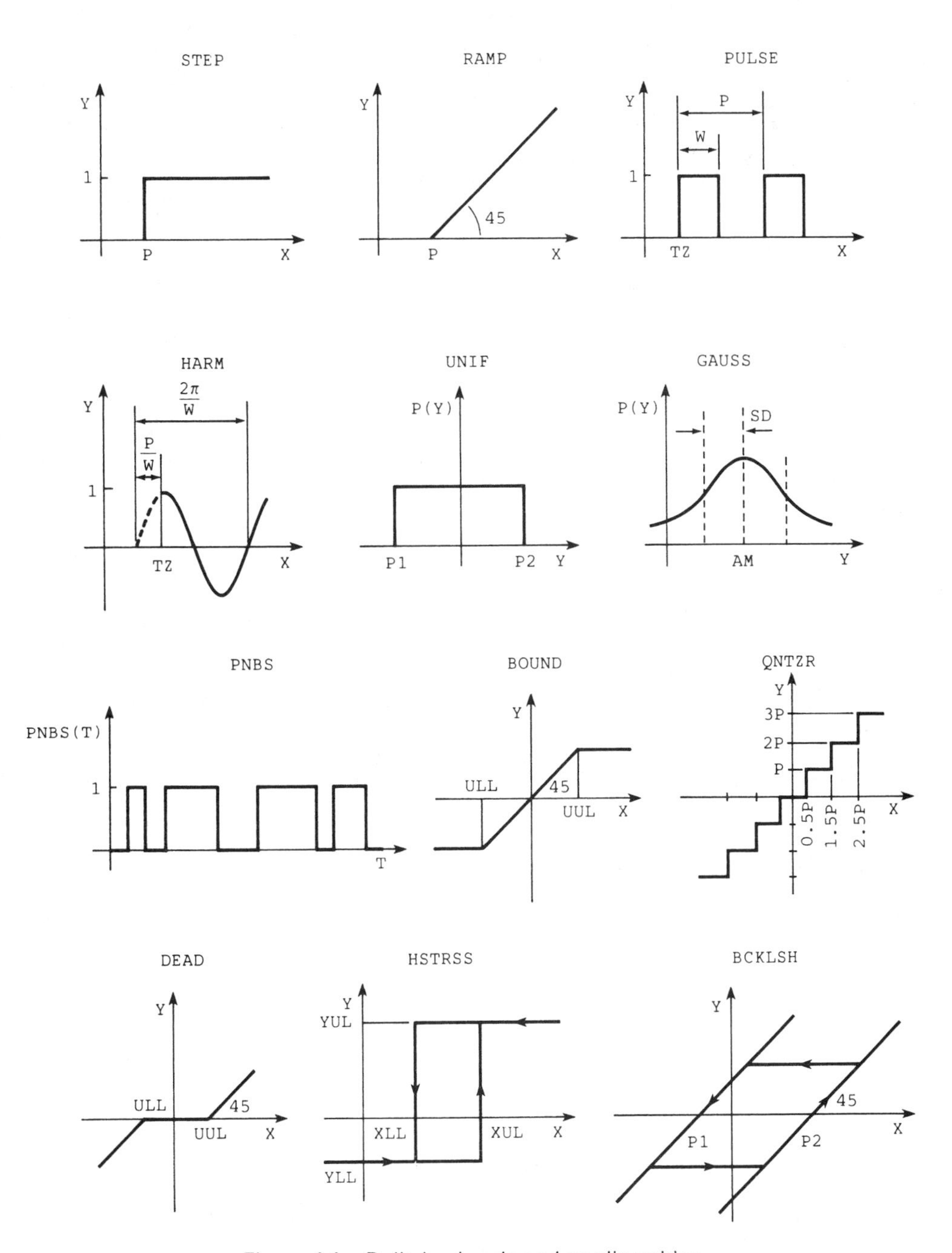


Figure 4.1 Built-in signals and nonlinearities.

form

```
var=NAME(arguments)
```

`var` ... name of the output variable; `NAME` ... name of the function; `arguments` ... constants, variables, arithmetic expressions.

### *Statements for the realization of discrete transfer functions*

Discrete transfer functions are used if hybrid systems, i.e. combinations of continuous and discrete systems, are simulated. Every transfer function $G(z)$ has a sampling unit on its input and a zero order hold on its output. The realization is shown in Fig. 4.2. PID is the discrete proportional integral differential controller with the transfer function

$$G(z) = \frac{q_0 + q_1 z^{-1} + q_2 z^{-2}}{1 - z^{-1}}$$

and is realized by calling the function

```
U=DPID(E,Q,W,TS)
```

`E` ... input variable; `Q` ... array of three variables determining the parameters $q_i$ of the discrete PID controller; `W` ... array of three variables for the `DPID` internal states; `TS` ... sampling time.

As well as the discrete PID controller, the general discrete transfer function `DTRAN`

$$G(z) = \frac{b_0 + b_1 z^{-1} + \cdots + b_n z^{-n}}{a_0 + a_1 z^{-1} + \cdots + a_m z^{-m}} \qquad a_0 \neq 0$$

the discrete delay (dead time) `DELAY`,

$$G(z) = z^{-d}$$

and the sample and hold element `SH` are implemented.

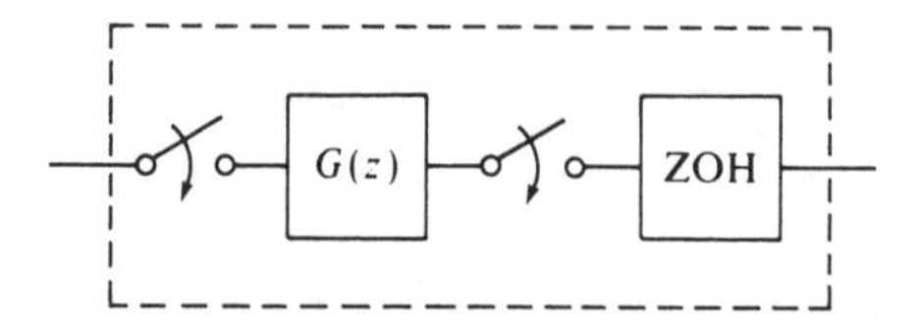


Figure 4.2 Discrete block realization.

**Output statements**

Output statements write simulation results on the screen or in the data file. The information in these statements can be changed interactively during simulation.

The `OUTPUT` statement causes the numeric output to be written on the screen every *n* communication intervals and has the form

```
OUTPUT [n,] variable [,variable], ...
```

The `PREPAR` statement stores the variables in the data file every *n* communication intervals and has the form

```
PREPAR [n,] variable,[variable,]...
```

### 4.2.2 The alternative language SIMNON

There are only a few simulation languages which are commercially attractive and which do not take into account the CSSL standard. One of them is the simulation language SIMNON (Elmqvist, 1975; Åström, 1985a). The language works as a compiler but as it does not compile into high level language modules (e.g. ACSL in FORTRAN) the time needed to obtain an executable program is very short. Because of this fact and due to the very powerful command language for interactive work, its interactiveness can be compared with that of the interpreter (usually block oriented) languages. SIMNON was developed at the Department of Automatic Control, Lund Institute of Technology.

**Model definition**

The model is described in a special model language. This language is simple and easy to learn and enables extensive error checking. The compiler works in parallel with an editor. When the compiler discovers an error the corresponding message is given and the user has an immediate opportunity to correct the erroneous line by certain editing commands.

SIMNON allows decomposition of the system into subsystems which can be described separately. This makes it possible to maintain a library of subsystems. The subsystems are not hierarchically organized but, nevertheless, are important steps towards modular modelling.

SIMNON allows description of both continuous subsystems and discrete (difference equation) subsystems. The sampling in discrete submodels need not be equidistant and need not be equal in all discrete subsystems. In order to allow connection between continuous subsystems and discrete subsystems, the outputs and states of the discrete subsystems must have defined values between sampling instants. So the outputs and states of discrete submodels are terminated by the zero order hold. The sampling instants for each discrete subsystem are specified by a special variable in the system description. This variable is updated at each sampling instant to contain the time for the next sampling. Thus after the activation of a discrete subsystem, it is apparent when the next sampling should be performed.

Continuous and discrete subsystems have the following structure:

```
CONTINUOUS SYSTEM ⟨name⟩ DISCRETE SYSTEM ⟨name⟩
declarations             declarations
assignments              assignments
END                      END
```

In the subsystem heading the type of the system and its name is defined.

The following declarations can be used:

```
INPUT  declaration of subsystem input variables
OUTPUT declaration of subsystem output variables
TIME   declaration of independent variable
STATE  declaration of states (for continuous and discrete subsystem)
DER    declaration of derivatives for continuous subsystem
NEW    declaration of predictions for discrete subsystem
TSAMP  declaration of sampling time in discrete subsystems
```

The system description can also contain parameters and auxiliary variables, which are not declared.

The parameters are assigned by statements of the following form:

```
⟨parameter⟩:⟨value⟩
```

The initial values of the state variables can be assigned in the same way. Variables are assigned as

```
⟨variable⟩=⟨expression⟩
```

This statement gives the SIMNON language its equation orientation. Outputs of the subsystems, derivatives and predictions are evaluated with such statements. Beside these statements the 'if-then- else' statement extends the modelling power of the SIMNON language significantly. INTEG statements are not used because the states and its derivatives are explicitly declared so there is no need to use them. The assignments are automatically sorted by the compiler into the appropriate calculation order.

Several standard functions are available in the package. These functions can be used in assignments in the subsystems or in the connecting system. The following standard functions are available:

| | | | |
|---|---|---|---|
| ABS(X) | absolute value | LOG(X) | logarithm base 10 |
| ATAN(X) | arctangent $(-\pi/2, \pi/2)$ | MAX(X,Y) | maximal value of $x$, $y$ |
| ATAN2(X,Y) | arctangent $(-\pi, \pi)$ | MIN(X,Y) | minimal value of $x$, $y$ |
| COS(X) | cosine function | MOD(X,Y) | the remainder when deriving $x$ by $y$ |
| EXP(X) | exponent function | SIGN(X) | sign of $x$ |
| INT(X) | integer part of $x$ | SIN(X) | sine function |
| LN(X) | natural logarithm | SQRT(X) | square root |
| | | TAN(X) | tangent |

Some standard systems of common interest are also included in the package. These are

| | |
|---|---|
| DELAY (continuous) | dead time |
| FUNC (continuous) | function generator |
| IFILE (discrete) | reads values from a file |
| LOGGER (discrete | samples and saves variables on a file |
| NOISE1 (discrete) | random noise generator, Gaussian or uniform |
| OPTA (discrete) | system for optimization |

The connections of the subsystems are defined in a connecting system, which defines how the inputs and outputs of the different subsystems are interconnected. The connecting system is a static system and may be time-varying but must not be dynamic. It has the following structure:

```
CONNECTING SYSTEM ⟨name⟩
declarations
connect section
END
```

The connect section contains assignment statements for connections of inputs and outputs of subsystems. The same variable names may be used in different subsystems. Therefore, the following notation is used in a connecting system to reference variables in the subsystems:

```
⟨variable⟩ [⟨subsystem⟩]
```

The right-hand side of the assignments may contain the STATE and OUTPUT variables, which are referenced in the same way.

### Interactive facilities

The core of the interactive program is a set of subroutines called Intrac. These routines handle the interaction between the user and the application package. For instance, Intrac handles the command decoding. The structure of each command is flexible. In some cases, arguments can be omitted and default values are then used.

The common situation when simulating a model is that the same sequence of commands is given several times. The user can then define a macro containing these commands. This macro is then used as a new command, possibly with different values of the arguments. The macros are defined using the commands for Intrac. Repetitive loops and control statements can be included in a macro.

### Example 4.1 Simulation of the prey and predator problem in SIMNON

To simulate the prey and predator problem with the simulation language SIMNON, we can use Eqn (2.6) directly as SIMNON is an equation oriented language. The

```
CONTINUOUS SYSTEM PREY_PREDATOR
''declarations
state RAB, FOX
der RABDOT, FOXDOT
''assignments
''model structure
RABDOT=A11*RAB-A12*RAB*FOX
FOXDOT=A21*RAB*FOX-A22*FOX
''model's constants
A11: 5
A12: 0.05
A21: 0.0004
A22: 0.2
''initial conditions
RAB: 520
FOX: 85
END
```

Figure 4.3 SIMNON program for the prey and predator problem.

correspondence between simulation program variables and physical variables is the same as in Example 3.1.

The model can be described as one SIMNON continuous system. The declarations of states and derivatives are followed by model structure equations. Finally, the model's constants and the initial conditions are defined. Fig. 4.3 represents the SIMNON simulation program for the same model constants and initial conditions as were given in Example 3.1.

When the model is defined it is compiled and simulated with the following commands:

```
syst PREY_PREDATOR
store RAB,FOX
simu 0, 10, 0.01
ashow RAB, FOX
```

The command `syst` compiles the model. With the command `store` the user selects variables to be stored during the simulation run. The command `simu` executes the simulation from the initial time 0 to the final time 10 with the initial calculation interval 0.01. The command `ashow` plots the stored values on the screen using an automatic scaling procedure.

To demonstrate the applicability of SIMNON subsystems, the model for the prey and predator problem can be realized with the aid of two continuous subsystems. The subsystem `PREY` includes the equation for rabbits and the subsystem `PREDATOR` includes the equation for foxes. Both subsystems communicate through the input and output variables. The corresponding model, which consists of two submodels, is shown in Fig. 4.4. Fig. 4.5 shows the SIMNON program, which includes two continuous systems and a connecting system. The variables `RAB` and `FOX` are the

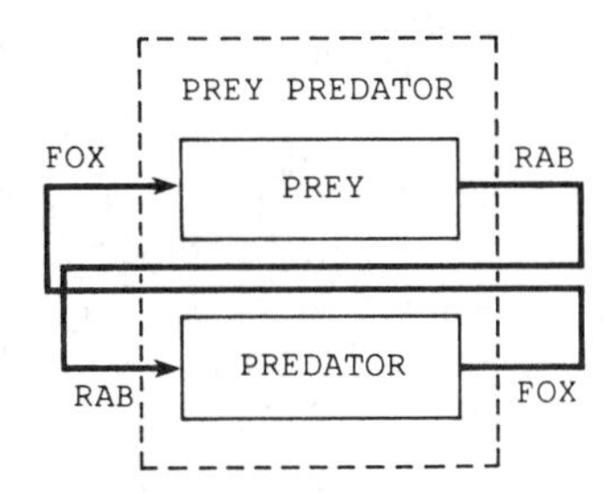


Figure 4.4 The prey and predator problem realized by two subsystems.

```
CONTINUOUS SYSTEM PREY          CONTINUOUS SYSTEM PREDATOR
state RAB                       state FOX
der RABDOT                      der FOXDOT
input FOX                       input RAB
RABDOT=A11*RAB-A12*RAB*FOX      FOXDOT=A21*RAB*FOX-A22*FOX
A11: 5                          A21: 0.0004
A12: 0.05                       A22: 0.2
RAB: 520                        FOX: 85
END                             END

CONNECTING SYSTEM PREY_PREDATOR
FOX [PREY]     =FOX [PREDATOR]
RAB [PREDATOR]=RAB [PREY]
END
```

Figure 4.5 SIMNON program for the prey and predator problem using subsystems.

outputs of the subsystems PRAY and PREDATOR. As these variables are the states, they must not be declared as outputs.

In order to simulate the model the following commands must be executed:

```
syst PREY, PREDATOR, PREY_PREDATOR
store RAB [PREY], FOX [PREDATOR]
simu 0, 10, 0.01
ashow RAB [PREY], FOX [PREDATOR]
```

The names of the subsystems and the connecting system should be defined as arguments of the compilation command `syst`. The variables must be declared with the appropriate subsystem names in brackets. □

## 4.3 EXAMPLE OF A BLOCK ORIENTED LANGUAGE

As mentioned in Section 3.3, interpreter oriented languages are almost always block oriented, so we often do not distinguish between the advantages and disadvantages of interpreter and block oriented languages which are described in the same section.

We shall therefore point out some main important features of block oriented languages:

- A block oriented language requires a description of the simulation model using low level simulation schemes or block diagrams.
- Blocks can usually be executed faster than equations.
- Modifications in the structure and parameters of the simulation model can be made interactively and the simulation can be rerun immediately.
- Block oriented languages are, from the point of the model definition, easier to learn than equation oriented languages and are therefore more convenient for beginners who simulate models with elementary blocks.

As an example of block oriented languages we shall briefly present the simulation language PSI (Interactive Simulation Program). The language was developed in the Laboratory for Control Engineering at the Delft University of Technology (van den Bosch, 1987). Therefore, PSI is written entirely in Fortran 77 and runs on mainframes as well as on PC computers.

PSI is an interpreter based, user friendly command language which enables highly interactive work with a simulation model.

### Model definition

The block diagram or the low level simulation scheme is used to describe the topological structure of the simulation model. Because the user cannot implement his own blocks, this disadvantage is reduced by a great number of blocks. Each block consists of one output, a maximum of three inputs and three parameters describing the relations between them. The model format coding is free, the symbolic names are used in the program.

The following standard blocks are available:

`ABS` absolute value
`ADC` analog-to-digital conversion
`BLN` boolean block
`BNG` bang-bang
`CON` constant
`DAC` digital analog conversion
`DIV` divider
`DPY` display block
`DRW` draw block
`DSP` dead space
`EXP` exponent function
`FFL` flip flop
`FIX` quantizer
`FNG` function generator
`GAI` gain
`HYS` hysteresis
`INC` mode controlled integrator
`INF` first order system
`INL` limited integrator
`INT` integrator
`LIM` limiter
`LOG` logarithm
`MAX` maximum relay
`MIN` minimum relay
`MUL` multiplier
`NOI` noise
`PDC` continuous PD controller
`PDZ` discrete PD controller
`PIC` continuous PI controller
`PIZ` discrete PI controller

| | | | |
|---|---|---|---|
| REL | relay | SUB | subtractor |
| SIN | sinus function | SUM | summer |
| SPL | sample & hold | TDE | delay time |
| SQT | square root | XXi | user defined block |
| STP | stop block | ZZZ | discrete delay |

In this review we omitted some blocks which are not so important. Function generators with equidistant points of one or two independent variables can be realized. Linear and quadratic interpolation between breakpoints is available. Some more complex dynamic blocks, such as first order lag, PD and PI controllers, as well as discrete PD and PI controllers, give the PSI wide applicability in the field of analysis and design of continuous and discrete control systems. Analog to digital and digital to analog conversion is also available in some versions. Several different integration blocks are implemented. Integration with the possibility of mode control enables simulation studies which are well known from hybrid techniques. Algebraic loops can also be handled with the simulation language PSI.

### Experimental features

The preceding text has described the facilities that are more or less characteristic for many block oriented languages. However, the following possibilities, which are based on multirun simulation studies (experimental features) and which cannot easily be implemented in block oriented languages are implemented in PSI:

- The comparison of variables between two or more simulation runs or the use of a variable from a previous run in the current run.
- Optimization experiment. A criterion is defined as a function of some parameters in simulation, and the algorithm (Hook–Jeves) must find such values of parameters that minimize or maximize the criterion function. The constraints in parameters can be included.

### Interactive facilities

Communications between the user and PSI is realized by means of a command language. Commands can be divided into several groups, some of which are as follows:

- simulation data definition statements (the definition of the topological structure, parameters, timing data and output representation);
- show commands (to show the actual values of a number of parameters on the screen);
- signal show commands (to study the signals stored in the memory);
- control commands (to control the simulation run);
- model save commands;

- function generator commands;
- optimization commands.

The fact that block oriented languages are more convenient for inexperienced users is not valid for the use of command language interaction. The command language has an advantage in comparison with menu oriented interaction only for skilled users who use the simulation tool frequently.

#### Example 4.2 Simulation of the prey and predator problem in PSI

In order to solve the prey and predator problem with the block oriented simulation language PSI, the digital simulation scheme must be used. The digital simulation scheme is in fact an extended simulation scheme, which was introduced in Section 2.1, because some additional blocks which are defined in a manual of a particular language are used. Let us redraw the simulation scheme in Fig. 2.7 and sign the outputs of all elements (blocks) with simulation program variables and constants. The correspondence with the names given in Fig. 2.7 is defined in Example 3.1. Due to the fact that all block outputs must be signed, some new variables have been introduced (`AUX1`, `AUX2`, `AUX3`, `AUX4`, `RABFOX`). Fig. 4.6 shows the appropriate digital simulation scheme.

To realize the scheme shown in Fig. 4.6 the following PSI blocks must be used: two integrators (`INT`), two summers (`SUM`), four gains (`GAI`) for model constant definitions and one multiplier (`MUL`). These blocks are defined in Fig. 4.7, where the

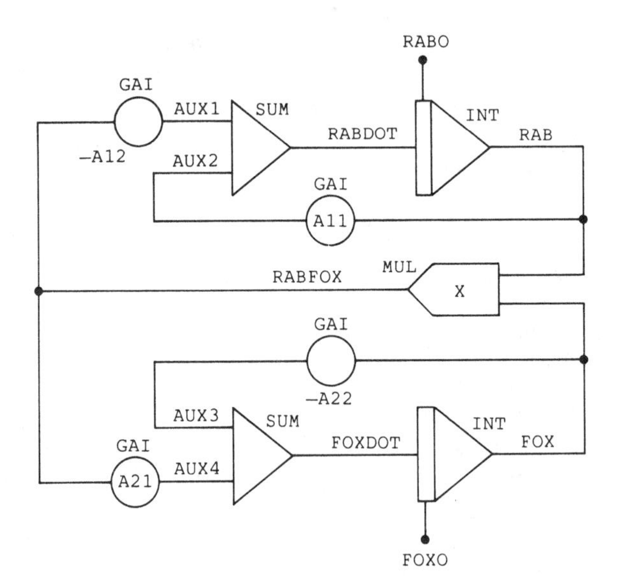


Figure 4.6 Simulation scheme.

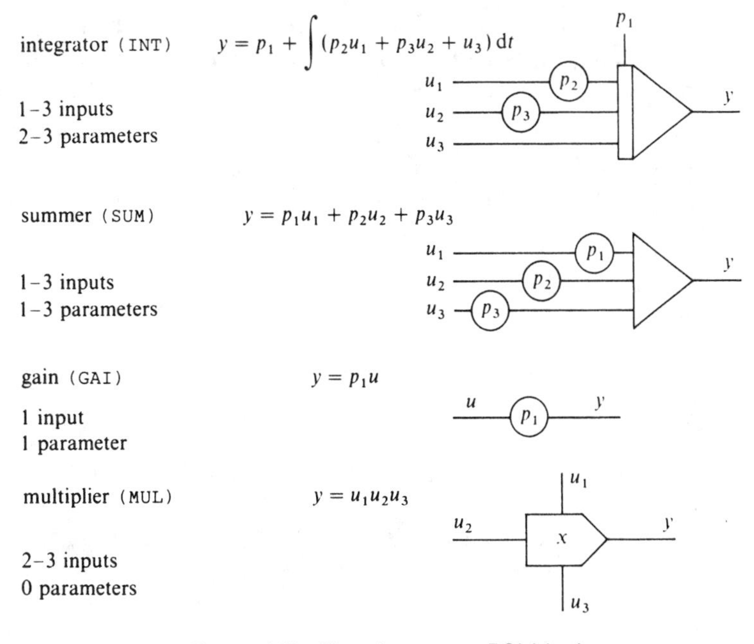


Figure 4.7 The elementary PSI blocks.

following notation is used: $y$ is the only output of each block, $u_i$ is the $i$th input and $p_i$ is the $i$th parameter of the block.

Using PSI command language the following items are defined:

- the structure of the model (configuration specification part);
- the model constants (parameter part);
- the simulation run control parameters.

The structure can be defined by typing the command B after starting PSI. Then all blocks are defined in an arbitrary order. Each block structure definition consists of: name of block output, type of block, names of block inputs

```
PSI*B
CONFIGURATION SPECIFICATION PART
Block, Type, Input 1, Input 2, Input 3
B*RABDOT, SUM, AUX1, AUX2
B*FOXDOT, SUM, AUX3, AUX4
B*RAB, INT, RABDOT
B*FOX, INT, FOXDOT
B*AUX1, GAI, RABFOX
B*AUX2, GAI, RAB
B*AUX3, GAI, FOX
B*AUX4, GAI, RABFOX
B*RABFOX, MUL, RAB, FOX
```

The model constants are defined in the parameter part by typing the command `P`. Each block is identified by its output variable name followed by the appropriate block parameters:

```
PSI*P
PARAMETER PART
Block, Par 1, Par 2, Par 3
P*RABDOT, 1, 1
P*FOXDOT, 1, 1
P*RAB, 520, 1
P*FOX, 85, 1
P*AUX1,-0.05
P*AUX2, 5
P*AUX3, -0.2
P*AUX4, 0.0004
```

By this definition the same model constants and initial conditions as were used in Example 3.1 are used.

Before starting the simulation we have to supply PSI with simulation run control parameters and output specifications. Several commands can be defined in one row:

```
PSI*T, O, R
Integration interval=0.01
Total time=10
Names of blocks to be shown=RAB, FOX
```

By these definitions the calculation interval `0.01` and the duration of simulation run `10` are used. After that the required responses (`RAB`, `FOX`) appear on the screen. □

## 4.4 THE USE OF GENERAL PURPOSE PROGRAMMING LANGUAGES

The concept of digital simulation systems shown in Fig. 3.3 can also be achieved with any general purpose program language. We begin here with BASIC (Beginners' All-purpose Symbolic Instruction Code) and with the prey and predator problem.

### Example 4.3 The prey and predator problem simulated in BASIC

According to the concept of digital simulation systems shown in Fig. 3.3, and in more detail in Fig. 3.16, the presimulation operations define the model data base and the simulation run control parameters. In our prey and predator problem, introduced in Problem 1.3 and already simulated with CSSL standard program in Example 3.1, the data base consists of four constants ($a_{11}$, $a_{12}$, $a_{21}$, $a_{22}$) and two initial conditions (initial sizes of both populations). The simulation run control parameters consist of the initial value of the independent variable, the calculation interval (independent variable increment) size and the simulation run duration.

There are several ways to use constants in a BASIC simulation program. The most straightforward way is to use them as numeric constants (numbers) in the model description part of the simulation program. However, if a constant appears several times in a model and the simulation run has to be repeated with different values for the constant, the user must change it at all its appearances. This is not only time consuming but may also be a source of error. Better programming practice is to give a name to the constant, i.e. to define the constant as a variable which is defined only once in the program and does not change its value. In BASIC the `DATA` and `READ` statements can be used for this purpose, but let us choose a clear constant definition using the standard BASIC assignment statement `LET`:

```
20 LET A11=5: LET A12=0.05: LET A21=0.0004: LET A22=0.2
```

Note that the same names as were used in Example 3.1 are used for constants, and several statements separated by colons are written on a single line. It must be mentioned that the operand `LET` can be omitted in almost all versions of BASIC. In the same way the initial and final values of the independent variable `T`, as well as the calculation interval `DT`, are defined:

```
40 LET T=0: LET DT=0.01: LET TFIN=10
```

The calculation interval `DT` corresponds (using CSSL'67 terminology) to the communication interval (usually denoted by `CINT`) divided by the number of steps inside it (usually denoted by `NSTEPS`) `DT=CINT/NSTEPS`. `DT` must be chosen according to the dynamic properties of the simulated model in order to avoid numerical problems and to satisfy the required accuracy. The numerical aspects of integration algorithms will be discussed in Section 5.1. Our choice of `DT` and `TFIN` corresponds to the choice of `CINT=0.01`, `NSTEPS=1` and `TFIN=10` in the CSSL standard program given in Example 3.1.

The next step in the presimulation operation is the initialization of states. In this step the states, i.e. the outputs of all memory blocks (integrator outputs in our case) must be initialized. Using the same names as were used in Example 3.1, this can be performed as follows:

```
60 LET RAB=520: LET FOX=85
```

Finally, the header of the results presentation, which appears only once per simulation run, is written in the presimulation operations:

```
80 PRINT "TIME      RAB      FOX"
```

The actual simulation run is performed according to Fig. 3.1 in a loop which is repeated until the simulation run termination condition is fulfilled. According to the basic digital simulation concept represented in Fig. 3.2 two operations are performed in the loop: evaluation of derivatives at moment $t$ and integration, which yields the outputs of all integrators for the moment $t + \Delta t$. Two other operations must be performed for complete simulation, namely the presentation of the results and the simulation run termination condition test. The scheduling of all four operations depends mainly on the integration algorithm, and the simplest possible scheduling is as follows:

1. *Evaluation of the derivatives:* the derivatives, i.e. the integrator inputs, are evaluated with standard BASIC assignment statements:

```
110 LET RABDOT=A11*RAB-A12*RAB*FOX
120 LET FOXDOT=A21*RAB*FOX-A22*FOX
```

   These two statements are written in a very similar way to the CSSL standard language statements in Example 3.1; however there is a principal difference concerning the order of the statements. The latter is arbitrary in CSSL standard language, while this is not the case in general purpose programming languages. More about this problem will be said later.
2. *Presentation of the results:* the simplest possible results presentation is to write all variables in which we are interested on the screen:

```
140 PRINT T,RAB,FOX
```

   Compare this statement with the `OUTPUT` statement of Example 3.1! There are two major differences in the argument lists:

   - The independent variable `T` is presented automatically in CSSL standard language; in BASIC it must be declared in the `PRINT` statement argument list.
   - The number `100` as the first argument in the CSSL standard language `OUTPUT` statement means that the output on the screen appears only every one-hundredth communication interval. In our BASIC program, output appears on the screen whenever the `PRINT` statement is executed. With chosen calculation interval (`DT=0.01`) and simulation run duration (`TFIN=10`) 1000 output lines are presented on the screen, which makes the interpretation of the results difficult.

   Let us program the CSSL standard language possibility of presenting the results only every $N$th communication interval in BASIC. First, with our previously obtained experience in good programming, we define $N$ as a constant in that part of the program in which constants are defined:

```
45 LET N=100.
```

   Then a counter must be introduced and initialized:

```
46 LET NCOUNT=0
```

Instead of a single print statement No. 140, the following sequence is programmed:

```
135 IF NCOUNT>0 GOTO 145
140 PRINT T,RAB,FOX
141 LET NCOUNT=N
145 LET NCOUNT=NCOUNT-1
```

The PRINT statement No. 140 is now executed only once in a hundred passes of the above sequence, i.e. ten times in a complete simulation run. This corresponds to the CSSL standard language output specifications.

3. *The simulation run termination condition test:* in the CSSL standard language program the simulation run is terminated when the independent variable exceeds the prescribed simulation run duration time. In BASIC this condition is programmed with the following conditional statement:

```
160 IF T>=TFIN GOTO 210
```

which, in the case of a fulfilled condition, causes the program to exit the loop and to terminate the simulation run.

4. *Integration:* the simplest integration algorithm is the Euler algorithm, defined by Eqn (3.3) (and more thoroughly described in Section 5.1), which is programmed in BASIC as follows:

```
180 LET T=T+DT:LET RAB=RAB+RABDOT*DT: LET FOX=FOX+FOXDOT*DT
```

After integration the whole procedure in the loop repeats, since the program flow is returned to item 1 by the BASIC program instruction:

```
200 GOTO 110
```

The complete program for the prey and predator problem in BASIC is now summarized, including the comments in Fig. 4.8. □

The realization of the derivative evaluations in the above prey and predator problem is straightforward: the right-hand sides of statements 110 and 120, defining the derivatives of both populations (RABDOT and FOXDOT respectively), contain only known constants (A11 to A22) and outputs of the memory (integrator) blocks. If several nonmemory blocks are chained, the sequence of block evaluation is important. Digital simulation languages embody a sorting algorithm which sorts the blocks according to the rule that each block can be evaluated only if the block inputs (variables on the right-hand side of the corresponding expression) are constants, outputs of memory blocks (including independent variable) or outputs of blocks already sorted by the sorting algorithm.

With the general purpose program languages the sorting procedure must be accomplished 'by hand' as shown in the following heating control example.

```
10 REM MODEL'S CONSTANTS DEFINITION
20 LET A11=5: LET A12=.05: LET A21=.0004: LET A22=0.2
30 REM INDEPENDENT VARIABLE INCREMENT AND
31 REM SIMULATION DURATION DEFINITION
40 LET DT=.01: LET TFIN=10
45 LET N=100
46 LET NCOUNT=0
50 REM INITIAL CONDITION DEFINITION
60 LET T=0: LET RAB=520: LET FOX=85
70 REM PRESENTATION OF THE RESULTS HEADER
80 PRINT " TIME     RAB     FOX"
90 REM BEGINNING OF THE LOOP
100 REM EVALUATION OF DERIVATIVES
110 LET RABDOT=A11*RAB-A12*RAB*FOX
120 LET FOXDOT=A21*RAB*FOX-A22*FOX
130 REM PRESENTATION OF RESULTS
135 IF NCOUNT>0 THEN GOTO 145
140 PRINT T,RAB,FOX
141 LET NCOUNT=N
145 LET NCOUNT=NCOUNT-1
150 REM SIMULATION TERMINATION CONDITION TEST
160 IF T>=TFIN THEN GOTO 210
170 REM INTEGRATION USING EULER ALGORITHM
180 LET T=T+DT:LET RAB=RAB+RABDOT*DT:LET FOX=FOX+FOXDOT*DT
190 REM END OF THE LOOP
200 GOTO 110
210 REM END OF THE SIMULATION PROGRAM
220 END
```

Figure 4.8 BASIC program for the prey and predator problem.

### Example 4.4 The heating control problem simulated in BASIC

The heating control problem described in Problem 1.2 and simulated with CSSL standard language in Example 3.2 consists of only one integrator, so its output (denoted by `TH`) and the independent variable (time denoted by `T`) are known at the beginning of the sorting process. So, first of all, the reference temperature and the absolute value of the room temperature are evaluated (the order of these two evaluations is not important since the path between both quantities is broken by the integrator). Then the control error, the output of the bang–bang controller (hysteresis function), the heater output signal, its delayed signal and, finally, the time derivative of the room temperature, must be evaluated in exactly this order (see block diagram in Fig. 1.5).

The simulation program in BASIC for the heating control problem is shown in Fig. 4.9. The program structure is very similar to the program structure of the prey and predator problem (Example 4.3) and does not need further explanation. Note that, again, the same variable names as in the CSSL standard program in Example 3.2 are used. There are two blocks, however, the function generator described

```
10 REM MODEL'S CONSTANTS DEFINITION
20 LET DELTAY=1: LET PMAX=5: LET GAIN=2: LET TIMCON=1
30 LET THE=15:LET TDELAY =.1: LET THWO=1
40 REM FUNCTION GENERATOR DEFINITION
50 DIM REFX(5): LET REFX(1)=0: LET REFX(2)=6:
51 LET REFX(3)=9: LET REFX(4)=15: LET REFX(5)=21
60 DIM REFY(5): LET REFY(1)=15: LET REFY(2)=20
61 LET REFY(3)=18: LET REFY(4)=20: LET REFY(5)=15
70 REM SIMULATION RUN PARAMETERS DEFINITION
90 LET DT=.02: LET TFIN=24: LET N=10
100 REM INITIALIZATION OF TIME DELAY INTERNAL MEMORY
110 LET INDEX=TDELAY/DT+1: DIM WORK(INDEX)
120 FOR I=1 TO INDEX: LET WORK(I)=0: NEXT I
130 REM INITIAL CONDITIONS DEFINITION
140 LET T=0: LET THW=THWO: LET U=0: LET NCOUNT=0
150 REM PRESENTATION OF THE RESULTS HEADER
160 PRINT " TIME     THR     P     TH"
170 REM BEGINNING OF THE LOOP
180 REM FUNCTION GENERATOR FOR THE REFERENCE TEMPERATURE
190 FOR I=1 TO 5: IF T>=REFX(I) THEN THR=REFY(I): NEXT I
210 LET TH=THW+THE: LET E=THR-TH: REM ERROR SIGNAL EVALUATION
230 REM HYSTERESIS:
240 IF E>DELTAY/2 THEN LET U=1
250 IF E<-DELTAY/2 THEN LET U=0
270 LET P=PMAX*U: REM HEATER GAIN
280 REM DELAY REALIZATION
290 FOR I=INDEX TO 2 STEP -1: LET WORK(I)=WORK(I-1): NEXT I
300 LET WORK(1)=P
310 LET PD=WORK(INDEX)
320 REM EVALUATION OF THE DERIVATIVE
330 LET THWD=(-1/TIMCON)*THW+GAIN/TIMCON*PD
340 REM PRESENTATION OF RESULTS
350 IF NCOUNTS>0 THEN GOTO 380
360 PRINT T,THR,P,TH: LET NCOUNT=N
380 LET NCOUNT=NCOUNT-1
390 REM SIMULATION TERMINATION CONDITION TEST
400 IF T>=TFIN THEN GOTO 450
420 LET T=T+DT: LET THW=THW+THWD*DT:REM INTEGRATION USING EULER ALGORITHM
440 GOTO 170: REM END OF THE LOOP
450 END
```

Figure 4.9 BASIC program for the heating control problem.

by Table 3.1 in Example 3.2 and the time delay, and these are now explained in more detail.

The function generator in BASIC is defined by two arrays, REFX and REFY, which define the breakpoints and function values, respectively, and are included in the presimulation operations:

```
50 DIM REFX(5): LET REFX(1)=0: LET REFX(2)=6
51 LET REFX(3)=9: LET REFX(4)=15: LET REFX(5)=21
60 DIM REFY(5): LET REFY(1)=15: LET REFY(2)=20
61 LET REFY(3)=18: LET REFY(4)=20: LET REFY(5)=15
```

This program sequence corresponds to the CSSL standard TABLE statement. The reference temperature as the function generator output is evaluated by the following program sequence:

```
190 FOR I=1 TO 5: IF T>=REFX(I) THEN THR=REFY(I): NEXT I
```

Note that the above program sequence is not a complete equivalent of the CSSL standard function generator because no interpolation is involved. But with step-shaped functions, as it is our reference temperature, it yields correct results.

The time delay is realized with an array WORK, which must have the dimension TDELAY/DT+1, which is denoted in the program as INDEX. The array WORK is defined and initialized with

```
110 LET INDEX=TDELAY/DT+1: DIM WORK(INDEX)
120 FOR I=1 TO INDEX: LET WORK(I)=0: NEXT I
```

At every passing of the loop the array WORK must be shifted, the time delay input is stored into its first element and the delayed variable is picked out at the position INDEX:

```
290 FOR I=INDEX TO 2 STEP -1: LET WORK(I)=WORK(I-1): NEXT I
300 LET WORK(1)=P
310 LET PD=WORK(INDEX)
```

Along with realization of the above time delay, the situation without time delay (TDELAY=0) can also be studied. □

It must be pointed out that sorting cannot always be accomplished. If the model consists of two or several nonmemory chained blocks and the output of the last block is fed back as the input of the first one, e.g.

```
LET X1=X2+...
LET X2=X1+...
```

sorting cannot be performed. Such model structures are called 'algebraic loops'. Section 5.2 deals with problems concerning algebraic loops.

With more sophisticated scheduling and integration algorithms, programs written in BASIC lose lucidity and in such cases languages which allow structured programming are preferable (e.g. PASCAL or FORTRAN). The program can be structured according to Figs 3.3 and 3.16: the main program and three subprograms – INTEG, DERIV and OUTPUT (subprograms are realized as procedures in PASCAL). Normally, the program is run several times with different data (e.g. initial conditions, model constants, etc.). In BASIC this is accomplished simply by changing the corresponding statements in the program and running it again and again. If the same method were used in PASCAL or FORTRAN, which are compilers, the source program would need to be recompiled each time, which may be a time consuming operation. Better programming practice is to include all the simulation parameters on a file, which is edited between simulation runs and read at the beginning of the program (in the presimulation operations). For our prey and predator problem such a file, named DATABASE, has the following form:

```
5      0.05  0.0004  0.2
0.01   10    100
0      520   85
```

In the first row are the model constants (A11, A12, A21 and A22), in the second are the simulation run control parameters (DT, TFIN and N) and in the third are the initial conditions of the independent variable ($T$) and both states (RAB and FOX respectively) are defined. Let us program the prey and predator problem first in PASCAL.

**Example 4.5 The prey and predator problem simulated in PASCAL**

The PASCAL program for the prey and predator problem is shown in Fig. 4.10 and uses the same variable names as the BASIC program (Fig. 4.8); and due to clear structured programming it is also understandable without comments. □

The heart of each simulation program is the integration subprogram, which usually performs operation scheduling along with integration – the DERIV subprogram activation for the evaluation of derivatives and the OUTPUT subprogram activation for presentation of the results.

The program loop in the INTEG subprogram is repeated until the termination condition is fulfilled. The use of professionally programmed integration subprograms, which are commercially available as procedures (in PASCAL) or subroutines (in FORTRAN) for digital simulation systems or other program packages is recommended. In general, the use of such subprograms involves four problems:

1. The commercially available subprogram 'does not know' the number of integrators in our program.
2. The commercially available subprogram 'does not know' the names of the variables to be integrated.
3. The commercially available subprogram 'does not know' the simulation run termination condition.
4. The commercially available subprogram 'does not know' the names of the derivatives evaluation and results presentation subprograms.

The first two problems are solved by the introduction of arrays for the integrator outputs and inputs and the use of them, together with the number of integrators, as the subprogram parameters.

The solution to the third problem is that the simulation run termination test is not evaluated in the integration subprogram but, rather, in either the derivatives or results presentation subprogram, which sends a simulation run termination notice (usually called a 'flag') to the integration subprogram. With more sophisticated

```
Program Prey_predator(input,output);
var t,dt,tfin,a11,a12,a21,a22: real;
  rab,fox,rabdot,foxdot: real;
  n,ncount: integer;
  data_base: text;
procedure deriv;
  (* defines the model structure, i.e. evaluates the derivatives *)
  begin
    rabdot: =a11*rab-a12*rab*fox; foxdot=a21*rab*fox-a22*fox;
  end; (* deriv *)
procedure output;
(* presents the simulation results ones in n activations *)
  begin
    if ncount<=0 then begin
                        writeln(t,rab,fox);ncount:=n;
                      end;
    ncount:=ncount-1;
  end; (*output *)
procedure integ;
  (* schedules operations and performs Euler integration *)
  begin
    repeat begin
             deriv; output;
             t:=t+dt;rab:=rab+rabdot*dt; fox:=fox+foxdot*dt;
           end
    until t>=tfin;
  end; (* integ *)
begin (* main program *)
  (* reads the model's constants and simulation run
  controlling data from the file *)
  assign(data_base,'database'); reset(data_base);
  readln(data_base, a11, a12, a21, a22);
  readln(data_base,dt,tfin,n);readln(data_base,t,rab,fox);
  ncount:=0; writeln('    TIME     RAB     FOX');
  integ;
end.
```

Figure 4.10 PASCAL program for the prey and predator problem.

integration algorithms the results presentation subprogram is called less frequently. So, for efficiency it is convenient to include the simulation termination test in it.

The fourth problem can be elegantly solved in FORTRAN, which allows the names of the called subprograms (subroutines) to be arguments in the integration subprogram reference. The actual names of the derivatives evaluation and results presentation subprograms must, however, appear in a preceding `EXTERNAL` statement. In PASCAL, either the source code of the integration subprogram (procedure) must be available or the default names must be used.

If arrays are introduced for integrator outputs and inputs the derivative evaluation subprogram may lose its lucidity since an array name with corresponding index must be used instead of a problem oriented name (e.g. in PASCAL `x[1]` and `xdot[1]` instead of RAB and RABDOT respectively). FORTRAN provides an elegant solution to this inconvenience also: the EQUIVALENCE statement. The EQUIVALENCE statement (e.g. EQUIVALENCE (X(1), RAB)) allows the same memory location to be referred either by first (e.g. X(1)) or second (e.g. RAB) name. Let us now program the prey and predator problem in FORTRAN.

**Example 4.6 The prey and predator problem simulated in FORTRAN**

We shall start with the integration subprogram and, in order to keep the continuity of prey and predator problem solutions, the integration subprogram will be given here with a simple Euler algorithm. Since FORTRAN has no global variables, the transfer of variables from one subroutine to another must be realized either by subprogram parameters or by COMMON statements. As only object code is usually available for commercially available integration subprograms, all variables must be transferred by subprogram arguments. The minimum set of integration subprogram arguments consists of initial time, calculation interval, integration process termination flag (all three are usually joined in a parameter vector), resulting vector of integrals, vector of derivatives, length of both vectors (i.e. number of integrators) and names of the derivatives evaluation and results presentation subprograms respectively. A very simple integration subprogram is shown in Fig. 4.11. The description of the above integration subprogram parameters is as follows:

- PRMT –an array with dimension greater than or equal to 3 with integration algorithm parameters: PRMT(1) initial value of the independent variable; PRMT(2) calculation interval; and PRMT(3) simulation run termination flag. All three parameters must be defined at the INTEG call; PRMT(3) must be zero and is later redefined to a nonzero value by the subroutine OUTPUT if the simulation run termination condition is fulfilled.
- X –an array of integrator output (states). At the INTEG call this must contain the initial values of all the integrators.
- XDOT –the array of derivatives (integrator inputs).
- NDIM –the number of integrators (states).
- DERIV –the name of the derivatives evaluation subprogram.
- OUTPUT –the name of the results presentation subprogram.

The INTEG subroutine is activated from the main program, which is a direct translation of the PASCAL main program of Example 4.5 and is shown in Fig. 4.12. The derivatives evaluation subprogram DERIV and the results presentation subprogram OUTPUT are shown in Figs. 4.13 and 4.14 respectively. The EQUIVALENCE statements in the derivatives evaluation and results presentation subprograms allow us to use our standard names (e.g. RAB, RABDOT) and to integrate the same variables as vectors in the integration subprogram (where, e.g., RAB and RABDOT are referred to as X(1) and XDOT(1) respectively). The model constants (A11 to A22), the simulation termination data (TFIN) and the results presentation data (N, NCOUNT) are passed to the derivatives evaluation (DERIV) and to the results presentation (OUTPUT) subprograms by the COMMON block.

The vector of derivatives and the vector of states are passed from the integration subprogram (INTEG) to the derivative evaluation subprogram (DERIV) and, vice versa, indirectly by INTEG arguments to the main program and from here by the

```
      SUBROUTINE INTEG(PRMT,X,XDOT,NDIM,DERIV,OUTPUT)
      DIMENSION PRMT(1),X(1),XDOT(1)
      T=PRMT(1)
   10 CALL DERIV(T)
      CALL OUTPUT(T,PRMT)
      IF(PRMT(3).NE.0) RETURN
      T=T+PRMT(2)
      DO 20 I=1,NDIM
   20 X(I)=X(I)+XDOT(I)*PRMT(2)
      GO TO 10
      END
```

Figure 4.11 The integration subprogram.

```
      PROGRAM PREY AND PREDATOR
      DIMENSION X(2),XDOT(2),PRMT(3)
      COMMON X,XDOT,A11,A12,A21,A22,TFIN,N,NCOUNT
      EXTERNAL DERIV,OUTPUT
      EQUIVALENCE (X(1),RAB),(X(2),FOX)
      EQUIVALENCE (PRMT(2),DT)
      OPEN(1,FILE='DATABASE',FORM='FORMATTED',STATUS='OLD')
      READ(1,*)A11,A12,A21,A22
      READ(1,*)DT,TFIN,N
      READ(1,*)T,RAB,FOX
      CLOSE(1)
      NCOUNT=0
      PRMT(1)=T
      PRMT(3)=0.
      CALL INTEG(PRMT,X,XDOT,2,DERIV,OUTPUT)
      STOP
      END
```

Figure 4.12 The FORTRAN main program.

```
      SUBROUTINE DERIV(T)
      DIMENSION X(2),XDOT(2)
      COMMON X,XDOT,A11,A12,A21,A22,TFIN,N,NCOUNT
      EQUIVALENCE (X(1),RAB),(X(2),FOX)
      EQUIVALENCE (XDOT(1),RABDOT),(XDOT(2),FOXDOT)
      RABDOT=A11*RAB-A12*RAB*FOX
      FOXDOT=A21*RAB*FOX-A22*FOX
      RETURN
      END
```

Figure 4.13 The DERIV subprogram.

```
      SUBROUTINE OUTPUT(T,PRMT)
      DIMENSION PRMT(1)
      DIMENSION X(2),XDOT(2)
      COMMON X,XDOT,A11,A12,A21,A22,TFIN,N,NCOUNT
      EQUIVALENCE (X(1),RAB),(X(2),FOX)
      IF (NCOUNT.LE.O) THEN
                        WRITE(**)T,RAB,FOX
                        NCOUNT=N
                        ENDIF
      NCOUNT=NCOUNT-1
      IF(T.GE.TFIN) PRMT(3)=1.
      RETURN
      END
```

Figure 4.14 The OUTPUT subprogram.

COMMON block to DERIV. The independent variable (T) and the simulation run parameter vector (PRMT) are passed among subprograms by means of their arguments. □

**Graphic possibilities**

According to Confucius, a picture says more than a thousand words and, indeed, the graphic presentation of results is generally preferred. Without doubt, graphics is a very attractive tool, but its programming is hardware dependent and requires considerable programming experience. For beginners, however, the use of commercially available graphic tools is recommended, where the results are written on a file and later processed by a graphics or data processing package (e.g. STATGRAPHIC, MSCHART, LOTUS, SIMPHONY, etc.). This method of data preparation corresponds to the PREPAR option of CSSL'67 syntax. Files for saving intermediate results must be opened in presimulation operations. For our prey and predator example in BASIC this may be accomplished by

```
55 OPEN "RESULTS" FOR OUTPUT AS #1
```

in PASCAL by

```
var results_file: text;
assign(results_file,'results')
rewrite(results_file);
```

and in FORTRAN by

```
OPEN(2,FILE='RESULTS',FORM='FORMATTED',STATUS='NEW')
```

During the course of the simulation the results must be written on the file, e.g. in BASIC, by

```
146 PRINT #1,T,RAB,FOX
```

Note that this statement will be executed at every communication interval which corresponds to the `PREPAR RAB,FOX` statement in Example 3.1. The corresponding PASCAL statement

```
writeln(results_file,t,rab,fox);
```

and FORTRAN statement

```
WRITE(2,*)T,RAB,FOX
```

must be included at appropriate points in the procedure `OUTPUT` and in the subprogram `OUTPUT` respectively. In postsimulation operations, files must be closed by

```
CLOSE #1
```

in BASIC, by

```
close (results_file)
```

in PASCAL, and by

```
CLOSE(2)
```

in FORTRAN respectively.

## 4.5 SIMULATION IN COMPUTER AIDED DESIGN PACKAGES

Simulation is inextricably involved in the design of systems of any kind, such as control systems, electronic circuits, digital filters, etc. Paper, pencil, ruler and simple simulation tools, used for the design of such systems a few years ago, have been replaced by powerful and user friendly packages running on a variety of computers including the PC. The use of computer aided design (CAD) packages is a necessity nowadays, especially for the design of complex systems. Simulation in specialized CAD packages is strongly problem oriented and is beyond the scope of this book which is devoted to the general approach to simulation. However, some ideas will be given on how to solve the examples, which represent the thread of this book, with computer aided control system design (CACSD) packages.

Most CACSD packages originated in the academic community and some of them became commercially available later. It is by no means the intention of this section to give a detailed overview of CACSD packages and so only some of them, well known in universities and related institutes, will be listed here, together with the references for more information: MATLAB – Moler (1982), MATLAB user's guide; MATRIXx – Shah *et al.* (1985); CTRL-C – Little *et al.* (1985); the set SIMNON, IDPACK, POLPACK and MODPACK – Åström (1985b); CADACS (formerly KEDDC) – Schmid (1985a, b); CYPROS – Tysso (1985); L-A-S – West *et al.* (1985); CLADP – Maciejowski and MacFarlane (1985); IMPACT – Rimvall and Cellier (1985); Rimvall and Cellier (1986); BLAISE – Delebeque and Steer (1985); Delebeque and Steer (1986); ANA – Šega *et al.* (1985).

CACSD packages are very efficient tools for the design of control systems. They include a set of preprogrammed functions for the special tasks of the control system concept, so that the user can concentrate on the problem to be solved rather than on the programming. User friendly interactive features mean that the design procedure can be repeated with different sets of parameters in order to achieve the most acceptable solution. Some of them possess graphics possibilities to render problem definition easier. So they are an ideal tool for solving the problems of control theory, where simulation plays an important role. By means of simulation the system represented in any form (transfer function, state space, block representation) is transformed into a time series, which tells the user how it behaves in the time domain. As we shall see in the Examples 4.7 and 4.8, a transfer function can be simulated in a single-line program which is a huge reduction of programming effort in comparison to general purpose simulation languages.

Since control systems are nothing but dynamic systems described by differential or difference equations, all systems represented in the same mathematical form (differential or difference equations) can be simulated by the simulation facilities of CACSD packages. The subsequent examples will illustrate this and MATLAB, being very popular in the academic community, will be used. In its original form, MATLAB is a package for solving problems in linear algebra. However, its matrix computation and numerical analysis capabilities provide a reliable foundation for applications in different areas of system theory. Collections of preprogrammed algorithms (the so-called `M` files) are available as TOOLBOXES. Only one function of the SIGNAL PROCESSING TOOLBOX and CONTROL TOOLBOX, respectively, will be used in our examples. MATLAB essentially works with only one kind of object, a rectangular numerical matrix which is surrounded by brackets. The elements of a row are separated by blanks or commas, the rows by semicolons.

### Example 4.7 Butterworth filter

The time response of the second order low pass Butterowrth filter with cut off frequency 50 Hz to an input signal (produced by a random number generator) has to be evaluated. The same filter has already been simulated with BASIC in

```
u=rand(100,1)
y=filter([0.06396 0.12792 0.03696],[1 -1.1638 0.4241],u);
k=1:1:100;
subplot(211);
plot(k,u,':',k,y,'-');
xlabel('k');
ylabel('y[_ _ _ _ _],u[.....]');
```

Figure 4.15 MATLAB program for the simulation of the Butterworth filter.

Example 2.6. Using a sampling frequency of 500 Hz the transfer function of the filter is

$$G(z) = \frac{Y(z)}{U(z)} = \frac{0.063\,96z^2 + 0.127\,92z + 0.063\,96}{z^2 - 1.1683 + 0.4241} \tag{4.1}$$

The *filter* function of the signal processing toolbox is an appropriate tool for the simulation of discrete transfer functions (and, of course, of all systems described by difference equations). Three arguments defining the numerator polynomial, the denominator polynomial and the input function must be specified. The fourth argument, which is optional, defines the initial conditions of the unit delays of the realization scheme in nested form. If it is omitted, all initial conditions are set to zero. Considering that in MATLAB the numerator and denominator polynomials are represented as two matrices with dimensions $1 \times 3$, i.e. as two rows, and that the *rand* function generates a random signal, the instruction for the simulation of the transfer function (4.1) can be written on one single line:

```
y=filter([0.06396 0.12792 0.06396],[1 -1.1638 0.4241],rand(100,1));
```

The arguments of the rand function define the size of the input signal. In our case this is a matrix, $100 \times 1$, i.e. a column of length 100. The duration of the simulation is defined by the length of the input signal, so in our case the simulation is terminated after the evaluation of 100 samples. The simulation results can be represented by the *plot* utility where the counter k is used for the time axis. Axes can be labelled as shown in Fig. 4.15, which gives the complete program for the simulation of the discrete transfer function and the representation of results. The simulation results, i.e. the input and output time series are shown in Fig. 4.16. □

**Example 4.8 Fibonacci series**

In this example the Fibonacci equation

$$y(k+2) = y(k+1) + y(k) \qquad y(1) = y(2) = 1 \tag{4.2}$$

will be solved by MATLAB. Recall that it has been solved in Example 2.5 by BASIC. As MATLAB has no facility for solving homogeneous difference or differential

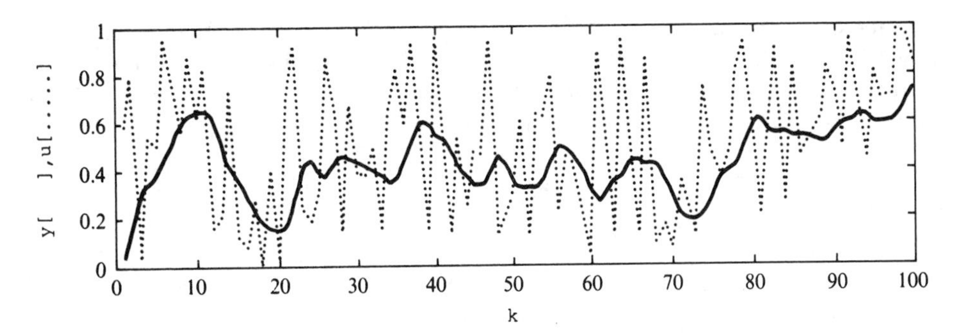


Figure 4.16 Input and output signals of the Butterworth filter.

equations, a trick must be used and Eqn (4.2) is written as the transfer function with zero numerator

$$G(z) = \frac{0}{z^2 - z - 1} \tag{4.3}$$

The initial conditions $y(1)$ and $y(2)$ are used as the initial conditions of the unit delay elements in the fourth argument of the *filter* function, which was explained in the preceding example. The input function is arbitrary; however its length determines the duration of the simulation. The MATLAB (SIGNAL PROCESSING TOOLBOX) statement for the evaluation of the Fibonacci numbers is thus

```
y=filter(0,[1 -1 -1],zeros(100,1),[1 1]);
```

□

**Example 4.9 Transfer function**

Continuous time transfer functions in MATLAB can be simulated in a similar way as discrete ones. The appropriate tool is the `lsim` function of the CONTROL TOOLBOX, which simulates continuous time linear systems with arbitrary inputs. When systems described by transfer functions are simulated, four arguments of the `lsim` function are used, representing the numerator and denominator polynomials, the input signal and the vector of the time axis. The time axis defines the time points at which the input signal is defined. The length of the time axis and the input signal must be the same and so the time axis also defines the time points in which the result, i.e. the transfer function output, is obtained. The response of the heating process with the transfer function

$$G(s) = \frac{2}{s+1} \tag{4.4}$$

to a step change of the heating power is simulated by the following instruction:

```
the=lsim(2,[1 1],[zeros(20,1);ones(81,1)],0:0.05:5)
```

The time axis is defined in 101 points ranging from 0 to 5 hours (with a resolution of 0.05 hours); the step function jumps from 0 to 1 after 20 samples (1 hour).

This is the easiest way to simulate linear continuous time systems in MATLAB. Another use of the `lsim` function will be shown in the next example. □

**Example 4.10 Car suspension system**

In this example a car suspension system will be simulated in MATLAB. The corresponding differential equation was derived in Example 1.1 and has the following form:

$$\dddot{y} + 70\ddot{y} + 300\dot{y} + 1000y = 0 \qquad y(0) = -0.05 \tag{4.5}$$

This equation is homogeneous so that the procedure given in Example 4.8 could be used (i.e. a transfer function with zero numerator). However, the `lsim` function has no capacity to introduce initial conditions, so Eqn (4.5) has to be transformed into the state space form and then the `lsim` function in the form appropriate for the simulation of systems described in state space can be used. The state space description of Eqn (4.5), which is obtained by introducing

$$x_1 = y \qquad x_2 = \dot{y} \qquad x_3 = \ddot{y} \tag{4.6}$$

yields

$$\begin{bmatrix} \dot{x}_1 \\ \dot{x}_2 \\ \dot{x}_3 \end{bmatrix} = \begin{bmatrix} 0 & 1 & 0 \\ 0 & 0 & 1 \\ -1000 & -300 & -70 \end{bmatrix} \cdot \begin{bmatrix} x_1 \\ x_2 \\ x_3 \end{bmatrix} \qquad \begin{bmatrix} x_1(0) \\ x_2(0) \\ x_3(0) \end{bmatrix} = \begin{bmatrix} -0.05 \\ 0 \\ 0 \end{bmatrix} \tag{4.7}$$

$$y = [1 \;\; 0 \;\; 0] \begin{bmatrix} x_1 \\ x_2 \\ x_3 \end{bmatrix} \tag{4.8}$$

The `lsim` function can be used for the simulation of linear systems described in state space form

$$\dot{x} = \mathbf{Ax} + \mathbf{b}u \qquad \mathbf{x}(0) = \mathbf{x}_0 \tag{4.9}$$

$$y = \mathbf{cx} + du \tag{4.10}$$

In this case seven arguments are used (the description of system $\mathbf{A}$, $\mathbf{b}$, $\mathbf{c}$ and $d$, the input function, the time axis and the initial conditions $\mathbf{x}_0$), and the car suspension system can be simulated by the following two instructions:

```
A=[0 1 0;0 0 1;-1000 -300 -70]
y=lsim(A,[0;0;0],[1 0 0],0,zeros(101,1),0:0.1:10,[-0.05;0;0]);
```

where the matrix `A` was defined in a separate line. □

**Example 4.11 Prey and predator problem**

The prey and pedator problem will illustrate how to simulate nonlinear systems in MATLAB. The problem was introduced in Problem 1.3 and simulated in Examples 3.1 and 4.3. It is described by two nonlinear differential equations:

$$\dot{x}_1 = 5x_1 - 0.05x_1x_2 \qquad x_1(0) = 520 \tag{4.11}$$

$$\dot{x}_2 = 0.0004x_1x_2 - 0.2x_2 \qquad x_2(0) = 85 \tag{4.12}$$

MATLAB has two functions for solving ordinary differential equations. These use second and third (`ode23`) or fourth and fifth (`ode45`) order Runge–Kutta formulas.

The nonlinear system must be represented in the state space, i.e. as a set of first order differential equations. Both functions require four arguments: a string variable with the name of the function for the evaluation of derivatives, the starting time of the integration, the finishing time of the integration and a column vector of initial conditions. Two optional arguments can be added for the definition of desired accuracy and for tracing the intermediate results. Both functions yield the time axis, i.e. the vector of time points in which the solution is obtained, and the solution itself in the form of a matrix where each column represents the solution in one time point. In our case the integration function can be invoked as follows:

```
[t,x]=ode45("deriv",0,10,[520;85])
```

The function `deriv`, which must be written by the user, must have two input arguments (the scalar `t` as the time and the column vector `x` as the state) and must generate the column vector of state derivatives. This function is the MATLAB counterpart of the `DERIV` subprogram used with general purpose programming languages described in Section 4.4. For our prey and predator problem the `deriv` function has the following form:

```
function xdot=deriv(t,x)
xdot(1)=5*x(1)-0.05*x(1)*x(2);
xdot(2)=0.0004*x(1)*x(2)-0.2*x(2);
```

The results can be presented by MATLAB's standard *plot* facility. □

## 4.6 ANALOG AND HYBRID SIMULATION SYSTEMS

Continuous systems were formerly simulated primarily on analog computers and thus the specifics of analog and hybrid computing will be discussed briefly in this section. Although it seemed initially that the application area of analog and hybrid computers would converge to zero it then became obvious that their use only decreased due to the increasing capabilities of digital computer configurations

(Breitenecker, 1989). However, some specialized areas still exist where analog and hybrid systems are needed. Besides, the important field of analog computers is surely educational since their properties enable very illustrative representation of the modelling and simulation approach. Finally, some principles, which originate in analog computation (e.g. scaling), can also be applied to other areas.

### 4.6.1 Basics of Analog Computing

First, let us summarize briefly the main properties and differences between digital and analog computers from the point of view of simulation.

*Digital computers.* This type of general purpose computer is available nowadays to almost everyone. Such devices, which originate from the abacus, use digits to express numbers (Jackson, 1960). The latter are generated by electrical pulses. Arithmetic and logical operations are performed sequentially. No problems occur in connection with memory, nor is there a vast amount of data storage (Neelamkavil, 1987). Many general and simulation languages are available. No direct mathematical integration is possible, while numerical integration may be problematic. However, for simulation purposes very high speeds of computing are required while real time simulations are questionable. Simulation execution time depends on the complexity of the simulated problem as well as on the required accuracy of simulation. More about modern hardware and software roles in simulation will be said in Chapter 6.

*Analog and hybrid computers.* These devices, which work on the principle of analogous relationships that exist between the physical quantities associated with a named device and those associated with a problem under study are called *analog computers.* Because their answers are obtained by measurements of some continuous quantity, such as the value of a voltage in an electronic analog computer, the latter is also called a continuous computer (Jackson, 1960). Analog computers probably originated from the development of the slide rule, where the mathematical operations were performed simply by adding or subtracting the distances on the frame and on the slider. Those distances were thus analogous to the problem variables. However, the most successful analog computers are electronic ones, which were developed from the special purpose machines (power network analysers, electronic analog simulators, etc.) to become, finally, general purpose computing systems. Due to parallelism in computation and because of the fact that particular mathematical operations are realized as analog electronic circuits, analog computers are to date the fastest simulation tool with a simulation speed that is independent of the model size. Here, highly interactive and illustrative simulation can be performed in compressed, real or expanded time. However, analog computers are expensive, especially the larger ones, because each operation requires its own computing element. Also, the capacity for program storage, documentation and representation of results are poor. Analog memory and function generation are also problematic. The relatively unusual kind of programming necessary must be noted (manual patching of particular components by cords), as well as the limited accuracy due to electronic element properties (Karba *et al.*, 1990). Also, the corresponding scaling which assures correlation between the problem and the computer variables is required in nearly all practical studies. However, the structure of the problem is apparent from the so-called analog simulation scheme, which represents correlation of analog elements and thus the program of the analog computer. Thus the results give a deeper insight into the behaviour of the system being studied. The output of the analog computer is displayed graphically in most cases and the solution of the given problem is determined mostly by several repetitions of the simulation run. *Hybrid systems* combine the advantages of digital and analog computers. In modern hybrid computation the programming is of a digital type. Some problems need extensive data transfer between the digital and analog parts, while the main disadvantage remains the high cost of such systems. Again more about such configurations will be said in Chapter 6.

**Basic analog computer operations**

The analog computer components, each of which performs a specific mathematical operation, can be classified into *linear and nonlinear* (*EAI Handbook*, 1967). The former perform the following mathematical operations:

- multiplication by a constant,
- inversion,
- algebraic summation,
- continuous integration,

while the latter perform the following operations:

- multiplication and division of variables,
- generation of arbitrary functions.

The above, supported by some others, which will be mentioned later, enable the simulation of complex mathematical models. Before giving some brief comments on analog components, let us introduce the *operational amplifier*, which is the basic element of the analog computer (*EAI Handbook*, 1967; Blum, 1969; and Tomović and Karplus, 1962). Any general purpose analog computer must have summing and integrating amplifiers as its most important components, and the direct coupled, inverting, high gain amplifier is the heart of such devices. The ideal operational amplifier must therefore have infinite gain $K$ (ratio of output to input voltage, typically of the order $10^8$), infinite bandwidth (functions with the same gain at all frequencies including zero), infinite input and zero output impedance, zero drift (zero output at zero input at all times) and zero phase shift. Note also, that such a device inverts the signal passing through it (the sign of the output is changed against that of the input). The input and output voltages are in general time functions which are measured with respect to ground. As the output voltage $e_o(t)$ is limited by the amplifier design in most cases (so-called 10 V machines) to a range of about $\pm 13$ V the input voltage $e_i(t)$ must remain less than 0.1 $\mu$V, taking into account the gain mentioned.

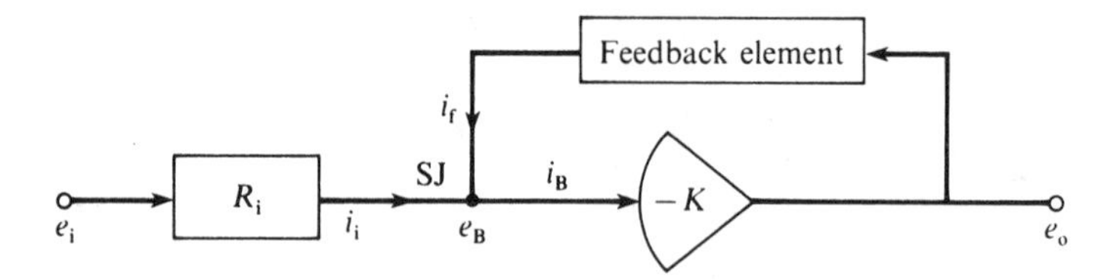


Figure 4.17 Operational amplifier.

This results in an input point, called *virtual ground* because it must remain near ground potential, which is used in the analysis of the network shown in Fig. 4.17.

Note that the symbol for the high gain amplifier is specific to the symbols of the general simulation scheme defined in Section 2.1.

Consider first the case where the feedback element in Fig. 4.17 is a pure resistor $R_f$. As $e_B$ is essentially zero (virtual ground), also $i_B \approx 0$. Kirchhoff's law for the summing junction (SJ) is therefore

$$i_i + i_f = 0 \tag{4.13}$$

and

$$\frac{e_i - e_B}{R_i} + \frac{e_o - e_B}{R_f} = 0 \tag{4.14}$$

From the relation

$$e_o = -ke_B$$

$e_B$ can be expressed as

$$e_B = -\frac{e_o}{k} \tag{4.15}$$

Combination of Eqns (4.14) and (4.15) gives

$$\frac{e_o}{R_f} = -\frac{e_i}{R_i} - \frac{e_o}{kR_i} - \frac{e_o}{kR_f} \tag{4.16}$$

Since $k \approx 10^8$ the last two terms of the right-hand side of Eqn (4.16) can be neglected giving, finally,

$$e_o = -\left(\frac{R_f}{R_i}\right) e_i \tag{4.17}$$

This relation means that the output of the amplifier is related to the input by a factor $R_f/R_i$ which can be unity or not. In the first case the analog inverter is realized, while in the second case a certain gain can be obtained. If the amplifier has several input resistors the summing operation is realized. A different effect can be achieved if the capacitor is used as the feedback element. In this case we have

$$\frac{e_i - e_B}{R_i} + C\frac{d}{dt}(e_o - e_B) = 0 \tag{4.18}$$

By substitution of Eqn (4.15) into Eqn (4.18), the following relation is obtained:

$$C\frac{de_o}{dt} = -\frac{e_i}{R_i} - \frac{e_o}{kR_i} - \frac{C}{k}\frac{de_o}{dt} \tag{4.19}$$

For the same reasons as in the preceding case, the last two terms on the right-hand side of Eqn (4.19) can be neglected yielding

$$\frac{de_o}{dt} = -\frac{e_i}{R_i C} \tag{4.20}$$

or, in integral form,

$$e_o = -\frac{1}{R_i C}\int e_i \, dt \tag{4.21}$$

Thus a device is obtained which has the output as the integral of the time varying input. The errors which can occur in connection with operational amplifiers can be the consequence of noninfinite amplifier gain, limited bandwidth, drift, parasitic capacitances, etc. (Tomović and Karplus, 1962). However, the reliability and accuracy of analog computers can be increased to some extent by the use of modern and quality electronic components.

Let us now give short descriptions of the basic analog computer components.

The *potentiometer* is the device for multiplication of a quantity with constant less than unity. The symbol for the potentiometer is shown in Fig. 4.18. Note that in the symbol the corresponding address of the component is usually written, which will also be the case for all other symbols. Without wishing to delve deeply into the electronic properties and analysis of the device, let us state (Blum, 1969) that the load of the potentiometer (the resistance to which the potentiometer is connected) influences the true setting of the potentiometer. Therefore, the latter must be set to the desired value under load. Any change of load changes the setting. General purpose analog computers have handset potentiometers; and potentiometers (sometimes called attenuators) which are set automatically to a given value in a special computer state where the potentiometers are already connected to their loads. The older type

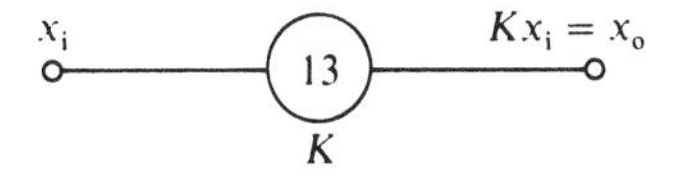


Figure 4.18 Symbol of the potentiometer.

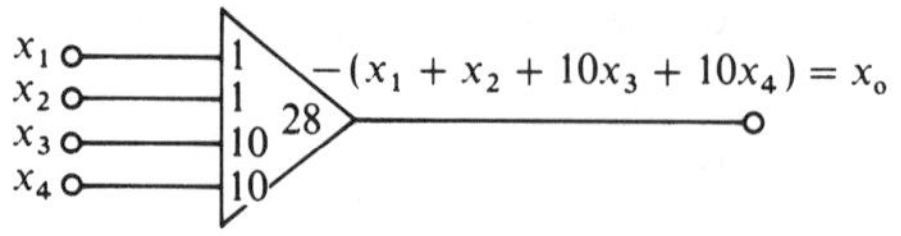

$x_1$, $x_2$, $x_3$, $x_4$ — 1, 1, 10, 10 — 28 — $-(x_1 + x_2 + 10x_3 + 10x_4) = x_o$

Figure 4.19 Symbol of the summer.

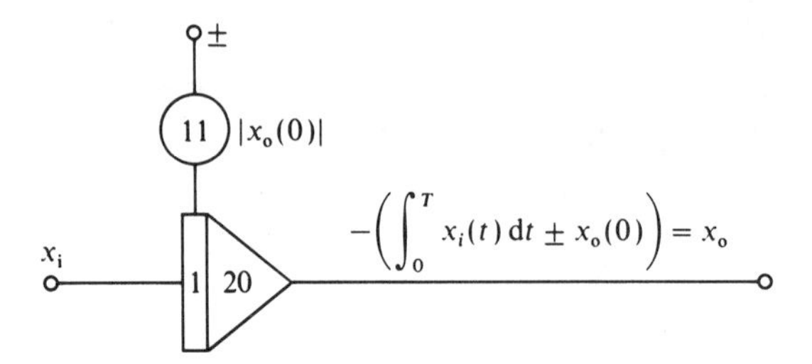


Figure 4.20 Symbol of the integrator.

of computer had so-called servopotentiometers (setting realized with correspondingly controlled servomotors) (EAI-580, 1968), while the newer types have digitally realized potentiometers which can realize positive or negative multiplication constants (special inverters are no longer necessary) (EAI-2000, 1982).

The *summer* is a device with several inputs, which gives as the output the negative sum of the inputs which according to the particular $R_f/R_i$ ratio, is weighted correspondingly. Its symbol is shown in Fig. 4.19 for one possible configuration of inputs. Note that the gain factors must be written in the symbol of the summer. The older analog machines had summers with four inputs in most cases (EAI-580, 1968), while the newer ones have a variety of possibilities (EAI-2000, 1982), namely have from four to seven inputs (some of them have the gain 10 and some the gain 1), which allow many combinations to achieve the desired gain.

The *integrator* is the main component of the analog computer, enabling fast and nonproblematic integration, which is the heart of simulation. Its symbol is shown in Fig. 4.20. Note that the integrator has an initial condition introduced through the potentiometer, whose input is connected to the reference voltage, usually $\pm 10$ V, and whose output is connected to a special point on the integrator circuit. The sign of the reference must be opposite to that of the integrator output for the defined positive initial condition, and equal to the sign of the integrator output for the defined negative initial condition. The configuration in Fig. 4.20 thus performs the following type of mathematical operation:

$$x_o(t) = -\left(\frac{1}{R_i C}\int_0^T x_i(t)\,dt + x_o(0)\right) \tag{4.22}$$

The fact that the integrator also changes the sign of the input must not be forgotten. Note that in simulation definite integrals are always solved, which means that the integration is performed in a certain prescribed time interval. Older types of analog machines had integrators with several inputs which performed summing and integration operations simultaneously (EAI-580, 1968), while the newer types mainly have integrators with only one input (EAI-2000, 1982). The factor $1/R_i C$ indicates that changes in input can cause the speed of integration to be increased or decreased, which can also be interpreted as a capacity for changing the integrator gain. General purpose analog computers have the ability to change factors to a greater or lesser extent and even to force each particular integrator to work in its specific regime. Also, the mode control for each particular integrator is enabled, which forces the integrator into different modes such as the mode of computing, the mode where the initial condition appears on the output of the integrator and the mode where the integrator holds the certain value of output.

Concerning the *nonlinear elements* let us say that these are more problematic from the analog computer point of view, with respect to electronic performance as well as to accuracy. The main element is the multiplier, which performs electronically. The performance of various analog computers differs, which is valid also for the symbols and patching. The operation of division can be performed by putting the multiplier in the feedback loop of the high gain amplifier. In a similar way, the element which produces the square root can be obtained by using the multiplier producing square of the variable as the feedback element of the high gain amplifier. Some specifics concerning sign change must be taken into account for any particular realization. General purpose analog computers sometimes also contain generators for specific functions such as sin, cos, arcsin, log, etc., as well as generators for the arbitrary function shapes. The latter devices perform linear interpolation between the points of the given function. Performance differs from that of diode function generators in older types (the so-called breakpoints of independent variables must be entered as well as the corresponding slopes to the next function value) to set the function generators digitally (where only the function values and the corresponding values of the independent variables are entered) in recent analog machines.

### 4.6.2 Analog Computer Programming

Preparation of the program for the analog computer is in fact so-called *analog simulation scheme* generation. It is closely related to the indirect approach for solving differential equations and the generation of the corresponding simulation scheme, which was discussed in detail in Section 2.1. The main differences are in the fact that the components which contain operational amplifier change the sign of the inputs and that some symbols which are specific to the type of computer used may be introduced. Let us try to illustrate the generation of an analog simulation scheme (often called an analog diagram) with the aid of examples.

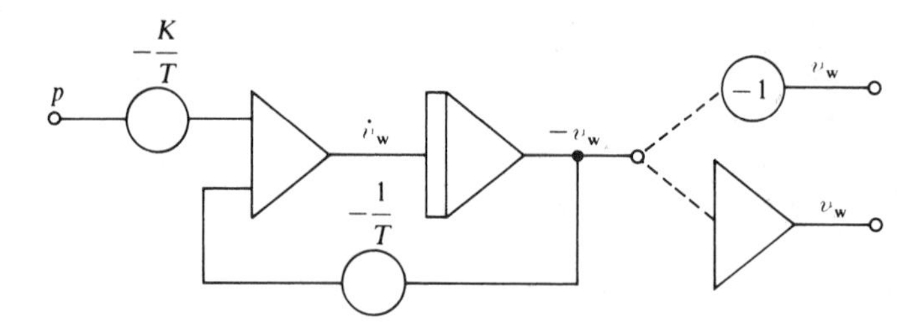


Figure 4.21 Analog simulation scheme for the heating control problem.

### Example 4.12 Heating control problem

The equation obtained for the indirect approach to the heating control problem (Example 2.1, Eqn (2.1)) should be rearranged into the form

$$\dot{v}_w = -\left(\frac{v_w}{T} - \frac{kp}{T}\right) \tag{4.23}$$

The negative sign in front of the right-hand side of Eqn (4.23) will be used to realize the sign change of the first summer in the analog simulation scheme which generates the highest derivative. Using a similar procedure to that used in Section 2.1, the analog simulation scheme in Fig. 4.21 is obtained.

We see that the sign is changed at the output of the integrator and that the terms generating the highest derivative have the sign which is valid inside the parentheses (the common minus sign is compensated by the summer). The result, $v_w$, which is obtained directly from Eqn (4.23), is negative and can be changed either by the potentiometer with value $-1$ or by the inverter. However, the best solution is to multiply the whole equation by a negative sign:

$$-\dot{v}_w = -\left(-\frac{v_w}{T} + \frac{kp}{T}\right)$$

obtaining a similar analog simulation scheme as that in Fig. 4.21 but with positive result and positive potentiometer $k/T$. The analog simulation scheme shown in Fig. 4.21 applies to the newer type of analog computer (potentiometer can change the sign, integrator with one input) and is also the same as that shown in Fig. 2.3, except that the sign of $v_w$ is negative here.

### Example 4.13 Car suspension system

In this example let us derive the analog diagram of a car suspension system for the older type of machines (summing integrators, servopotentiometers, inverters, etc.).

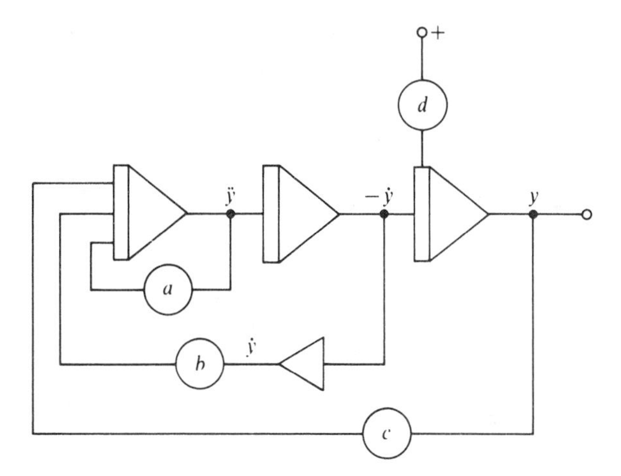


Figure 4.22 Analog simulation scheme for the car suspension system.

Thus Eqn (2.5) must be rearranged into the form

$$\frac{\mathrm{d}}{dt}(\ddot{y}) = -(a\ddot{y} + b\dot{y} + cy) \qquad y(0) = -d \tag{4.24}$$

The analog simulation scheme shown in Fig. 4.22 can thus be derived. Note that, due to the negative sign of the defined initial condition, the sign of the reference voltage at the input of the corresponding potentiometer is equal to the sign of its integrator output. For a sign change of $\dot{y}$ the inverter is used as well as the summing integrator for the highest derivative. As can be seen from Eqn (4.24), it is of lower order, which means that the time response for $\dddot{y}$ is not attainable. □

**Example 4.14 Methods for transfer function simulation**

Deriving the analog simulation scheme for the nested form method of transfer function simulation given in Section 2.1 is very easy as applied to the general simulation scheme in Fig. 2.8. For the newer analog machines with integrators having only one input, the realization of the subsidiary variables (Eqn (2.16)) is exactly the same as that for the general simulation scheme due to the fact that the summer and integrator are always used in pairs, thus giving no sign change. As the output is generated using Eqn (2.17), the only difference between the analog and general simulation schemes for the example discussed, and of course for the corresponding type of analog computer, is in the sign of $Y$ which is negative in analog simulation, and in the signs of the feedback potentiometers which are now positive as shown in Fig. 4.23.

The situation is slightly different in the case of the partitioned form method where some signs of potentiometers must be in accordance with the sign changes of

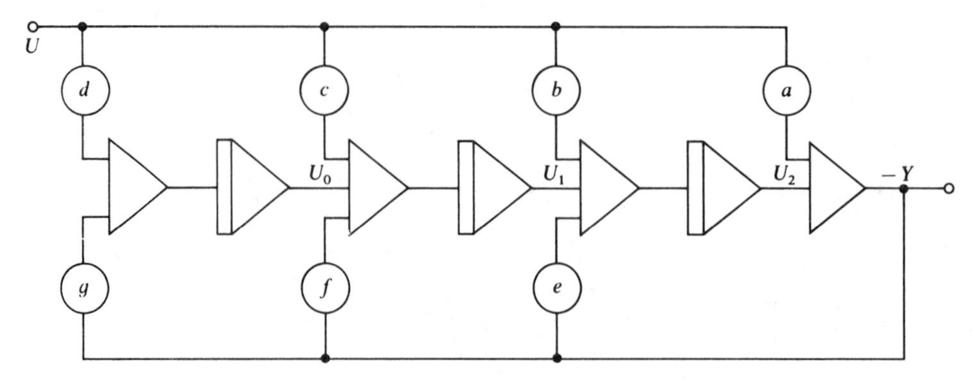


Figure 4.23 Analog simulation scheme for the nested form of transfer function simulation.

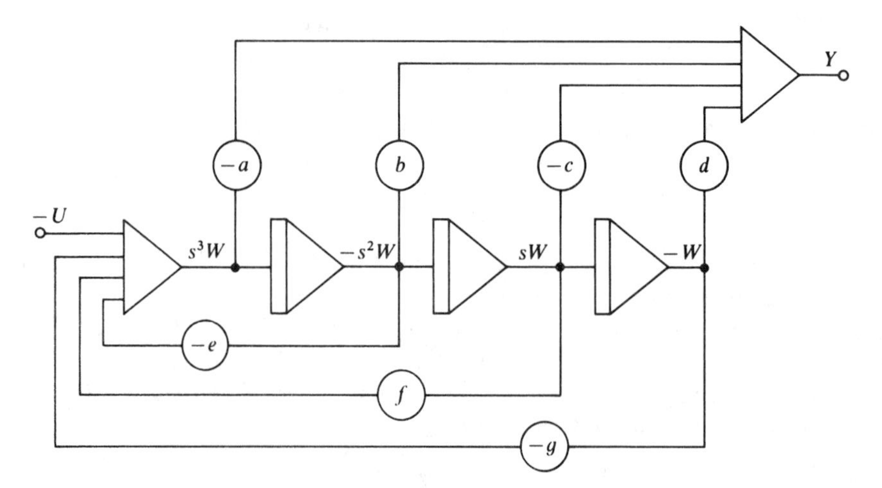


Figure 4.24 Analog simulation scheme for the partitioned form of the transfer function simulation.

the derivatives of $W$, which is a consequence of rearranging Eqn (2.20):

$$s^2W = -(es^2W + fsW + gW - U) \tag{4.25}$$

as well as Eqn (2.21):

$$Y = -(-as^3W - bs^2W - csW - dW) \tag{4.26}$$

The modified simulation scheme from Fig. 2.9, representing the analog simulation scheme, is shown in Fig. 4.24. □

The examples discussed show that the main differences between simulation and analog simulation schemes originate from the fact that the analog components containing the operational amplifier change the sign of the component output. However, the analog schemes obtained are still not the required representation of the analog program. As mentioned earlier, the problem variables must be adapted to the analog computer and converted into so-called *computer variables* by the magnitude and time scaling procedure, which will be discussed in the next subsection. The so-called *scaled analog simulation scheme* is finally obtained, containing the computer variables and the components in which their addresses are written. Together with the table containing the values of the potentiometers and their gain factors, it represents all necessary information for the analog program, which means patching the defined electronic components into the overall structure. If the addresses of the elements are chosen appropriately the structure of the problem is clearly visible on the computer patch panel, which increases the illustrativeness of simulation.

The scaled analog simulation scheme for the car suspension system is illustrated in Fig. 4.25, together with the table of potentiometers (Table 4.2). Here the scaling potentiometers (20, 21, 22) are chosen to be digitally set devices, while those representing the problem parameters are handset potentiometers enabling the efficient study of parameter influence on system behaviour. As can be seen in Fig. 4.25, the computer variables are given in square brackets representing the ratio between the problem variable and its maximum value. From this ratio the transformation can be made between voltage and problem variable quantities on an analog computer. All the elements in the scaled analog simulation scheme are addressed, defining uniquely the computer components which will be connected into the overall structure. The table of potentiometers (Table 4.2) gives the exact values and signs of particular potentiometers, as well as the gain factors of the inputs to which they are connected. Note that the signs of the potentiometers are also marked in the table of potentiometers (for certain types of computers, otherwise inverters are used). So a simple visual check of the scheme is possible. For the negative feedback loops, which are the most

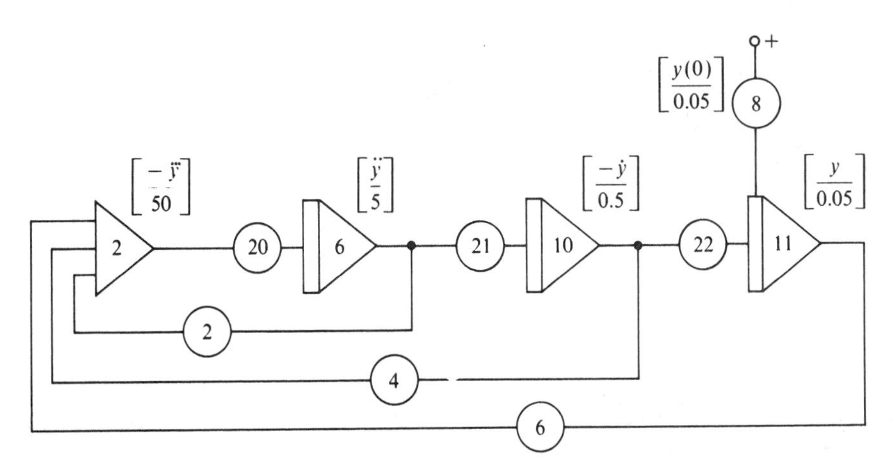


Figure 4.25 Scaled analog simulation scheme for the car suspension system.

Table 4.2 Table of potentiometers

| *Address* | *Meaning* | *Value* | *Gain* |
|---|---|---|---|
| 2 | $-\frac{k_1 + k_2}{f}$ | $-0.7$ | 10 |
| 4 | $\frac{k_2}{m}$ | 0.3 | 10 |
| 6 | $-\frac{k_1 k_2}{mf}$ | $-1.0$ | 1 |
| 8 | $\frac{y_0}{y_{max}}$ | 1.0 | / |
| 20 | $\frac{\dddot{y}_{max}}{n\ddot{y}_{max}}$ | 1.0 | 1 |
| 21 | $\frac{\ddot{y}_{max}}{n\dot{y}_{max}}$ | 1.0 | 1 |
| 22 | $\frac{\dot{y}_{max}}{n y_{max}}$ | 1.0 | 1 |

usual, an odd number of sign changes must occur (caused either by the potentiometer or by the operational amplifier in some component).

### 4.6.3 Magnitude and Time Scaling

When discussing basic methodologies for solving problems with simulation in Chapter 2, and the concepts of simulation systems tools in Chapter 3, we were not concerned with the magnitudes of the problem variables and the corresponding units. The range of problem variables may vary considerably, e.g. when simulating crystal structures the distances are measured in $10^{-10}$ m, while when simulating planetary movements they are measured in $10^{10}$ m. Also, the problem dynamics (speed of variation) may be quite different, ranging from microseconds (when simulating, e.g., the movement of an electron in a magnetic field) to thousands of years (when simulating, e.g., radioactive decay).

On the other hand, analog computers usually operate with voltages in the range $\pm 10$ V and with time dynamics which must be adjusted to electrical and mechanical characteristics (frequency range) of computing units and output devices. In order to be able to simulate problems on an analog computer, the appropriate magnitude and time scaling must be done. As shown in Example 4.17, the same procedures can also be applied in a dedicated analog hardware design. Magnitude and time scaling is in general not required in digital simulation. However, it is necessary if the problem magnitudes and/or time constants are in the vicinity of or exceed the minimum or maximum absolute real numbers which can be represented on the digital computer. Magnitude scaling is also required if algorithms are implemented on dedicated digital hardware using integer or fixed point arithmetic.

As mentioned, magnitude and time scaling are needed on analog–hybrid computers in order to establish the correspondence between problem and computer variables. This correspondence cannot be direct (one to one), which can be seen by the following example: If a problem variable, e.g., $x_1$ in our prey and predator example, has its value $x_1 = 700$ rabbits, then the amplifier whose output corresponds to $x_1$ would have to output 700 V, which is impossible. Another reason why the simulation schemes as represented in Chapter 2.1 cannot be run in their original form lies in the fact that the potentiometer setting is limited to 1 and the analog unit gains are at most 10 or 100 (depending on the hardware). If a constant in the problem equation turns out be be, e.g., $c = 1000$ in our car suspension system, it is also clear that it is impossible to obtain such a constant with a potentiometer and an appropriate unit gain.

On the other hand, problems also arise if an amplifier output or a potentiometer setting is too small. For example, on high accuracy computers potentiometers can be set with an accuracy of $10^{-4}$, which yields a relative error of 0.02% when the potentiometer is set to 0.5 and a relative error of 10% when it is set to 0.001. Setting the potentiometer to 0.000 05 is not possible at all. It is obvious that the amplifier noise becomes dominant if the amplifier output drops below the amplifier noise level. Similar effects can be observed when using integer arithmetic for simulation. The solutions of the above described problems are magnitude and time scaling.

Magnitude scaling is an algebraic operation which is performed on a system of differential equations in order to transform the problem variables to corresponding computer variables. There are several scaling methods, but only the most systematic and consistent method of normalized variables will be described here.

The idea of normalized variables is to divide the problem variable $x$ by its maximum absolute value $x_{max}$ yielding the normalized problem variable $[x/x_{max}]$. Modern analog–hybrid computers also deal with normalized variables, which are obtained by dividing the amplifier output by the reference voltage $e_{max}$, yielding normalized computer variables $[e/e_{max}]$.

The normalized problem and computer variables are dimensionless quantities within the range $\pm 1$. So the most straightforward way to use normalized variables is to make the problem normalized variable directly proportional to the computer normalized variable

$$\left[\frac{x}{x_{max}}\right] = \left[\frac{e}{e_{max}}\right] \tag{4.27}$$

The actual scaling factor which relates the problem physical quantity to the amplifier output voltage is obtained by division of the computer reference voltage by the maximum value of the problem variable:

$$e = \left(\frac{e_{max}}{x_{max}}\right) x \tag{4.28}$$

However, the user will never need to be concerned about the scale factor. The simulation results obtained by the standard output devices of analog–hybrid computers are in normalized variables, and in order to obtain physical variables the normalized variables have to be multiplied by the corresponding maximum values.

The use of normalized variables is equivalent to the use of an old analog voltmeter with a scale ranging from 0 to 1 and with different voltage ranges. A readout of 0.8 using a 5 V range results in a voltage of 4 V, while the same readout using a 100 V range results in 80 V.

The problem of scaling consists mainly of the problem of determining the maximum values of problem variables. If the estimated maximum value of a problem variable is smaller than its true maximum value, then the corresponding computer variable is greater than 1, which forces the corresponding amplifier to saturation. The simulation results are of course useless. If, on the other hand, the estimated maximum value of a problem variable exceeds its true value considerably (by a factor greater than 10), the amount of amplifier noise becomes significant. The simulation results are inaccurate. The same effects are observed when using integer arithmetic: underestimated and considerably overestimated maximum values result in overflows and increased quantization noise respectively.

## Estimation of maximum values

Several methods can be used for the estimation of maximum values of problem variables:

1. The basic method is via knowledge of the physical background.
2. When dealing with the $n$th order differential equation

$$a_0 x^{(n)} + a_1 x^{(n-1)} + \cdots + a_n x = f(t) \tag{4.29}$$

   or corresponding representation in state space, some speculative methods, which try to estimate the maximum values without solving the equation, can be used.

   According to the *equal-coefficient rule* (Jackson, 1960), the maximum values of $x$ and its derivatives are chosen in such a way that all coefficients in the scaled equation are approximately of the same magnitude. A restriction exists, however, which implies that the assumed maxima constitute a monotonic set. It means that the values of the coefficients continually increase (decrease) for each derivative in turn.

Another possibility for the estimation of maximum values of variables is based on an examination of time constants and *natural frequencies* of the system, described by Eqn (4.29). For first and second order unforced ($f(t) = 0$) and stable systems

$$a_0\dot{x} + a_1 x = 0 \tag{4.30}$$

and

$$a_0\ddot{x} + a_1\dot{x} + a_2 x = 0 \tag{4.31}$$

respectively, it can be easily shown that the estimated maximum values of the variable $x$ and its derivatives form a geometric progression. For the first order system (Eqn (4.30)) the solution is

$$x(t) = x(0)\exp\left(-\frac{a_1}{a_0}t\right) = x_{max}\exp\left(-\frac{a_1}{a_0}t\right) \tag{4.32}$$

Obviously, for stable first order systems the maximum value equals the initial value ($x_{max} = x(0)$). For second order systems the worse case estimate is obtained for undamped systems ($a_1 = 0$, i.e. damping neglected) with solution

$$x(t) = x_{max}\sin\left(\sqrt{\frac{a_2}{a_0}}t + \varphi\right) \tag{4.33}$$

yielding

$$\dot{x}_{max} = \sqrt{\frac{a_2}{a_0}}x_{max} \tag{4.34}$$

and

$$\ddot{x}_{max} = \sqrt{\frac{a_2}{a_0}}\dot{x}_{max} = \frac{a_2}{a_0}x_{max} \tag{4.35}$$

Analogously for higher order unforced and stable systems an 'average natural frequency' (Carlson *et al.*, 1967)

$$\bar{\omega} = \sqrt[n]{\frac{a_n}{a_0}} \tag{4.36}$$

is introduced. This frequency is actually the geometric mean of the absolute values of all characteristic equation roots and is the ratio of estimated maximum value of a derivative and its preceding derivative:

$$\bar{\omega} = \frac{\dot{x}_{max}}{x_{max}} = \frac{\ddot{x}_{max}}{\dot{x}_{max}} = \cdots = \frac{x_{max}^{(n)}}{x_{max}^{(n-1)}} \tag{4.37}$$

giving

$$\dot{x}_{max} = \bar{\omega} x_{max} \tag{4.38}$$

$$\ddot{x}_{max} = \bar{\omega} \dot{x}_{max} = \bar{\omega}^2 x_{max} \tag{4.39}$$

$$\vdots$$

$$x_{max}^{(n)} = \bar{\omega} x_{max}^{(n-1)} = \bar{\omega}^n x_{max} \tag{4.40}$$

Calculation of the maximum value of the highest derivative from the original equation using maximum values of the lower derivatives should be used in the case where all maxima occur at the same time. Both methods yield only the ratio between the maximum values of a variable and its derivatives. In order to establish actual maximum values, one has to estimate the maximum value of the variable or of one derivative. For unforced stable systems with all but one zero initial condition, the maximum value of the variable or derivative with nonzero initial condition is equal to the initial condition. For constant forcing function $f(t) = c$ the steady state value of $x$ is $c/a_n$. With all zero initial conditions and an assumption of 100% overshoot, $2c/a_n$ is a good guess for the maximum value of $x$. Here, the estimates for other maximum values, using the equal coefficient rule, can be obtained from the relation $x_{max}^{(i)} = c/a_{(n-i)}$. If the forcing function and several nonzero initial conditions are present, the maximum value estimate is obtained by the corresponding combination of particular maximum value estimates.

It must be pointed out again that both methods are speculative and that real maximum absolute values may differ from the estimated ones as shown in Example 4.15. The same principles can also be applied to nonlinear systems with corresponding approximations and linearizations, as shown in Example 4.16.

3. Digital simulation is used to determine maximum values in advanced hybrid systems.
4. Maximum values can be obtained by an initial simulation run with conservative scaling.

If the problem turns out to be poorly scaled, the scaling procedure has to be repeated. In this case the (inaccurate) results of the poorly scaled simulation run are used for a better estimation of maximum values.

## Magnitude scaling

Technically, the magnitude scaling procedure is performed as follows:

1. The original (nonscaled) equation is written for each block on the simulation scheme.
2. Problem variables are substituted with computer ones but in such a way that the equations remain valid, i.e. the problem variables are divided and

multiplied by the corresponding maximum values. The nonscaled equation is rearranged in such a way that normalized variables, potentiometer settings and input gains are shown explicitly. The resulting scaled equation has the following typical form:

$$[\textit{normalized output}] = -f\{\textit{potentiometer setting})$$
$$\cdot \textit{gain} \cdot [\textit{normalized input}]\} \qquad (4.41)$$

where $f$ is the function symbol of the block. For the integrator the function symbol $\int\{\ \}\,dt$ + (*initial condition*) is usually replaced by the time derivative operator $d/dt$ applied on the left-hand side of the equation and a separate equation is written for the initial condition:

$$[\textit{normalized output}] = -\{(\textit{potentiometer setting})\cdot[\pm 1]\} \qquad (4.42)$$

If the problem parameters (coefficients of the differential equation) vary within a certain range it is good practice to declare the corresponding potentiometer settings algebraically, as shown in Example 4.15.

In the magnitude scaling procedure given above it was supposed that the negative minimum value lies in the same magnitude range as its maximum value and so the whole range of the scaled variable is used. If this is not the case, which is true in particular when equations describe nonlinear systems (e.g. populations in the prey and predator problem cannot be negative) the normalized variable must be proportional to the deviation of the problem variable from its nominal value, also called the 'operating point'. The same result is, however, obtained by reformulation of the problem in terms of deviations from the operating point, as shown in Example 4.16.

**Time scaling**

The magnitude scaling described above was concerned with dependent variables of the problem. However, it is equally important to establish the correspondence between problem and computer independent variables. The independent variable on all analog–hybrid computers is time. The rate of computer variable change must be consistent with the dynamic properties of the computer and its peripheral devices. Also, the solution must take place within a reasonable amount of time. The problem independnet variable may not be time (it may be length, for instance) and a scaling factor must be introduced. Even if the problem independent variable is time, the dynamic range of the problem may be quite different from the dynamic range of the computer. For instance the problem may take hours (or millions of years) to come to an end, which would not be practical. The procedure that establishes the correspondence between problem and computer independent variables is called time scaling and is defined by

$$\tau = nt \qquad (4.43)$$

where $\tau$ is the computer independent variable (time, usually in seconds), $t$ is the problem independent variable and $n$ is the time scale factor. If the problem independent variable is time (in seconds), $n$ is a dimensionless quantity, otherwise its dimension is seconds divided by the problem independent variable unit. In the first case there are three possibilities: if $n > 1$ the solution on the computer is slower than the original problem, if $n < 1$ the computer simulation is faster than the original problem, and for $n = 1$ real time simulation is performed. When simulated on analog–hybrid computers, chemical, controlled thermonuclear, thermal, pharmacokinetical and socio-economical processes are in most cases speeded up, while electrical processes require a slowdown. Mechanical processes are usually simulated in real time or close to it ($n$ in the vicinity of 1).

How does the time scaling influence the scaled equations? Examination of all computer components indicates that the only time dependent element is the integrator. So all magnitude scaled integrator equations have to be changed in such a way that the operator $\mathrm{d}/\mathrm{d}t$ appearing on the left-hand side of them is replaced by the operator $n \cdot \mathrm{d}/\mathrm{d}\tau$, yielding a typical scaled equation of the form

$$\frac{\mathrm{d}}{\mathrm{d}\tau}[\textit{normalized output}] = -\{(\textit{potentiometer setting}) \cdot \textit{gain} \cdot [\textit{normalized input}]\} \tag{4.44}$$

The time scale factor $n$ thus influences (divides) the potentiometer settings (and gains) of all integrator inputs, while the equation for the integrator initial condition remains unchanged.

A good choice of $n$ is where the majority of integrator coefficients (potentiometer settings multiplied by gains) are in the range between 0.1 and 10. This cannot always be accomplished of course. In a correct magnitude scaled problem, if the ratio between the largest and smallest integrator coefficient is greater than 1000, an indication of the so-called stiff problem is given. The problem of stiffness cannot be reduced by the magnitude or time scaling.

### Example 4.15 Scaling of the car suspension system

The purpose of this example is to scale the equations for our car suspension problem. The corresponding simulation scheme and the analog simulation scheme were shown in Figs 2.5 and 4.22, respectively, while the equation is

$$\dddot{y} + 70\ddot{y} + 300\dot{y} + 1000y = 0 \qquad y(0) = -0.05 \tag{4.45}$$

Our next step is estimation of the maximum values. A damped transition is expected, so it is obvious that the maximum absolute value of $y$ is 0.05, i.e. the absolute value of the initial condition. So Eqn (4.45) is written in the form

$$\dddot{y}_{\max}\left[\frac{\dddot{y}}{\dddot{y}_{\max}}\right] + 70\ddot{y}_{\max}\left[\frac{\ddot{y}}{\ddot{y}_{\max}}\right] + 300\dot{y}_{\max}\left[\frac{\dot{y}}{\dot{y}_{\max}}\right] + 50\left[\frac{y}{0.05}\right] = 0 \tag{4.46}$$

Table 4.3 Maximum values of car suspension problem variables

| | *Estimated maximum values* | | | |
|---|---|---|---|---|
| | *Equal coefficient rule* | *Average natural frequency method* | *True values* | *Values for rescaling* |
| $y_{max}$ | 0.05 | 0.05 | 0.05 | 0.05 |
| $\dot{y}_{max}$ | 0.1666 | 0.50 | 0.101 | 0.125 |
| $\ddot{y}_{max}$ | 0.71 | 5.0 | 0.49 | 0.50 |
| $\dddot{y}_{max}$ | 50.0 | 50.0 | 50.0 | 50.0 |

and since the coefficient preceding $[y/0.05]$ is 50, the maximum values $\dddot{y}_{max}$, $\ddot{y}_{max}$ and $\dot{y}_{max}$ are chosen in such a way that all coefficients in Eqn (4.46) are 50 according to the equal coefficient rule. The resulting estimates are summarized in the first column of Table 4.3.

With the average natural frequency rule, the average natural frequency is

$$\bar{\omega} = \sqrt[3]{1000} = 10 \tag{4.47}$$

and the corresponding maximum values are summarized in the second column of Table 4.3.

For scaling, the more conservative estimates (the second column of Table 4.3) are chosen yielding the following equation for the first block (summer):

$$50\left[\frac{-\dddot{y}}{50}\right] = \left\{-70\cdot 5\cdot\left[\frac{\ddot{y}}{5}\right] + 300\cdot 0.5\cdot\left[\frac{-\dot{y}}{0.5}\right] - 1000\cdot 0.05\cdot\left[\frac{y}{0.05}\right]\right\} \tag{4.48}$$

and, in the final scaled form,

$$\left[\frac{-\dddot{y}}{50}\right] = -\left\{(-0.7)\cdot 10\cdot\left[\frac{\ddot{y}}{5}\right] + (0.3)\cdot 10\cdot\left[\frac{-\dot{y}}{0.5}\right] + (-1.)\cdot 1\cdot\left[\frac{y}{0.05}\right]\right\} \tag{4.49}$$

The scaled equations for all three integrators are as follows:

$$\frac{d}{d\tau}\left[\frac{\ddot{y}}{5}\right] = -\left\{\frac{10}{n}\left[\frac{-\dddot{y}}{50}\right]\right\} \tag{4.50}$$

$$\frac{d}{d\tau}\left[\frac{-\dot{y}}{0.5}\right] = -\left\{\frac{10}{n}\left[\frac{\ddot{y}}{5}\right]\right\} \tag{4.51}$$

$$\frac{d}{d\tau}\left[\frac{y}{0.05}\right] = -\left\{\frac{10}{n}\left[\frac{-\dot{y}}{0.5}\right]\right\} \qquad \left[\frac{y(0)}{0.05}\right] = -\{(1.)\cdot[+1]\} \tag{4.52}$$

With chosen time scale factor $n = 10$, all integrator coefficients become 1.

The results of the simulation are shown in Fig. 4.26 and from it the true absolute maximum values are obtained and shown in the third column of Table 4.3.

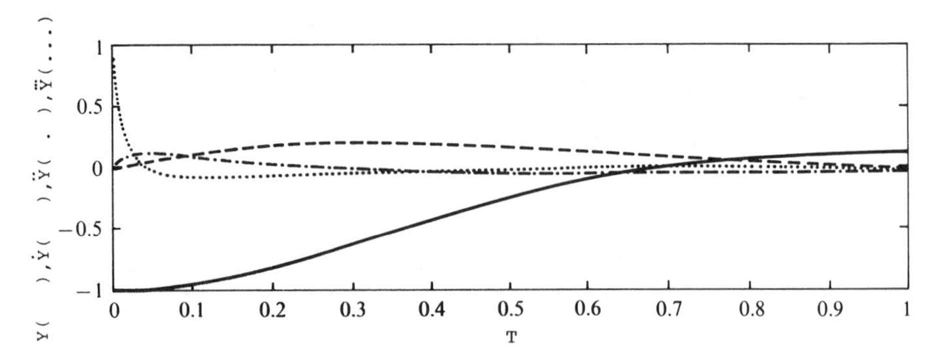


Figure 4.26 Scaled car suspension problem results.

Finally, the problem is rescaled with maximum values shown in the fourth columns of Table 4.3, yielding (with $n = 10$) the following scaled equations:

$$\left[\frac{-\dddot{y}}{50}\right] = -\left\{\left(\frac{-a}{100}\right)\cdot\left[\frac{\ddot{y}}{0.5}\right] + \left(\frac{b}{400}\right)\cdot\left[\frac{-\dot{y}}{0.125}\right] + \left(\frac{-c}{1000}\right)\cdot\left[\frac{y}{0.05}\right]\right\} \tag{4.53}$$

$$\frac{d}{d\tau}\left[\frac{\ddot{y}}{0.5}\right] = -\left\{(1.0)\cdot 10\cdot\left[\frac{-\dddot{y}}{50}\right]\right\} \tag{4.54}$$

$$\frac{d}{d\tau}\left[\frac{-\dot{y}}{0.125}\right] = -\left\{0.4)\cdot\left[\frac{\ddot{y}}{0.5}\right]\right\} \tag{4.55}$$

$$\frac{d}{d\tau}\left[\frac{y}{0.05}\right] = -\left\{(0.25)\cdot\left[\frac{-\dot{y}}{0.125}\right]\right\} \qquad \left[\frac{y(0)}{0.05}\right] = -\{(1.)\cdot[+1]\} \tag{4.56}$$

The scaled equation form of Eqn (4.53) is especially suitable if several simulation runs are executed with various parameters. The choice of $n = 10$, however, is not the only possibility. The choice of $n = 25$, for instance, would also give the integrator coefficients in the suggested range. Sometimes the choice can be dependent upon the real problem characteristics. □

### Example 4.16 Scaling of the prey and predator example

The purpose of this example is to show the maximum value estimation and scaling procedure for nonlinear systems (prey and predator problem):

$$\begin{aligned} \dot{x}_1 &= 5x_1 - 0.05x_1x_2 \qquad & x_1(0) &= 520 \\ \dot{x}_2 &= 0.0004x_1x_2 - 0.2x_2 \qquad & x_2(0) &= 85 \end{aligned} \tag{4.57}$$

First the operation point of the problem is determined. A good choice of operating point is the stationary value $(\bar{x}_1, \bar{x}_2)$, which is defined by

$$\dot{x}_1 = \dot{x}_2 = 0 \tag{4.58}$$

yielding

$$\bar{x}_1 = 500 \qquad \bar{x}_2 = 100 \tag{4.59}$$

Now new problem variables are introduced as deviations from the operation point:

$$\tilde{x}_1 = x_1 - \bar{x}_1 = x_1 - 500 \tag{4.60}$$

$$\tilde{x}_2 = x_2 - \bar{x}_2 = x_2 - 100 \tag{4.61}$$

and the system (4.57) becomes

$$\left.\begin{array}{ll} \dot{\tilde{x}}_1 = -0.05\tilde{x}_1\tilde{x}_2 - 25\tilde{x}_2 & \tilde{x}_1(0) = 20 \\ \dot{\tilde{x}}_2 = 0.0005\tilde{x}_1\tilde{x}_2 + 0.04\tilde{x}_1 & \tilde{x}_2(0) = -15 \end{array}\right\} \tag{4.62}$$

Our next step is the estimation of maximum values. It is expected that the magnitude of populations will oscillate around the stationary point and since populations cannot be negative, the favourable absolute maximum values of $\tilde{x}_1$ and $\tilde{x}_2$ are estimated to be 500 and 100 respectively.

The maximum values of $\dot{\tilde{x}}_1$ and $\dot{\tilde{x}}_2$ will again be estimated by two speculative methods applied on nonlinear systems, which makes them even more speculative. With the equal coefficients rule $\dot{\tilde{x}}_{1\max}$ and $\dot{\tilde{x}}_{2\max}$ are 2500 and 20, respectively, which can easily be shown if the maximum values of $\tilde{x}_1$ and $\tilde{x}_2$ are introduced into the system (4.62).

Using the average natural frequency rule the problem must first be linearized around the stationary value ($\tilde{x}_1 = \tilde{x}_2 = 0$). In our case this is done by neglecting the first term on the right-hand side of both equations in Eqn (4.62). From the remaining system, the second order linear equations

$$\left.\begin{array}{l} \ddot{\tilde{x}}_1 + \tilde{x}_1 = 0 \\ \ddot{\tilde{x}}_2 + \tilde{x}_2 = 0 \end{array}\right\} \tag{4.63}$$

are obtained with the natural frequency $\bar{\omega} = 1$. The estimated maximum values of $\dot{\tilde{x}}_1$ and $\dot{\tilde{x}}_2$ are thus $\dot{\tilde{x}}_{1\max} = \tilde{x}_{1\max} = 500$ and $\dot{\tilde{x}}_{2\max} = \tilde{x}_{2\max} = 100$. For scaling, more conservative values ($\dot{\tilde{x}}_{1\max} = 2500$ and $\dot{\tilde{x}}_{2\max} = 100$) are used, yielding the scaled equations

$$\left[\frac{\dot{\tilde{x}}_1}{2500}\right] = -\left\{\left[\frac{\tilde{x}_1}{500}\right]\cdot\left[\frac{\tilde{x}_2}{100}\right] + \left[\frac{\tilde{x}_2}{100}\right]\right\} \tag{4.64}$$

$$\left[\frac{-\dot{\tilde{x}}_2}{100}\right] = -\left\{(0.2)\cdot\left[\frac{\tilde{x}_1}{500}\right]\cdot\left[\frac{\tilde{x}_2}{100}\right] + (0.2)\cdot\left[\frac{\tilde{x}_1}{500}\right]\right\} \tag{4.65}$$

$$\frac{\mathrm{d}}{\mathrm{d}t}\left[\frac{\tilde{x}_1}{500}\right] = -\left\{(0.5)\cdot 10\cdot\left[\frac{-\dot{\tilde{x}}_1}{2500}\right]\right\} \quad \left[\frac{\tilde{x}_1(0)}{500}\right] = -\{(0.04)\cdot[-1]\} \tag{4.66}$$

$$\frac{\mathrm{d}}{\mathrm{d}t}\left[\frac{\tilde{x}_2}{100}\right] = -\left\{\left[\frac{-\dot{\tilde{x}}_2}{100}\right]\right\} \quad \left[\frac{\tilde{x}_2(0)}{100}\right] = -\{(0.15)\cdot[+1]\} \tag{4.67}$$

Note, that here the output of the multiplier is scaled to the product of the maximum values of both factors.

All the coefficients in the above equations lie within the range 0.1 to 10, so no time scaling is considered. Using simulation, the real maximum values $\tilde{x}_1 = 507$, $\tilde{x}_2 = 16.7$, $\tilde{\dot{x}}_1 = 520$ and $\tilde{\dot{x}} = 20.5$ are obtained. □

The scaling procedure described can also be used for the design of analog hardware, as shown in the next example.

**Example 4.17 Application of scaling procedure to a dedicated analog hardware design**

The purpose of this example is to show the applicability of the scaling procedure to the design of analog filters. Although digital filtering is preferable nowadays, analog filters must be used, e.g. at the input of an analog–digital converter in order to prevent aliasing. In terms of the simulation and control theory literature, an analog filter is a realization of a transfer function. A second order Chebyshev low pass filter with maximal flatness in the passband, stopband ripple 30 dB down from the peak in the passband and cut off frequency 50 Hz has the transfer function

$$G(s) = \frac{0.031\,62s^2 + 6242}{s^2 + 110s + 6242} \tag{4.68}$$

The corresponding simulation scheme in 'nested form' is obtained as described in Chapter 2.1 and is shown in Fig. 4.27.

The range of coefficients in Fig. 4.27 indicates the necessity of scaling. Let us first estimate the corresponding maximum values. Usually, no attenuation or amplification is desired in the passband, so the maximum values of the filter input $u$ and its output $y$ are the same and, without loss of generality, they will be assumed to be 1. It must be pointed out that such choice of maximum values may cause same problems with transient responses or with filters having ripple in the passband. The

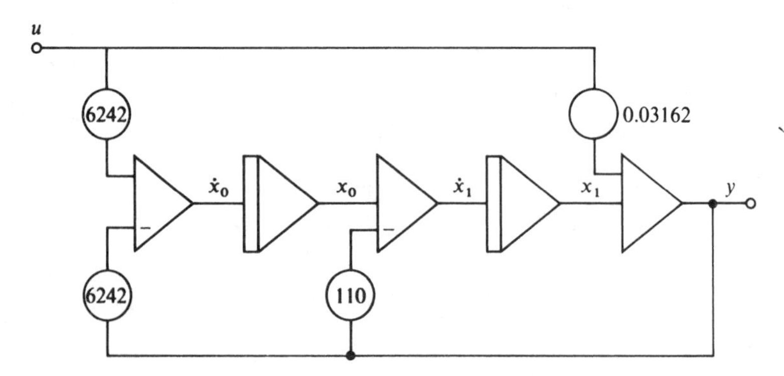


Figure 4.27 Simulation scheme for the Chebyshev filter.

solution of such problems depends on the complete system, which includes our filter and is not of interest to us. The above discussion is only a justification for not using the worst case '100% overshoot rule', which would give double the maximum absolute value of $y$. Maximum values of other quantities in Fig. 4.27 will now be estimated by the following two methods:

1. Using the 'worst case rule' and average natural frequency $\bar{\omega} = \sqrt{6242} = 79$, the maximum values of $\dot{x}_0$ and $x_0$ are estimated to be

$$\dot{x}_0 = |6242 \cdot u_{max}| + |6242 \cdot y_{max}| = 12\,494 \tag{4.69}$$

and

$$x_{0max} = \frac{\dot{x}_{0max}}{\bar{\omega}} = 158 \tag{4.70}$$

respectively. Note that this relation for $\dot{x}_0$ is derived from the simulation scheme in Fig. 4.27. The same procedure could now be applied for the estimation of $\dot{x}_{1max}$ and $x_{1max}$ but these estimates, obtained by proceeding with the simulation scheme from left to right, would be too pessimistic. Instead of this we apply the procedure of deriving maximum values from right to left in the simulation scheme and obtain

$$x_{1max} = |y_{max}| + |0.031\,62 \cdot \mathrm{u}_{max}| = 1.031\,62 \tag{4.71}$$

and

$$\dot{x}_{1max} = \bar{\omega} x_{1max} = 81.5 \tag{4.72}$$

All estimated maximum values are summarized in the first column of Table 4.4.

2. If steady state behaviour only is of interest, the amplitude/frequency response method can be used. The amplitude/frequency response (Saucedo and Schiring, 1968; Terrel, 1988) yields the attenuation of a transfer function (filter) for a steady state sine wave as a function of frequency and is obtained by evaluation of the expression $|G(s)|$ for $s = j\omega$, $j$ being the imaginary unit and $\omega$ the radial frequency. The idea of the amplitude/frequency response method is to find out the transfer functions from input to all interesting variables and to evaluate the maximum value of the corresponding amplitude/frequency response. Applying this method to our filter, the maximum values for $\dot{x}_{0max}$, $x_{0max}$, $\dot{x}_{1max}$ and $x_1$ are summarized in the second column of Table 4.4. It must be pointed out that the amplitude/frequency response evaluation is a complex mathematical operation and thus the justification of its usage may sometimes be questionable.

Observing columns 1 and 2 in Table 4.4 it can be seen that the average natural frequency rule gives more pessimistic estimates, but since steady state responses are of major interest in filtering, the second column of Table 4.4, obtained by the amplitude/frequency method, will be used for scaling. The application of the scaling

Table 4.4 Maximum values of the Chebyshev filter

| *Variable* | *Average natural frequency rule* | *Amplitude/frequency response method* |
|---|---|---|
| $u$ | 1.00 | 1.00 |
| $\dot{x}_0$ | 12 484 | 7870 |
| $x_0$ | 158.0 | 114.0 |
| $\dot{x}_1$ | 81.37 | 55.00 |
| $x_1$ | 1.03 | 0.9684 |
| $y$ | 1.00 | 1.00 |

procedure yields

$$\left[\frac{-\dot{x}_0}{7870}\right] = -\left\{(0.7931)\cdot\left[\frac{u}{1}\right] + (0.7931)\cdot\left[\frac{-y}{1}\right]\right\} \tag{4.73}$$

$$\frac{\mathrm{d}}{\mathrm{d}t}\left[\frac{x_0}{114}\right] = -\left\{(0.6904)\cdot 100\cdot\left[\frac{-\dot{x}_0}{7870}\right]\right\} \tag{4.74}$$

$$\left[\frac{-\dot{x}_1}{55}\right] = -\left\{(0.2073)\cdot 10\cdot\left[\frac{x_0}{114}\right] + (0.2)\cdot 10\cdot\left[\frac{-y}{1}\right]\right\} \tag{4.75}$$

$$\frac{\mathrm{d}}{\mathrm{d}t}\left[\frac{x_1}{0.9684}\right] = -\left\{(0.5679)\cdot 100\cdot\left[\frac{-\dot{x}_1}{55}\right]\right\} \tag{4.76}$$

$$\left[\frac{-y}{1}\right] = -\left\{(0.031\,62)\cdot\left[\frac{u}{1}\right] + (0.9684)\cdot\left[\frac{x_1}{0.9684}\right]\right\} \tag{4.77}$$

As our filter is supposed to operate in real time, no time scaling is performed. Actually, the time scaling of the filter with $n=30$ (making the time response thirty times slower) corresponds to the scaling of the frequency response characteristic for the same factor (decreasing the cut off frequency from 30 Hz to 1 Hz).

If the filter is built as a dedicated analog hardware filter (operational amplifier) economy has to be taken into account and integrators and summers are joined into one element, as described in Section 4.6.1 and shown in Fig. 4.28. In this case the

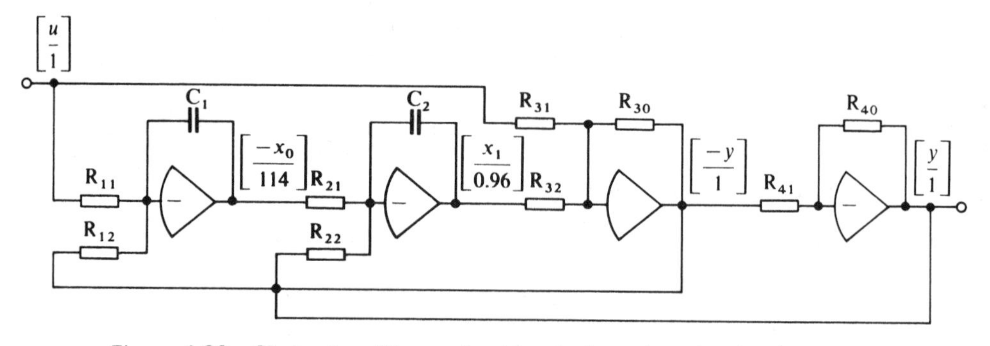


Figure 4.28 Chebyshev filter realized by dedicated analog hardware.

Table 4.5 Elements for the Chebyshev filter

| *Element* | *Summing integrator* 1 | *Summing integrator* 2 | *Summer* | *Inverter* |
|---|---|---|---|---|
| Feedback element | $C_1 = 1\ \mu F$ | $C_2 = 1\ \mu F$ | $R_{30} = 1\ k\Omega$ | $R_{40} = 10\ k\Omega$ |
| Input1 coefficient | 54.75 | 117.7 | 0.031 62 | 1 |
| Input1 element | $R_{11} = 18.26\ k\Omega$ | $R_{21} = 8.4966\ k\Omega$ | $R_{31} = 31.63\ k\Omega$ | $R_{41} = 10\ k\Omega$ |
| Input2 coefficient | 54.75 | 113.6 | 0.9684 | — |
| Input2 element | $R_{12} = 18.26\ k\Omega$ | $R_{22} = 8.803\ k\Omega$ | $R_{32} = 1.033\ k\Omega$ | — |

scaled equations for elements performing summations and integrations are

$$\frac{d}{dt}\left[-\frac{x_0}{114}\right] = -\left\{54.75 \cdot \left[\frac{u}{1}\right] + 54.75 \cdot \left[\frac{-y}{1}\right]\right\} \tag{4.78}$$

$$\frac{d}{dt}\left[\frac{x_1}{0.9684}\right] = -\left\{117.7 \cdot \left[\frac{-x_0}{114}\right] + 113.6 \cdot \left[\frac{y}{1}\right]\right\} \tag{4.79}$$

where the potentiometer settings and unit input gains are not shown explicitly but are joined in a single number as an element coefficient. Element coefficients are realized by an appropriate choice of capacitors and resistors as shown in Table 4.5. □

### 4.6.4 Static Test

Magnitude and time scaling influence the accuracy of the simulation results. Besides this, the reliability of simulation depends mostly on the elimination of faults. It will be supposed that the mathematical model used in simulation is valid, i.e. we shall not deal with modelling faults. The sources of faults can be quite heterogeneous, e.g. wrong scaling or patching on the analog computer and typewriting faults in digital simulation. Hardware defects are sources of faulty results when using analog–hybrid computers. The static test is a systematic procedure for detecting and eliminating all kinds of simulation faults. It is used more frequently in analog–hybrid simulation; its principles, however, can also be used for eliminating errors in incorrectly typed digital simulation programs, as shown in Example 4.18.

The static test or static check is a systematic fault detection and elimination procedure, the idea of which is to assume arbitrary values for the outputs of all integrators, to calculate the outputs of all blocks and the inputs (derivatives) of all integrators from the original problem equations, to calculate them from the scaled block scheme and to measure them from the patch panel on the computer. Any disagreement indicates an error. The systematic procedure is as follows:

(a) Choose arbitrary values for the outputs of all integrators in such a way that no overloads and no extremely small outputs occur in the scaled equations.

(b) Calculate the output of every block and the derivative of every integrator in the simulation scheme from the original equations.

(c) Using the scaled simulation scheme, potentiometer settings and assumed scaled integrator outputs from (a), calculate on the scaled block diagram the output of every block and the derivative of every integrator.
(d) Compare the calculated values from (b) and (c). Any disagreement indicates an error in programming or scaling procedure. The source of error can be located by working backwards through the block diagram.
(e) Measure the outputs of every block and the derivative of every integrator.
(f) Compare the read-outs of (e) with the values obtained in (b) or (c). Any disagreement indicates an error in patching or a computer component malfunction. The error source can be located as described in (d).

The principles of the static test may also be used for detecting and eliminating typewriting faults in digital simulation, as shown in the following example.

**Example 4.18 Typing error elimination by static test principles**

Suppose the following program was typed in for the prey and predator problem:

```
PROGRAM PREY AND PREDATOR
CONSTANT A11=5,A12=0.05,A21=0.0004,A22=0.2
CONSTANT RABO=520,FOXO=85
RABDOT=A11*RAB–A12*RAB*FOX
FOXDOT=A12*RAB*FOX–A22*FOX
RAB=INTEG(RABDOT,RABO)
FOX=INTEG(FOXDOT,FOXO)
CONSTANT TFIN=10
TERMT(T.GE.TFIN)
CINTERVAL CI=0.01
OUTPUT 100, RABDOT,RAB,FOXDOT,FOX
END
```

The simulation results for T=0 are obtained as follows:

```
T    RABDOT  RAB    FOXDOT  FOX
0.   390.    520.   2193.   85.
```

Using the initial condition data the derivatives RABDOT and FOXDOT, calculated from original Eqns (1.25)–(1.28) are 390 and 0.68 respectively. Disagreement in FOXDOT indicates an error in the corresponding expression and by inspecting it a misprint, A12, instead of A21 is detected.

### 4.6.5 Working with Analog Computers

Here we comment briefly on working with general purpose analog computers, but only in appropriate depth. Due to the very different types of machines, it would be impossible to go into details but, on the other hand, basic ideas on the organization and capabilities of analog computers must be presented.

Analog programs given in the form of scaled analog simulation schemes and tables of potentiometers are realized on the computer patch panel. There, the corresponding inputs and outputs of the electronic circuits, realizing particular mathematical operations, are attainable. The panel is organized so that the elements (inputs, outputs, etc.) are clearly visible. Each element is correspondingly addressed. Patch panels are changeable in most computers, enabling at least minimal saving of analog programs. The connections between particular elements are realized with the aid of suitable cords. The structure of the analog computer is modular so that different combinations can be included in the computer. The older types have some form of command panel, while for the newer configurations the commands are entered through the keyboard and terminal, which are PC-like. The outputs of the simulation are recorded on digital voltmeters, oscilloscopes and plotters. The main modes of an analog computer are *operate*, where integrators and other elements operate to produce the dynamic solution of a patched problem; *initial conditions*, where on the outputs of the integrators the initial conditions appear; and *hold*, where the integrators hold the level of the outputs in the moment of mode onset. Very important is the state of waiting for the beginning of the simulation runs where the potentiometers, already loaded, can be set to their prescribed values. Also, the state for checking the program and computer components is available. With the aid of the appropriate timer the simulation runs can be automatically repeated, which is called repetitive operation. The speed of analog computation enables the model time responses to be permanently visible on the oscilloscope almost immediately (a few thousand simulation runs per second). The capacity for repetitive computation increases the illustrative nature of simulation and promotes understanding of model and, indirectly, of the behaviour of the process. The influence of every model parameter change (for instance, as caused by handset potentiometer setting) is evident immediately. With the aid of corresponding commands a variety of information can be obtained, such as the status and voltage of particular elements and possible overloads. Also, the duration of the computer modes can be defined for the whole device or for particular integrators (the duration of the operation period defines the observation interval, i.e. the duration of the simulation run). The computer can thus work in real time or in computer time which may be compressed or expanded. The speed of computing can also be defined by the selection of an appropriate master time scale factor whose change does not influence the time scaling (the response can, for instance, be observed on the oscilloscope using faster computing, while it can be plotted only by selecting the factor for slower computing – thus no time rescaling is necessary).

Finally, let us briefly summarize the working cycle on an analog computer when a simulation study is performed for a given problem.

- With the aid of the mathematical model prepared for the indirect approach to differential equations solving, the simulation scheme which is correspondingly modified to the analog simulation scheme is generated.

- The magnitude and time scaling procedure gives, as the result, the scaled analog simulation scheme with the table of potentiometers, which represents all the information necessary for the analog program performance.
- On the basis of the scaled analog simulation scheme and the table of potentiometers the analog program is patched and the so-called *static test* procedure is performed to find the mistakes in programming, patching and possible failure of the computer components used.
- The simulation runs, in most cases repetitive, are then performed. The possibility of the need for rescaling must also be checked. The recorded or displayed results are analysed in order to make a decision about the further course of the study. As the parameter changes and model structure modifications can easily be made during the simulation, and as their influence on the model behaviour is visible immediately, the derivation of an acceptable model is illustrative and relatively short. The correct documentation and interpretation of the results are also of extreme importance.

As can be seen, the working procedure on an analog computer is quite compatible with the modelling and simulation approach cycle.

### 4.6.6 Analog–Hybrid Computing

As mentioned earlier, the traditional distinction between analog and digital computers is that the former are parallel while the latter are serial in nature. Thus if a problem requires fifty multiplications, the analog computer would use fifty multipliers, while the corresponding digital program would use the same arithmetic unit fifty times in succession. To blur this distinction to some extent the digital logic elements which, like analog ones, operate simultaneously are added in analog computers and such a configuration is called an *analog–hybrid computer* (or parallel hybrid computer) (EAI Basics, 1968).

The logic elements may have their terminations on the analog patch panel, or a separate panel may be provided solely for logic elements. Such extension of analog computers enables the logic decisions to be included in the programming which expand their capabilities significantly. More complex procedures and experiments with models can be performed, such as optimization, curve fitting, etc. As well as pure logical elements, a group of components exists which enable the mixture of logical and analog signals in the sense that they are controlled by logical signals or produce logical output for analog input signals. Let us discuss, again in an appropriate depth, the elements for the combination of logical and analog signals as well as pure logical elements. The former are as follows:

- *The comparator*. This is an electronic device which, according to the relation of two analog inputs, generates true and false logical output. The symbol is shown in Fig. 4.29. In order to make a decision as to when the input signals are nearly equal, a suitable hysteresis is built into the comparator. There are many special requirements regarding the signs of inputs for circuits in different types of computer. Also, the capacity for logical control of the comparator operation is often included (it can be put in a kind of HOLD mode).

- *Switches and relays.* All kinds of electronic switches, as well as relays in older types of computers, can be found. They all operate on the same principle. According to the logical control signal, the output is either equal to the first or to the second input. Also, realization with zero output in the case of low logical signal is sometimes possible. One of the possible symbols is shown in Fig. 4.30.
- *The track-store unit.* This unit represents the analog memory. Its operation can be explained if it is supposed to be the combination of summer and the integrator which can be put to HOLD operation. In the state *track* the device acts exactly like the analog summer, whereas in the state *store* it acts like the integrator which is put to the HOLD operation. So in track mode the unit tracks the negative sum of the analog inputs, while in store mode it holds the value of the output. The logic control signal changes between track and store states. As shown in Fig. 4.31, the unit has analog and logic IC (initial condition) inputs so that the state IC can also be achieved.

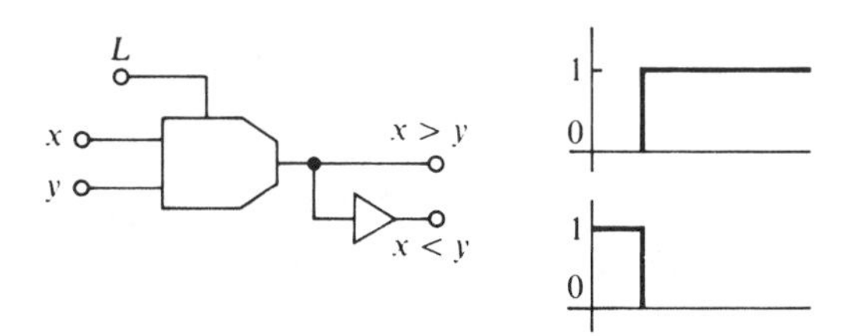


Figure 4.29 Symbol for the comparator.

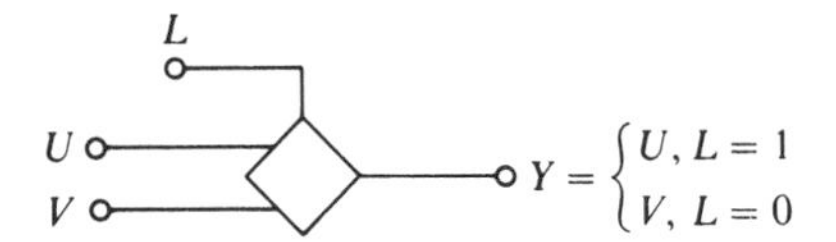


Figure 4.30 One of the possible realizations of the switch and its symbol.

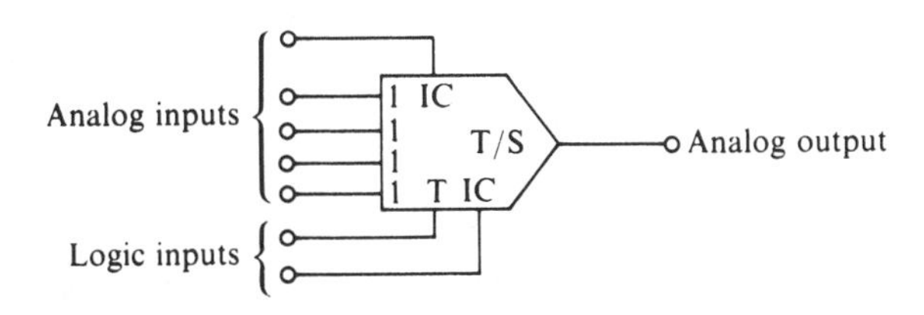


Figure 4.31 Symbol for the track–store unit.

True logical components are for the particular types of computers very different in their performance as well as in their symbols, and thus the latter are sometimes omitted in our overview. The main components are as follows:

- *Logical gates*. Most computers include only AND gates with true and false outputs and two to four inputs. So all logic operations can be obtained using De Morgan's theorem. The symbol is shown in Fig. 4.32.
- *Flip-flops*. The most important fact which has to be mentioned when introducing flip-flops is that they have memory (they are in some sense analogous to integrators where output also depends upon past history). They are synchronized with the clock pulses of suitable and changeable periods in order to resolve uncertainties caused by delays in signal propagation. The basic idea of any synchronous system is the separation of *decision* and *action*. Thus the two processes are not allowed to occur simultaneously. Flip-flops first decide what to do and then do it, but in precisely defined moments. As seen in Fig. 4.33, they have two inputs and logic control input, enabling termination of operation as well as true and false output.
- *The logic differentiator*. This device generates one clock period pulse at the instant when the input changes its value from 0 to 1 (leading edge differentiator). Falling edge and bipolar differentiators can also be incorporated, using the corresponding combination of flip-flop and gate. The symbol shown in Fig. 4.34 indicates that the logic differentiator has one input as well as true and false output.
- *The general purpose register*. This device contains four to eight flip-flops which can be used collectively or individually. General purpose registers are of very different types and capabilities. They are used as binary counters as well as

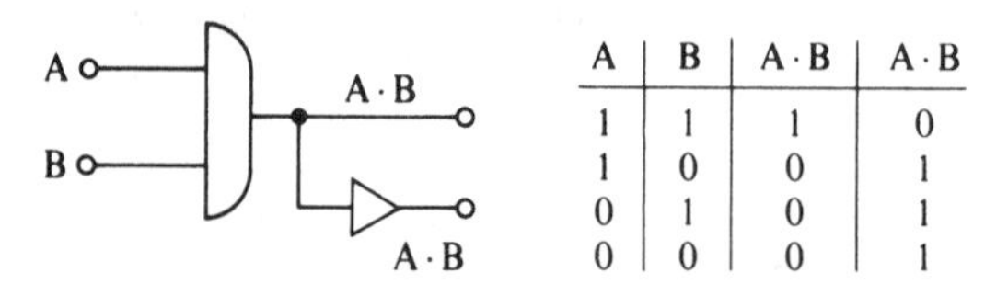


| A | B | A·B | $\overline{A \cdot B}$ |
|---|---|---|---|
| 1 | 1 | 1 | 0 |
| 1 | 0 | 0 | 1 |
| 0 | 1 | 0 | 1 |
| 0 | 0 | 0 | 1 |

Figure 4.32 Symbol for the AND gate and state transition table.

| | | $Q_N = 0$ | | $Q_N = 1$ | |
|---|---|---|---|---|---|
| $S$ | $R$ | $Q_{N+1}$ | $\bar{Q}_{N+1}$ | $Q_{N+1}$ | $\bar{Q}_{N+1}$ |
| 0 | 0 | 0 | 1 | 1 | 0 |
| 0 | 1 | 0 | 1 | 0 | 1 |
| 1 | 0 | 1 | 0 | 1 | 0 |
| 1 | 1 | 1 | 0 | 0 | 1 |

Figure 4.33 One of the possible symbols for the flip-flop with state transition table.

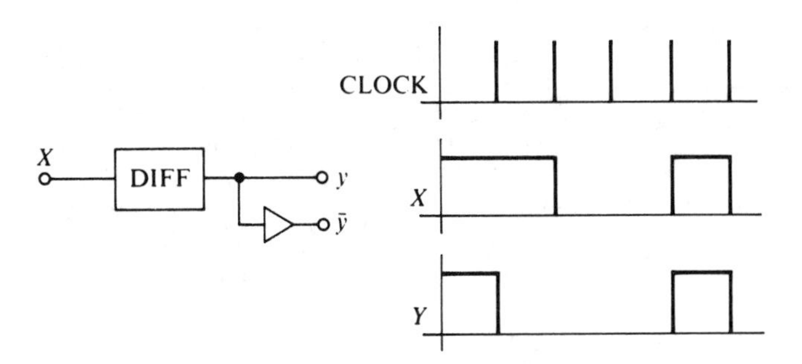


Figure 4.34 Symbol for the logic differentiator with its timing diagram.

shift and ring shift registers. Various inputs enable the entering of shift signals, the clearing of information, the loading of information from the patch panel or manually, the logic control of register operation, etc.

- *Counter*. This device represents the capability for decimal counting from some defined value down to zero. It counts the input pulses and generates a certain output signal. It can be set or reset by command or from the patch panel.

Finally, let us note the existence of logical modes, which are analogous to the analog modes, as well as all kinds of logical component status indicators, several pulse sources, the capability of manual setting of the components, etc.

### 4.6.7 Hybrid Computer Systems

The development of hybrid computer systems was initiated as a result of the complementary characteristic of analog and digital computers. Modern general purpose hybrid systems are of three major types: the analog computer, the digital computer and the linkage system (Bekey and Karplus, 1968). A great majority employ analog and digital parts which can function independently or in the hybrid mode. In decisions regarding hybrid system configurations a great deal of information must be taken into account, especially if the system is used for some special task in industry (analog and digital characteristics and size, software, input–output devices, speed of computation, etc.). The linkage systems were much more problematic for the older configurations where, besides analog–digital and digital–analog converters, specialized hybrid interfaces were needed to enable the digital control of some analog computer elements (servopotentiometers, function generators, etc.). In the more modern hybrid systems where the analog part is microprocessor controlled this linkage is nonproblematic. Two kinds of system can be distinguished.

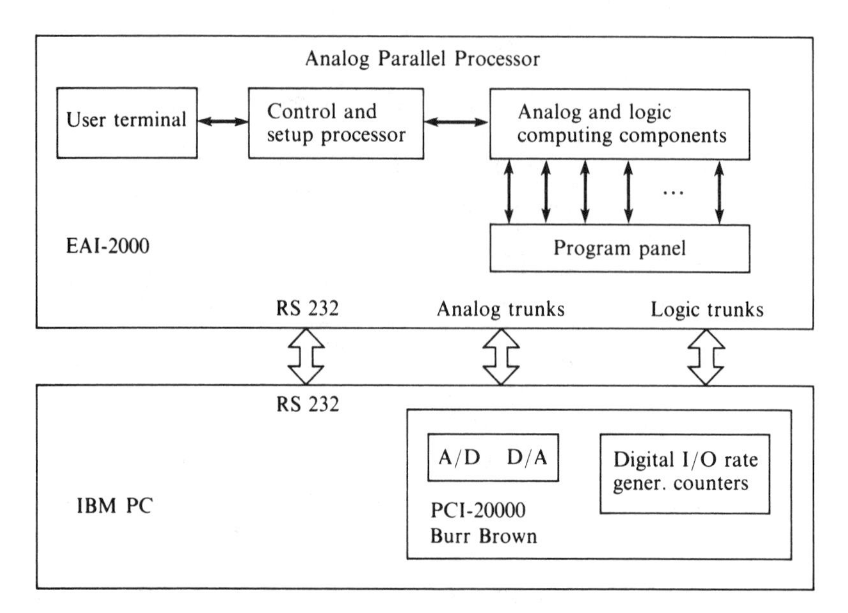


Figure 4.35 The possibility of nonstandard hybrid configuration.

The first kind is intended for non-time-critical operation (EAI-2000, 1982). The digital part is connected through the serial data communication channel (RS 232 port) to the control and setup interface microprocessor of the analog part, enabling bidirectional data flow. Therefore, the digital computer can be programmed to perform automatically all the procedural tasks involved in the setup, checkout and operation of the analog part. The data associated with the modelled problem cycle between the computers through A/D and D/A conversions. Such configurations, which can also be nonstandard in the sense that the digital part and the conversions can be chosen by the user of the system, are very powerful tools for simulating certain classes of problem (e.g. simulating computer control of continuous processes). They also support the operation of the analog part because they enable its efficient setup as well as the documentation of results in numeric and graphic form (with the aid of the corresponding software on the digital part). One of the most important areas of use for such systems (which can also be used in smaller configurations) is, in our opinion, education. No simulation tool is as illustrative in the teaching of modelling, simulation and control in several branches of science as is analog or hybrid configuration. One possible combination is shown in Fig. 4.35.

Hybrid systems for time critical operations are ranged in the second group. They include the standard parallel interface to special hybrid components (control lines, interrupt lines, high speed mode control, DAM channels, multiplexed ADC channels, etc.). They provide the high speed bidirectional connections required for tasks in some special areas where hybrid systems now represent the only simulation tool which can solve the problem.

Finally, let us list some simulation tasks and simulated process characteristics for which hybrid simulation is needed:

- high-speed simulation;
- real time, or faster, simulation;
- control system design, evaluation and testing;
- man-in-the-loop simulation (direct interaction between user and simulation);
- hardware-in-the-loop simulation (computer connected to the pilot or real plant, real controller connected to the computer);
- need for equal simulation speed for small and large models;
- complex, time variable, nonlinear processes;
- presence of discontinuities (depending on states and on the time);
- highly stiff systems (large differences in the constants);
- processes described by highly coupled sets of differential, algebraic and boolean equations.

More about advanced hybrid systems will be said in Section 6.2.

## 4.7 PROBLEMS

### Solved problems

#### Problem 4.1

Simulate the modified model of the car suspension (Example 2.2) defined by Eqn (2.34) and the simulation scheme in Fig. 2.18. An additional nonlinearity of the car suspension system which reduces heavy oscillations is included in the model by the following inequalities:

$$y_2 - y_1 > y_0 \qquad k_2 = k_{21} \tag{4.80}$$

$$y_2 - y_1 \leqslant y_0 \qquad k_2 = k_{20} \tag{4.81}$$

The following constants are used:

$m_1 = 250$ kg $\qquad m_2 = 20$ kg
$k_{20} = 5*10^4$ N/m $\qquad k_{21} = 10^6$ N/m
$f = 1000$ Ns/m $\qquad y_0 = 0.08$ m
$z = 0.1$ m $\qquad k_1 = 10^4$ N/m

The simulation run duration should be 2.5 s and the communication interval is 0.005 s.

The simulation program which shows the PROCEDURAL feature of the CSSL standard language for the simulation of additional nonlinearity is shown in Fig. 4.36, while the results of the simulation are shown in Fig. 4.37.

```
CONSTANT M1=250,M2=20
CONSTANT K1=10000,K20=5.E+04,K21=1.E+06
CONSTANT F=1000,TFIN=2.5,Z=0.1,Y0=0.08
Y1DD=-K1/M1*DIF-F/M1*DIFD
Y2DD=F/M2*DIFD+K2/M2*ZMY2+K1/M2*DIF
DIFD=Y1D-Y2D
DIF=Y1-Y2
ZMY2=Z-Y2
Y1D=INTEG(Y1DD,0)
Y2D=INTEG(Y2DD,0)
Y1=INTEG(Y1D,0)
Y2=INTEG(Y2D,0)
PROCEDURAL(K2=Y1,Y2)
   K2=K20
   IF((Y2-Y1).LE.Y0)GO TO 10
   K2=K21
10 CONTINUE
END
ERRTAG IERR
HDR SIMULATION OF CAR SUSPENSION SYSTEM
PREPAR Y1,Y2
OUTPUT 20,Y1,Y2
TERMT T.GE.TFIN
CINTERVAL CI=0.005
END
```

Figure 4.36 Simulation program for the car suspension system with nonlinearity.

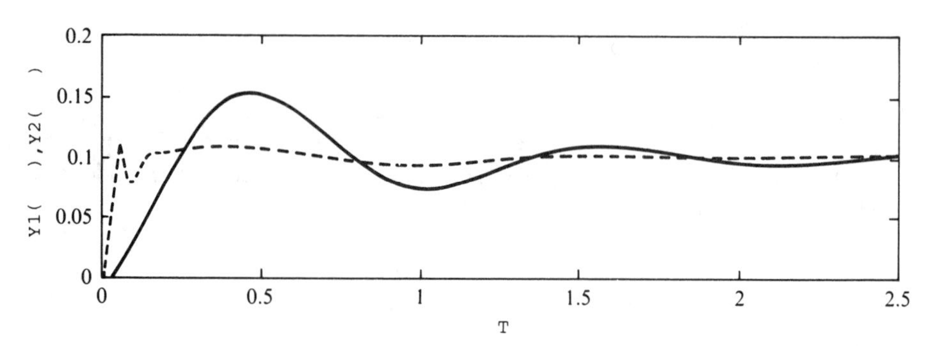


Figure 4.37 The time responses of car body ($y_1$) and wheel ($y_2$) displacements.

**Problem 4.2**

Solve Mathieu's equation, which is introduced in Problem 2.3 and defined by Eqn (2.43), with the following constants and initial conditions:

$$a = 6.25 \qquad \omega = 2 \qquad \nu = 1.5 \qquad y_0 = 6 \qquad \dot{y}_0 = 0$$

The duration of the simulation run is 40 s and the communication interval is 0.1 s.

```
CONSTANT A=6.25,OMEGA=2
CONSTANT THETA=1.5
CONSTANT Y0=6.,YD0=0
CONSTANT TFIN=40
YDD=-A*Y+2*THETA*COS(OMEGA*T)*Y
YD=INTEG(YDD,YD0)
Y=INTEG(YD,Y0)
TERMT T.GT.TFIN
CINTERVAL CI=0.1
ERRTAG IERR
OUTPUT 5,Y,YD
PREPAR Y,YD
END
```

Figure 4.38 The program for Mathieu's equation.

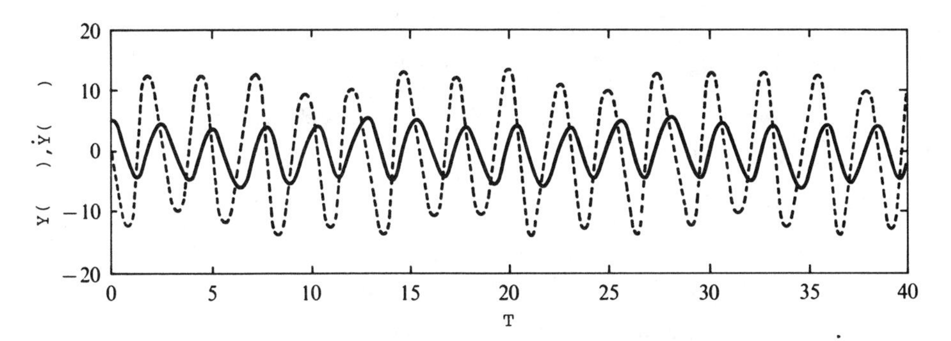


Figure 4.39 The results of Mathieu's equation.

The simulation program is shown in Fig. 4.38 and the results of the simulation are presented in Fig. 4.39.

## Problem 4.3

Simulate the Lorentz chaotic system defined in the form:

$$\dot{x} = \sigma(y - x) \tag{4.82}$$

$$\dot{y} = (1 + \lambda - z)x - y \tag{4.83}$$

$$\dot{z} = xy - bz \tag{4.84}$$

with constants $\sigma = 10, \lambda = 24, b = 2$ and initial conditions $x(0) = y(0) = z(0) = 0.1$.

This problem arises from the study of turbulent convection in fluids, with chaotic behaviour for chosen constants. The chaotic behaviour is indicated by an exotic steady state response, which is neither a constant nor a periodical limit cycle but a so-called chaotic attractor. Thus chaotic systems are sometimes defined as nonlinear deterministic systems with random behaviour. Chaos, also called strange behaviour, is currently one of the most exciting topics in nonlinear systems research. Many sophisticated phenomena can be investigated and analysed only with the aid of simulation.

Fig. 4.40 shows the simulation program. The results of the simulation in the phase plane $y = y(x)$ are shown in Fig. 4.41.

```
''Lorentz attractor''
''-------------------
PROGRAM LORENTZ
''
'' constants''
CONSTANT sigma=10.,b=2.,lambda=24.
CONSTANT TFIN=60.
CONSTANT X0=0.1,Y0=0.1,Z0=0.1
''
''structure representative statements''
XD=sigma*(Y-X)
YD=(1+lambda-Z)*X-Y
ZD=X*Y-b*Z
X=INTEG(XD,X0)
Y=INTEG(YD,Y0)
Z=INTEG(ZD,Z0)
''
''output and control statements''
PREPAR X,Y,Z
OUTPUT X,Y,Z
HDR SIMULATION OF LORENTZ ATTRACTOR
TERMT(T.GE.TFIN)
CINTERVAL CI=0.01
END
```

Figure 4.40 Program for the simulation of the Lorentz chaotic system.

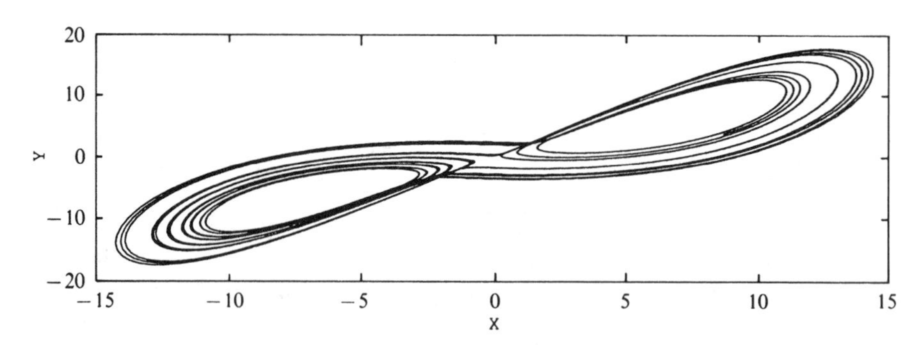


Figure 4.41 Lorentz attractor in the $(x, y)$ plane.

**Problem 4.4**

Find the Poincaré map of the Duffing chaotic system, described by the differential equations

$$\dot{x} = y \tag{4.85}$$

$$\dot{y} = x(c - x^2) - ay + b \cos \omega t \tag{4.86}$$

with constants $a = 0.05$, $b = 7.5$, $c = 0$, $\omega = 1$ with zero initial conditions.

Nonautonomous nonlinear systems, which are forced by a cosine function, can have periodic, quasi-periodic or chaotic stationary behaviour. The latter can be confirmed by the Poincaré map. It is realized by the stroboscopic sampling of the phase plane trajectory in multiples of the forcing period. These samples form the Poincaré limit set. In the case of chaotic steady state they do not lie on a simple geometric object in the phase space, as in the case of periodic or quasi-periodic solutions. The shapes of Poincaré maps are specific and do not depend upon the initial conditions.

As the forcing period in our example is $2\pi$, the communication interval 6.283 184 is chosen to obtain one sampling in each period. Fig. 4.42 shows the simulation program. The appropriate Poincaré set (also called the Cantor set) is shown in Fig. 4.43. The complex shape confirms the chaotic behaviour.

```
''Duffing chaotical system''
''----------------------------------------
PROGRAM DUF
''
''constants''
CONSTANT a=0.5,b=7.5,c=0.,w=1.
CONSTANT TFIN=20.E+03,X0=0.,Y0=0.
''
''structure representative statements''
XD=Y
YD=X*(c-X*X)-a*Y+b*COS(w*T)
X=INTEG(XD,X0)
Y=INTEG(YD,Y0)
''
''control and output statements''
PREPAR X,Y
OUTPUT X,Y
TERMT (T.GE.TFIN)
NSTEPS NST=10
CINTERVAL CI=6.283184
END
```

Figure 4.42 Program for the simulation of the Duffing system.

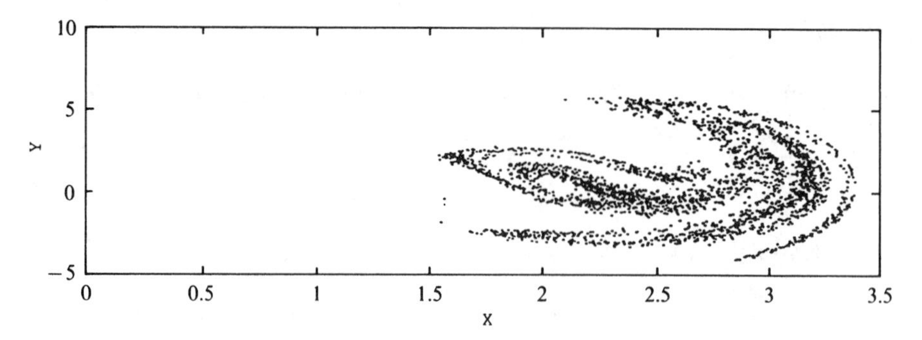


Figure 4.43 Poincaré map of the Duffing system.

**Problem 4.5**

Simulate the digital control system with the continuous process and discrete PID controller. The block diagram is shown in Fig. 4.44 and the transfer functions are

$$G_R(z) = \frac{q_0 + q_1 z^{-1} + q_2 z^{-2}}{1 - z^{-1}} \tag{4.87}$$

$$q_0 = 2.332 \qquad q_1 = -3.074 \qquad q_2 = 1.105 \tag{4.88}$$

$$G_P(s) = \frac{1 - 4s}{(1 + 4s)(1 + 10s)} \tag{4.89}$$

The sampling time is $T_s = 4$ s, the observation period is 60 s and the communication interval is 0.5 s. For realization of the continuous transfer function (4.89), use the simulation scheme shown in Fig. 2.22, Problem 2.4.

Fig. 4.45 illustrates the program in the simulation language SIMCOS and the simulation results are shown in Fig. 4.46.

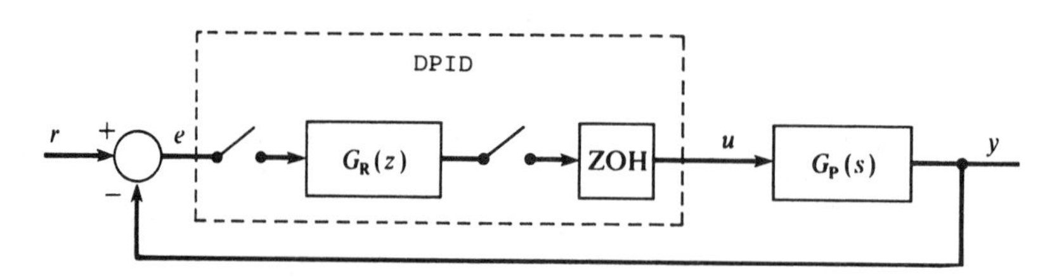


Figure 4.44 The block of the digital control system.

```
ARRAY Q(3),WORK(3)
CONSTANT Q=2.332,-3.074,1.105
CONSTANT WORK=0.,0.,0.
CONSTANT R=1.
CONSTANT TFIN=60.,TS=4.
''
''     ****************************************
''     *         PROCESS TRANSFER FUNCTION      *
''     ****************************************
''
W1D=0.25*U-0.25*W1
W=-4*W1D+W1
YD=0.1*W-0.1*Y
W1=INTEG(W1D,0.)
Y=INTEG(YD,0.)
''
''     ****************************************
''     *         DISCRETE PID CONTROLLER        *
''     ****************************************
''
U=DPID(E,Q,WORK,TS)
''
''     ****************************************
''
E=R-Y
HDR DISCRETE PID CONTROL SYSTEM
ERRTAG IERR
CINTERVAL CI=0.1
NSTEPS NST=2
TERMT(T.GT.TFIN)
OUTPUT 5,E,U,Y
PREPAR R,E,U,Y
END
```

Figure 4.45 The simulation program for the digital control system.

## Exercises

### Problem 4.6

Solve Problems 2.6 and 2.7 using your favourite digital simulation language.

### Problem 4.7

Solve Problem 2.8 using your favourite digital simulation language. Use amplitude 1 as the input signal.

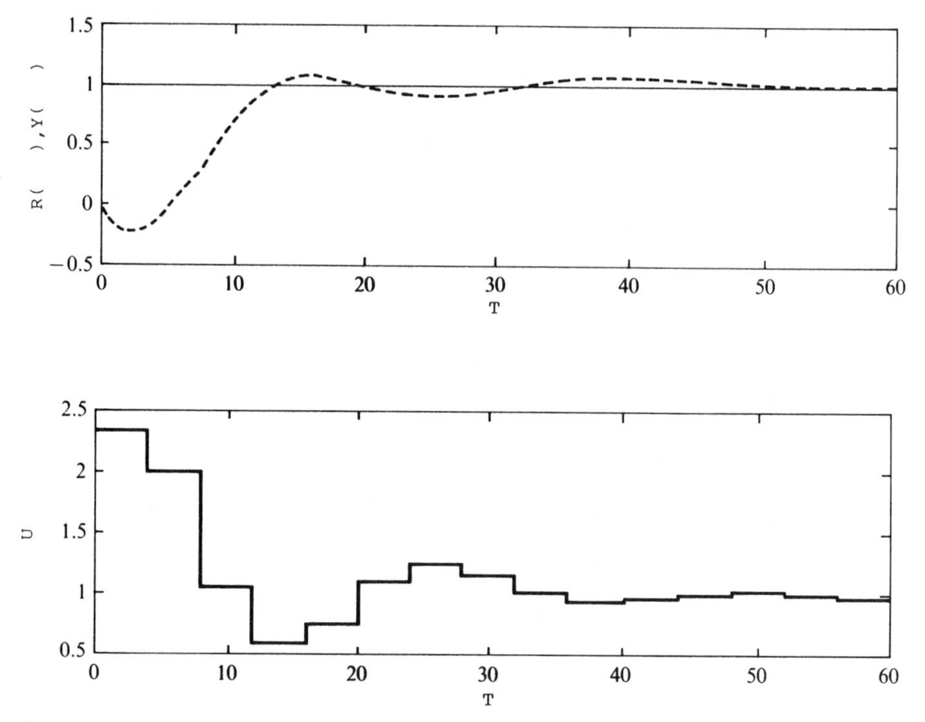


Figure 4.46 The control ($u$) and the controlled ($y$) variables of the digital control system.

## Problem 4.8

Solve Problem 2.9 using your favourite simulation language. Use sine wave signals with different frequencies as inputs and observe the results.

## Problem 4.9

A chaotic model due to Rössler is defined by the following equations:

$$\dot{x} = -y - z \tag{4.90}$$

$$\dot{y} = x + ay \tag{4.91}$$

$$\dot{z} = b + z(x - c) \tag{4.92}$$

Use constants $b = 2$ and $c = 4$. Simulate the model in your favourite simulation language and show that the system has limit cycle with period $T$ for $a = 0.30$ and with period $2T$ for $a = 0.35$ and a chaotic attractor for $a = 0.398$.

**Problem 4.10**

Chua circuits are described by the following equations:

$$\dot{x} = \alpha(y - h(x)) \tag{4.93}$$

$$\dot{y} = x - y + z \tag{4.94}$$

$$\dot{z} = -\beta y \tag{4.95}$$

$$h(x) = \begin{cases} bx + a - b & x \geqslant 1 \\ ax & |x| < 1 \\ bx - a + b & x \leqslant -1 \end{cases} \tag{4.96}$$

Use the constants $\beta = 14\frac{2}{7}$, $a = -\frac{1}{7}$, $b = \frac{2}{7}$. Simulate the model using an appropriate simulation language and show that for $\alpha = 8$ the steady state is a limit cycle with period $T$, for $\alpha = 8.2$ the steady state is a limit cycle with period $2T$, and for $\alpha = 8.5$ the steady state is the chaotic attractor.

## BIBLIOGRAPHY

ABURDENE, M.F. (1988), *Computer Simulation of Dynamic Systems*, Wm C. Brown Publishers, Dubuque, Iowa.

ÅSTRÖM, K.J. (1985a), *A Simnon Tutorial*, Department of Automatic Control, Lund Institute of Technology, CODEN: LUTFD2/(TFRT-3168)/(1982), Sweden.

ÅSTRÖM, K.J. (1985b), 'Computer aided tools for control system design', in M.J. Jamshidi and C.J. Herget (eds), *Computer-aided Control System Engineering*, North-Holland, Amsterdam, pp. 3–40.

BAKER, N.J.C. and P.J. SMART (1983), 'The SYSMOD simulation language', *Proceedings of the 1st European Simulation Congress*, Aachen, W. Germany, pp. 281–6.

BAKER, N.J.C. and P.J. SMART (1985), 'Elements of the SYSMOD simulation language', *Proceedings of the 11th IMACS World Congress*, Oslo, Norway, vol. 2.

BAUSCH-GALL, I. (1987), 'Continuous system simulation languages', *Veh. Syst. Dyn. (Netherlands)*, **16**, 347–66.

BEKEY, G.A. and W.J. KARPLUS (1968), *Hybrid Computation*, John Wiley, NY.

BLUM, J.J. (1969), *Introduction to Analog Computation*, Harcourt, Brace & World, Inc., NY.

BOSCH, P.P.J. VAN DEN and A.J.W. VAN DEN BOOM (1985), 'Industrial applications in the Netherlands of computer-aided design packages for control systems; trends and practice', *Preprints of the 3rd IFAC/IFIP International Symposium CADCE '85*, Copenhagen, Denmark, pp. 58–61.

BOSCH, P.P.J. VAN DEN (1987), *Simulation Program PSI* (manual), Delft University of Technology.

BREITENECKER, F. (1989), 'Need for hybrid simulation?', *Proceedings of the 3rd European Simulation Congress* (D. Murray-Smith, J. Stephenson and R.N. Zobel, eds), Edinburgh, pp. 421–5.

BRUIJN, M.A. DE and F.P.J. SOPPERS (1986), 'The Delft parallel processor and software for continuous system simulation', *Proceedings of the 2nd European Simulation Congress*, Antwerp, Belgium, pp. 777–83.

CARLSON, A., G. HANNAUER, T. CAREY and P.J. HOLSBERG (1967), *Handbook of Analog Computation*, Electronic Associates, Inc., Princeton, NJ.

CELLIER, F.E. (1979), Combined Continuous/Discrete System Simulation by Use of Digital Computers: Techniques and Tools, PhD, Swiss Federal Institute of Technology, Zürrich.

CELLIER, F.E. (1983), 'Simulation software – today and tomorrow', *Proceedings of the IMACS Conference*, Nantes, France.

CROSBIE, R.E. (1984), 'Simulation on microcomputer – experiences with ISIM', *Proceedings of the 1984 Summer Computer Simulation Conference*, Boston, Massachusetts, USA, **1**, 13–17.

CROSBIE, R.E. (1986), 'Developments in ESL and others for the 1990's, *Proceedings of the 1986 Summer Computer Simulation Conference*, USA, pp. 313–14.

CROSBIE, R.E. and F.E. CELLIER (1982), 'Progress in simulation language standards', *An Activity Report for Technical Committee TC3 on Simulation Software, Proceedings of the 10th IMACS World Congress*, Montreal, Canada, **1**, pp. 411–2.

CROSBIE, R.E., S. JAVEY, J.S. HAY and J.G. PEARCE (1985), 'ESL – a new continuous system simulation language', *Simulation*, **44** (5), 242–6.

CROSBIE, R.E. and J.L. HAY (1985), 'Applications of ESL', *Proceedings of the 11th IMACS World Congress*, Oslo, Norway, **3**, 193–6.

DELEBEQUE, F. and S. STEER (1985), 'The interactive system BLAISE for control engineering', *Prep. 31st IFAC Int. Symp. CADCE'85*, Copenhagen, pp. 44–6.

DELEBEQUE, F. and S. STEER (1986), 'Some remarks about the design of an interactive CACSD package: The BLAISE experience', in K.L. Lineberry (ed.), *Proc. IEEE 31st Symp. on Computer Aided Control System Design*, NY, pp. 48–51.

DENCE, T.P. (1980), *The FORTRAN Cookbook*, Tab Books, Blue Ridge Summit, PA.

DIVJAK, S. (1975), Synthesis of Time Optimal Digital Simulation System for Continuous Dynamical Systems, PhD thesis, Faculty of Electrical and Computer Engineering, University of Ljubljana.

*EAI Handbook of Analog Computation* (A. Carlson, G. Hannauer, T. Carey and P.J. Holsburg, eds, 1967), Electronic Associates, Inc., West Long Branch, NJ.

*EAI Basics of Parallel Hybrid Computers* (G. Hannauer, ed., 1968), Electronic Associates, Inc., West Long Branch, NJ.

*EAI-580 Analog–Hybrid Computing System* – Reference Handbook (1968), West Long Branch, NJ.

*EAI-2000 Analog Reference Handbook* (1982), West Long Branch, NJ.

ELMQVIST, H. (1975), *SIMNON, an Interactive Simulation Program for Nonlinear Systems*, users manual, Department of Automatic Control, Lund Institute of Technology, Sweden, Report 7502.

ELMQVIST, H. (1977), 'SIMNON: an interactive simulation program for nonlinear system', *Proceedings Simulation '77*, Montreux.

FASOL, K.H and K. DIEKMAN (1990), *Simulation in der Regelungstechnik*, Springer-Verlag, Heidelberg.

FAVRAU, R.R., S.A. MURTHA and G.R. MARR (1984), 'Desk top simulation – a state of the art perspective', *Proceedings of the 1984 Summer Computer Simulation Conference*, Boston, Massachusetts, USA, **1**, 3–8.

GALBRAITH, R. (1982), *Professional Programming Techniques: Starting with the basics*, Blue Ridge Summit, Tab Books.

GILOI, W.K. (1975), *Principles of Continuous System Simulation*, B.G. Teubner, Stuttgart.

GOFORTH, R.R. and R.M. CRISP (1988), 'Simulation with microcomputers: an overview', *Access, a journal of microcomputer applications*, **7** (1), 4–14.

GRIERSON, W.O. (1986), 'Perspectives in simulation hardware and software architecture', *Modeling, Identification and Control* (Norwegian research bulletin), **6** (4), 249–55.

GULLAND, W.G. (1973), 'Continuous system simulation', *IEE Conference on Computer Aided Control System Design*, Cambridge, UK, pp. 186–92.

HALLIN, H.J. and S.A.R. HEPNER (1984), 'Solving benchmark problems with PSCSP and other simulation languages' *Proceedings of the 1984 Summer Computer Simulation Conference*, Boston, Massachusetts, USA, **1**, 3–8.

HAY, J.L. and R.E. CROSBIE (1981), *Outline Proposal for a New Standard for Continuous System Simulation Language (CSSL 81)*, Computer Simulation Centre, Department of Electrical Engineering, University of Salford.

HAY, J.L. and R.E. CROSBIE (1984), 'ISIM – a simulation language for microprocessors', *Simulation*, September 1984, pp. 133–6.

HEISERMANN, D. (1981), *PASCAL*, Tab Books, Blue Ridge Summit, PA.

HUBER, R.M. and A. GUASCH (1985), 'Towards a specification of a structure for continuous system simulation languages', *Proceedings of the 11th IMACS World Congress*, Oslo, Norway, pp. 109–13.

JACKSON, A.S. (1960), *Analog Computation*, McGraw-Hill, London.

JENSEN, K. and N. WIRTH (1978), *PASCAL – User Manual and Report*, 2nd edn, Springer, Berlin.

KARBA, R., B. ZUPANČIČ, F. BREMŠAK, A. MRHAR and S. PRIMOŽIČ (1990), 'Simulation tools in pharmacokinetic modelling', *Acta Pharm., Jugosl.*, **40**, 247–62.

KLEINERT, W., D. SOLAR and F. BERGER (1983), 'Status report on TU Vienna's hybrid time sharing system', *Proceedings of the 1st European Simulation Congress*, Aachen, W. Germany, pp. 193–200.

KLEINERT, W., M. GRAFF, R. KARBA and B. ZUPANČIČ (1988), 'Simulation einer Destillationskolonne – Modellierung mit SIMCOS und Vergleich der Ergebnisse von ACSL und SIMSTAR Simulationen', *Proceedings of the 5th Symposium Simulationstechnik*, Aachen, W. Germany, pp. 254–9.

KORN, G.A. (1983a), 'A wish for simulation-language specifications', *Simulation*, January, p. 83.

KORN, G.A. (1983b), 'Direct-executing languages for interactive simulation and computer-aided experiments', *Proceedings of the 1st European Simulation Congress*, Aachen, W. Germany, pp. 225–33.

KORN, G.A. and J.V. WAIT (1978), *Digital Continuous System Simulation*, Prentice Hall.

LANDAUER, J.P. (1988), *EAI STARTRAN Environment*, users' manual, Electronic Associates, Inc., West Long Branch, NJ.

LIPSCHUTZ, S. and A. POE (1978), *Theory and Problems of Programming with FORTRAN*, Schaums Outline Series, NY.

LITTLE, J.N., A. EMAMI-NAEINNI and S.N. BANGERT (1985), 'CTRLC-C and matrix environments for computer aided design of control systems', V.M. Jamshidi and C.J. Herget (eds), *Computer-Aided Systems Engineering*, North-Holland, Amsterdam, pp. 111–24.

MACIEJOWSKI, J.M. and A.G.J. MACFARLANE (1985), 'The Cambridge linear analysis and design programs', in M.J. Jamshidi and C.J. Herget (eds), *Computer-aided Control System Engineering*, North-Holland, Amsterdam, pp. 125–37.

MITCHEL & GAUTHIER, ASSOC. (1981), *ACSL: Advanced Continuous Simulation Language* (user guide/reference manual).

MOLER, C. (1982), *MATLAB User's Guide*, Department of Computer Science, University of New Mexico, Albuquerque.

MOLNAR, I. (1983), 'Some problems in research of general simulation systems', *Proceedings of the 1st European Simulation Congress*, Aachen, W. Germany, pp. 88–92.

NEELAMKAVIL, F. (1987), *Computer Simulation and Modelling*, John Wiley, NY.

NILSEN, N.R. (1984), *The CSSL IV Simulation Language* (reference manual). Simulation Service, Chatsworth, California, USA.

NILSEN, N.R. (1985), 'Recent advances in CSSL IV', *Proceedings of the 11th IMACS World Congress*, Oslo, Norway, **3**, 101–3.

ÖREN, T.I. and B.P. ZIEGLER (1979), 'Concepts for advanced simulation methodologies', *Simulation*, **32** (3), 69–82.

*PC – MATLAB™ for MS-DOS personal computers*, The MathWorks Inc., South Natick, MA (1989).

PETKOV, P.Hr. and N.D. CHRISTOV (1985), 'SYSLAB: An interactive system for analysis and design of linear multivariable systems', *Prep. 31st IFAC Int. Symp., CADCE'85*, Copenhagen, pp. 140–4.

RIMVALL, M. and F.E. CELLIER (1985), 'A structural approach to CACSD' in M.J. Jamshidi and C.J. Herget (eds), *Computer-aided Control System Engineering*, North-Holland, Amsterdam, pp. 149–58.

RIMVALL, M. AND F.E. CELLIER (1986), 'Evaluation and perspectives of simulation languages following the CSSL standard', *Modeling, Identification and Control* (Norwegian research bulletin), **6** (4), 181–99.

SAUCEDO, R. and E.E. SCHIRING (1986), *Introduction to Continuous and Digital Control Systems*, Macmillan Applied System Science, NY.

SCHMID, CHR. (1985a), 'KEDDC – A computer-aided analysis and design package for control systems', in M.J. Jamshidi and C.J. Herget (eds), *Computer-aided Control System Engineering*, North-Holland, Amsterdam, pp. 159–80.

SCHMID, CHR. (1985b), 'Technique and tools of CACSD', *Prep. 4th IFAC Symp. on Computer-aided Design in Control Systems, CACSD'88*, Beijing, pp. 67–75.

SCHMIDT, B. (1986), 'Classification of simulation software', *Syst. Anal. Model. Simul.* (Benelux journal), **3** (2), 133–40.

SCHMIDT, G. (1980), *Simulationstechnik*, R. Oldenbourg, Munich.

SHAH, M.J. (1988), *Engineering simulation: Tools and applications using the IBM PC family*, Prentice Hall, Englewood Cliffs, NJ.

SHAH, S.C., M.A. FLOYD and L.L. LECHAMN (1985), 'MATRIXx: Control design and model building CAE capabilities', in M.J. Jamshidi and C.J. Herget (eds), *Computer-aided Control System Engineering*, North-Holland, Amsterdam, pp. 181–207.

STRAUSS, J.C. (1967), 'The SCi continuous system simulation language', *Simulation*, **9**, pp. 281–303.

SYN, W.M. and H. DOST (1985), 'On the dynamic simulation language (DSL/VS) and its use in the IBM corporation', *Proceedings of the 11th IMACS World Congress*, Oslo, Norway, **3**, 115–18.

ŠEGA, M., S. STRMČNIK, R. KARBA and D. MATKO (1985), 'Interactive program package ANA for system analysis and control design', *Prep. 3rd IFAC Int. Symp., CADCE'85*, Copenhagen, pp. 95–8.

TERREL, J.T. (1988), *Introduction to Digital Filters*, Macmillan Education, Houndmills.

TOMOVIĆ, R. and W.J. KARPLUS (1962), *High Speed Analog Computers*, John Wiley, NY.

*TUTSIM User's Manual for IBM PC Computers* (1983), Twente University of Technology.

TYSSO, A. (1985), 'Practical CAD tools for mathematical modelling', *Prep. 31st International Symp., CADCE'85*, Copenhagen, pp. 360–4.

WELSH, J. (1979), *Introduction to PASCAL*, Prentice Hall, Englewood Cliffs, NJ.

WEST, P.J., S.P. BINGULAC and W.R. PERKINS (1985), 'L-A-S: A computer aided control system design language', in M.J. Jamshidi and C.J. Herget (eds), *Computer-aided Control System Engineering*, North-Holland, Amsterdam, pp. 243–61.

ZIEGLER, B.P. (1976), *Theory of Modelling and Simulation*, John Wiley & Sons, Inc., NY.

ZUPANČIČ, B., D. MATKO, M. ŠEGA and P. TRAMTE (1986), 'Simulation in the program package ANA', *Proceedings of the 2nd European Simulation Congress*, Antwerp, Belgium, pp. 314–18.

ZUPANČIČ, B. (1989a), 'Digital Simulation Language Synthesis for the Computer Aided Control System Design', PhD, Faculty of Electrical and Computer Engineering, University of Ljubljana, Ljubljana.

ZUPANČIČ, B. (1989b), *SIMCOS – the Language for Continuous and Discrete Systems Simulation*, users' manual, Faculty of Electrical and Computer Engineering, University of Ljubljana, Ljubljana.

ZUPANČIČ, B., D. MATKO, R. KARBA, M. ATANASIJEVIĆ and Z. ŠEHIĆ (1991), 'Extensions of the simulation language SIMCOS towards continuous – discrete complex experimentation system', *Preprints of the IFAC Symposium, CADCS'91*, University of Wales, Swansea, UK, pp. 351–6.

# 5

# Further Procedures Concerning Reliable Simulation Results

Numerical integration methods represent the heart of each digital simulation system. For the simulation of simple problems, users usually do not need much knowledge of integration algorithms. However, if a problem is numerically more complicated, it is important for the user to have basic knowledge of integration algorithms and to appreciate the advantages and disadvantages of a particular integration algorithm. So we will give a brief survey of integration methods which are closely related to the physical background of problems and not to the numerical behaviour of algorithms.

Algebraic loops present another numerical problem. They rarely appear in simulation studies if the problems are modelled properly. However, these problems are frequently met by novice programmers and modellers. Thus the principles of algebraic loop solving will be discussed briefly.

With some basic knowledge of integration algorithms and algebraic loop solving, the results will be much more reliable and a lot of computer time will be spared.

## 5.1 NUMERICAL INTEGRATION METHODS

The simulation software that is currently available usually offers a comprehensive selection of different integration algorithms. Nevertheless, a user does not usually know which method is best for his particular problem. Sometimes he is not aware of this problem at all because the simulation tool itself chooses the default integration algorithm. It usually works quite well for fairly uncomplicated problems. Thus experience has shown that average and less-skilled users always use the default integration algorithm, which is in most cases the Runge–Kutta algorithm of fourth order. The reason for the user's choice is the fact that he is not able to make a decision on something he does not really understand. So he simply ignores the possibility of integration algorithm selection and after some time he even forgets that the tool provides him with this facility. The best solution to this problem would be the integration algorithm, which would be able to handle all kinds of problems with the same efficiency. Because it is unrealistic to expect the user to be able to choose algorithms a great deal of work has gone into producing the kind of expert systems which help the user to choose the best integration algorithm for a particular simulation application.

The following references may provide users with an understanding of numerical problems: Hairer and Wanner (1990); Gustaffson (1990); Hairer and Lubich (1988); Gustaffson (1988); Press *et al.* (1986); Korn and Wait (1978); Hall and Watt (1976); Lapidus and Seinfeld (1971); and Gear (1971).

### 5.1.1 General Form of Numerical Integration Formulae

In the late eighteenth century, when Euler developed his algorithm, the numerical integration of ordinary differential equations was an important part of mathematics. We shall restrict ourselves here to a selection of methods which are useful either for a large class of problems or for their pedagogical value in tracing the state of the art.

Since real models are usually described with higher order differential equations, these equations must be transformed into a system of first order equations by the appropriate definition of new variables. In this way the problem can be described as the initial value problem with vector differential equation

$$\dot{\mathbf{x}} = \mathbf{f}(t, \mathbf{x}) \tag{5.1}$$

with initial condition $\mathbf{x}(t_0) = \mathbf{x}_0$, where $\mathbf{x}$ is the state vector and $\mathbf{f}$ is the derivative function. The simulation means that Eqn (5.1) must be integrated from the initial time $t_0$ to the final time $t_{max}$. Using a numerical integration procedure the simulation time is divided into several calculation intervals. The calculation interval $h$ is supposed to be constant so the solution must be evaluated at points $t_k = t_0 + k \cdot h$, $k = 0, 1, 2, \ldots, k_{max}$. The accurate solution at time $t_{k+1}$ can be written as

$$\mathbf{x}(t_{k+1}) = \mathbf{x}(t_0) + \int_{t_0}^{t_{k+1}} \mathbf{f}(t, \mathbf{x})\, dt = \mathbf{x}(t_k) + \int_{t_k}^{t_{k+1}} \mathbf{f}(t, \mathbf{x})\, dt \tag{5.2}$$

With the discretization of Eqn (5.2) we obtain

$$\mathbf{x}_{k+1} = \mathbf{x}_k + \mathbf{I}_k \tag{5.3}$$

where $\mathbf{I}_k$ approximates to the integral

$$\mathbf{I}_k \approx \int_{t_k}^{t_{k+1}} \mathbf{f}(t, \mathbf{x})\, dt \tag{5.4}$$

Because of this approximation $\mathbf{x}_{k+1}$ is only an approximation of the true value $\mathbf{x}(t_{k+1})$. So an error appears in each calculation interval (local error) and it can even be accumulated during the simulation run (global error).

There are many integration algorithms that solve Eqn (5.4). They can be divided into the following:

- single-step methods (implicit and explicit);
- multistep methods (implicit and explicit);
- extrapolation methods;
- methods for stiff systems.

### 5.1.2 Types of Numerical Integration Error

Two types of errors commonly appear when an integration procedure is performed by a numerical algorithm on a digital computer: the numerical approximation (truncation) error due to the finite order of the integration procedure, and the roundoff error due to the finite precision of the computer.

#### Numerical approximation (truncation) errors

We shall use the term 'numerical approximation (truncation) error' for errors due to the inherent limitations of the integration algorithm. These errors would arise regardless of the digital computer precision (number of bits in the computer word) used for integration implementation. With infinite computer precision, a given integration algorithm would produce a local per calculation interval error

$$\mathbf{e}_{k+1} = \mathbf{x}_{k+1} - \left(\mathbf{x}_k + \int_{t_k}^{t_{k+1}} \mathbf{f}(t, \mathbf{x})\, dt\right) = \mathbf{I}_k - \int_{t_k}^{t_{k+1}} \mathbf{f}(t, \mathbf{x})\, dt \qquad (5.5)$$

for a solution $\mathbf{x}_{k+1}$ started at $t_k$ (accurate value at $t_k$). Often, $\mathbf{e}_{k+1}$ can be estimated as the solution proceeds. However, a more important quantity is the total or global approximation error

$$\mathbf{e}_{T,k+1} = \mathbf{x}_{k+1} - \left(\mathbf{x}_0 + \int_{t_0}^{t_{k+1}} \mathbf{f}(t, \mathbf{x})\, dt\right) \qquad (5.6)$$

for a solution started at $t = t_0$. Usually, the global error is larger than the local error and in some cases may grow without bound as time increases. But the global error cannot be estimated during simulation. We can usually estimate the local error; however, a bounded local error does not assure a bounded global error. Nevertheless, it does represent the only possibility of error control during simulation. It can be shown that for the majority of integration methods the local truncation error is proportional to the $m + 1$ power of the calculation interval

$$\mathbf{e}_{k+1} \propto h^{m+1} \qquad (5.7)$$

where $h$ is the calculation interval and $m$ is the order of an integration method. So for a more accurate result the calculation interval must be reduced or the order of the integration algorithm must be increased.

#### Roundoff errors

In practice, integration algorithms are implemented by computer arithmetic with finite precision (finite word length, number of bits). This leads to a second class of errors, usually called 'roundoff errors'. Roundoff errors accumulate and become

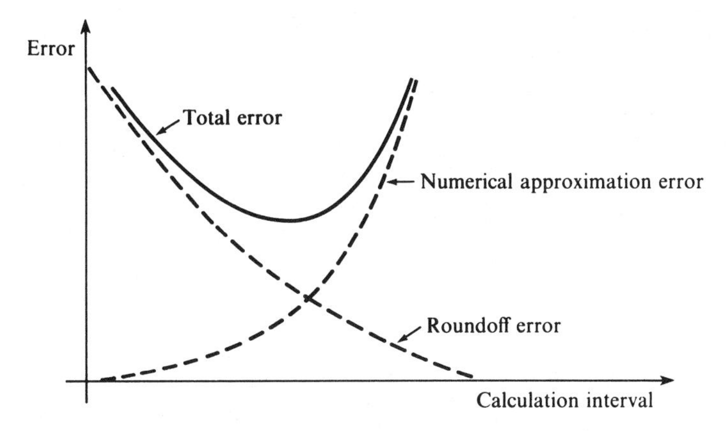


Figure 5.1 The influence of the calculation interval on the total error.

increasingly serious with increasing integration time $t_{\max} - t_0$, with increasing order of the integration method $m$ (more computations) and also, unfortunately, with decreasing calculation interval $h$, since a smaller $h$ means more calculation intervals for given $t_{\max} - t_0$.

Roundoff error accumulation is difficult to predict accurately. With true rounding in computer arithmetic, the errors may be assumed to be uniformly distributed between plus and minus $\frac{1}{2}10^{-n}$, where $n$ is the number of decimal digits. Using this assumption the roundoff error can be estimated with the expression (Korn and Wait, 1978)

$$\mathbf{e}_{T,k+1} \approx \frac{10^{-n}}{2} \sqrt[2]{\frac{m(t_{\max} - t_0)}{12h}} \tag{5.8}$$

Fig. 5.1 shows the influence of the calculation interval $h$ on the numerical approximation and roundoff errors and on the total error, which is the sum of both. There exists an ideal calculation interval $h_{\text{opt}}$ which minimizes the total error. Of course, it is not easy to evaluate $h_{\text{opt}}$ as it depends on the system of differential equations $(\mathbf{f}(t, \mathbf{x}))$, on the integration algorithm and on the computer or its arithmetics.

### 5.1.3 Single-step Integration Methods

A single-step integration method prescribes the numerical technique for calculating the approximate solution at $t_{k+1}$ in terms of the value at the previous step, i.e. at $t_k$. The general formula is obtained by a Taylor expansion series of Eqn (5.2) in the neighbourhood of the point $\mathbf{x}(t_k)$:

$$\mathbf{x}(t_k + h) = \mathbf{x}(t_k) + h\dot{\mathbf{x}}(t_k) + \frac{h^2}{2!}\ddot{\mathbf{x}}(t_k) + \frac{h^3}{3!}\dddot{\mathbf{x}}(t_k) + \cdots \tag{5.9}$$

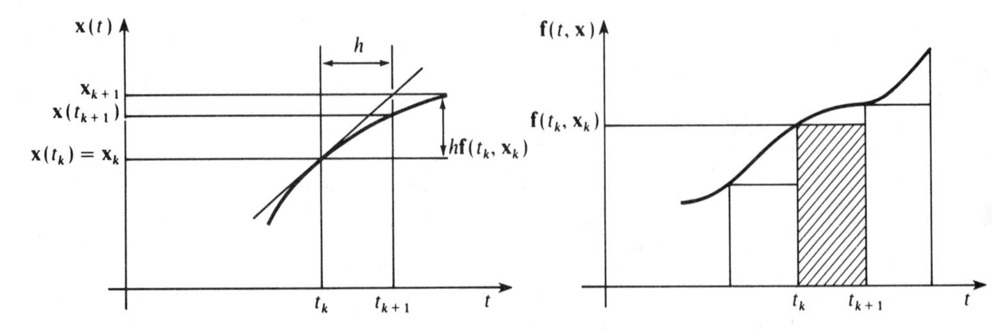


Figure 5.2 Graphic representation of the Euler integration algorithm.

A single-step method is of order $m$ if $m+1$ terms in the Taylor series are used. As higher order terms are neglected, the local error can be approximated by the equation

$$\mathbf{e}(t_k) \approx \frac{h^{m+1}}{(m+1)!}\mathbf{x}^{(m+1)}(t_k) \tag{5.10}$$

If only two terms are used, the elementary single-step algorithm known as the Euler algorithm (first order) is obtained:

$$\mathbf{x}(t_{k+1}) \approx \mathbf{x}(t_k) + h\dot{\mathbf{x}}(t_k) \tag{5.11}$$

or

$$\mathbf{x}_{k+1} = \mathbf{x}_k + h\mathbf{f}(t_k, \mathbf{x}_k) \tag{5.12}$$

Eqn (5.12) represents the Euler integration algorithm. It can be clearly represented in a graphical way (Fig. 5.2). As $\mathbf{x}(t_k) = \mathbf{x}_k$ we assume that the result at point $t_k$ is accurate or that point $t_k$ is the initial integration time. Due to the local truncation error, which is proportional to the square of the calculation interval, the approximate solution $\mathbf{x}_{k+1}$ differs from the true solution $\mathbf{x}(t_{k+1})$.

The Euler integration algorithm is intuitively clear and needs only one derivative calculation in one calculation interval. But it can approximate a differential equation solution within acceptable errors only for very small calculation intervals $h$. The resulting large number of calculation intervals increases the computing effort and the roundoff error due to the finite precision of the computer, so that practical simulation studies require better integration algorithms. To obtain better accuracy, several terms must be used in Taylor expansion series. But this necessitates higher order derivatives which cannot easily be evaluated. Runge was the first to point out that it is possible to avoid successive differentiation in a Taylor series while preserving accuracy. He replaced the part containing high order derivatives in the Taylor series with a part which contained undetermined coefficients and several derivative evaluations within the interval $(t_k, \mathbf{x}_k)$ and $(t_{k+1}, \mathbf{x}_{k+1})$. In other words, the derivatives in the Taylor series were bypassed by $\mathbf{f}(t, \mathbf{x})$ being evaluated several times within the interval

Table 5.1 Runge–Kutta classical method coefficients

|  |  |  |  |  |  |
|---|---|---|---|---|---|
| 0 |  |  |  |  | $\frac{1}{6}$ |
| $\frac{1}{2}$ | $\frac{1}{2}$ |  |  |  | $\frac{1}{3}$ |
| $\frac{1}{2}$ | 0 | $\frac{1}{2}$ |  |  | $\frac{1}{3}$ |
| 1 | 0 | 0 | 1 |  | $\frac{1}{6}$ |

mentioned. Thus we set up the general single-step equation

$$\mathbf{x}_{k+1} = \mathbf{x}_k + \sum_{i=1}^{v} w_i \mathbf{k}_i \tag{5.13}$$

where $w_i$ are the weighting coefficients to be determined, $v$ is the number of derivative evaluations in the calculation interval and $\mathbf{k}_i$ satisfies the explicit sequence

$$\mathbf{k}_i = h\mathbf{f}\left( t_k + c_i h, \mathbf{x}_k + \sum_{j=1}^{i-1} a_{ij}\mathbf{k}_j \right) \qquad i = 1, 2, \ldots, v \tag{5.14}$$

As $\mathbf{k}_i$ in this equation can be evaluated recursively, these methods are called explicit single-step integration methods, also well known as Runge–Kutta methods. The order of the method $m$ is not directly evident from Eqns (5.13) and (5.14), but it is the same as the order of the Taylor prototype series. Deriving Eqns (5.13) and (5.14) for an appropriate order $m$ it is possible to obtain a large number of formulae by a suitable choice of some free parameters. For $1 \leqslant m \leqslant 4$ the number of derivative evaluations $v$ per calculation interval is the same as the order, but for methods with $m > 4$ the number of derivative evaluations is always greater than the order.

The reader interested in developing coefficients $c_i$, $a_{ij}$ and $w_i$ is advised to refer to the literature, e.g. Gear (1971) and Press *et al.* (1986). Here only the single-step explicit integration methods can be presented in the condensed form developed by Butcher (Butcher, 1964). This is the array form of the coefficients in Eqns (5.13) and (5.14):

$$\begin{array}{c|cccc|c}
0 & & & & & w_1 \\
c_2 & a_{21} & & & & w_2 \\
c_3 & a_{31} & a_{32} & & & w_3 \\
\vdots & \vdots & \vdots & \ddots & & \vdots \\
c_v & a_{v1} & a_{v2} & \cdots & a_{vv-1} & w_v
\end{array} \qquad \mathbf{c}|\mathbf{A}|\mathbf{w}$$

The typical single-step explicit methods (Runge–Kutta) are: Euler (1, 1), improved Euler (2, 2), Heun (2, 2), Nystrom (3, 3), Heun (3, 3), classical Runge–Kutta (4, 4), England (4, 4), Runge–Kutta–Fehlberg (4, 5), Runge–Kutta–Fehlberg (5, 6). The numbers in brackets $(m, v)$ denote the order $(m)$ and the number of derivative evaluations $(v)$. The coefficients of the classical Runge–Kutta method are presented in condensed array form as shown in Table 5.1.

Besides single-step explicit methods, implicit ones are also well known. They can be obtained from Eqns (5.13) and (5.14) if the upper limit in the summation in Eqn (5.14) is changed from $i-1$ to $v$:

$$\mathbf{x}_{k+1} = \mathbf{x}_k + \sum_{i=1}^{v} w_i \mathbf{k}_i \tag{5.15}$$

$$\mathbf{k}_i = h\mathbf{f}\left( t_k + c_i h, \mathbf{x}_k + \sum_{j=1}^{v} a_{ij}\mathbf{k}_j \right) \qquad i = 1, 2, \ldots, v \tag{5.16}$$

The matrix **A** in the condensed Butcher form is not triangular but has elements in the right upper part. The main problem appears because the coefficients $\mathbf{k}_i$ depend upon the coefficients $\mathbf{k}_j$ (for $j = 1, 2, \ldots, v$) and the recursive evaluation is no longer possible. So the main disadvantage of the implicit method is that an iterative procedure is required to solve the equations for the coefficients $\mathbf{k}_i$. In comparison with explicit methods they have higher orders for the same number of derivative evaluations, which results in smaller local error. They also have better stability properties. Gauss (2, 1), Gauss (4, 2), Radau (5, 3), Milne (4, 3) and Lobatto (6, 4) are representatives of these methods.

Semiimplicit forms are intermediate between the explicit and implicit formulae. The motivation behind this approach is based upon obtaining stable methods as well as maintaining computational efficiency, i.e. avoiding the necessity to iterate. But instead of iterations the Jacobian matrix

$$\mathbf{J} = \frac{\partial \mathbf{f}(t, \mathbf{x})}{\partial \mathbf{x}} \tag{5.17}$$

and its inversion must be evaluated to calculate each $\mathbf{k}_i$. The procedure demands less computational effort than in the case of implicit iterations, although $v$ evaluations of the Jacobian matrix and $v$ matrix inversions must be performed in each calculation interval. Two Rosenbrock methods are well-known representatives of the semiimplicit methods.

### Estimation of the local error

It is known that the local truncation error is proportional to $h^{m+1}$, where $h$ is the calculation interval and $m$ is the order of a method. To obtain reliable results this error must be evaluated. The common procedure demands integration on the interval $2h$ with two calculation intervals: $h$ and $2h$. It is possible to evaluate the local truncation error from the difference of both solutions. This error can be added to the solution using the calculation interval $h$ whereby the improved solution is obtained. For the fourth order Runge–Kutta method the approximation of the local error is

$$\mathbf{e}_{k,h} \approx \frac{1}{15}(\mathbf{x}_{k,h} - \mathbf{x}_{k,2h}) \tag{5.18}$$

where $\mathbf{x}_{k,h}$ and $\mathbf{x}_{k,2h}$ are the solutions obtained with calculation intervals $h$ and $2h$ respectively. The improved solution is then

$$\mathbf{x}_{k,h}^{*} = \mathbf{x}_{k,h} + \mathbf{e}_{k,h} = \frac{1}{15}(16\mathbf{x}_{k,h} - \mathbf{x}_{k,2h}) \tag{5.19}$$

The above procedure requires a great deal of additional calculation effort.

Another approach to the estimation of the local truncation error uses two single-step methods of orders, $m - 1$ and $m$, and takes the difference between both results. Computation is minimized by a corresponding choice of Runge–Kutta coefficients so that some intermediate results are common to both formulae. Runge–Kutta–Merson and Runge–Kutta–Fehlberg are well-known representatives of this approach. Fehlberg calculated such coefficients that, for the fifth order evaluation, only one additional derivative evaluation is needed in comparison with fourth order evaluation.

**Stability of single-step methods**

It is well known that a continuous system is stable if the eigenvalues of the Jacobian matrix

$$\mathbf{J} = \frac{\partial \mathbf{f}(t, \mathbf{x})}{\partial \mathbf{x}} \tag{5.20}$$

which are obtained by the equation

$$\det(\lambda_i \mathbf{I} - \mathbf{J}) = 0 \tag{5.21}$$

have negative real parts. Numerical integration methods transform a differential equation system into a system of difference equations. The latter can be unstable (and thus amplify small numerical errors during simulation) even if the original differential equation system is completely stable. Let us consider the explicit Euler algorithm as applied to a simple differential equation:

$$\left.\begin{aligned} \frac{dx}{dt} &= f(t, x) = \lambda x \\ x_{k+1} &= x_k + hf(t_k, x_k) = x_k + h\lambda x_k = x_k(1 + \lambda h) \end{aligned}\right\} \tag{5.22}$$

As $\lambda$ is the eigenvalue of the continuous system, it is stable for every $\lambda < 0$ because the well-known solution is $x(t) = x(0)\exp(-\lambda t)$. The difference equation solution $x_k = x(0)(1 + \lambda h)^k$ decays for

$$|1 + \lambda h| < 0 \tag{5.23}$$

representing the equation of the unit circle with centre $-1$ in the $\lambda h$ plane. The inside of this circle represents the stability of the Euler integration algorithm.

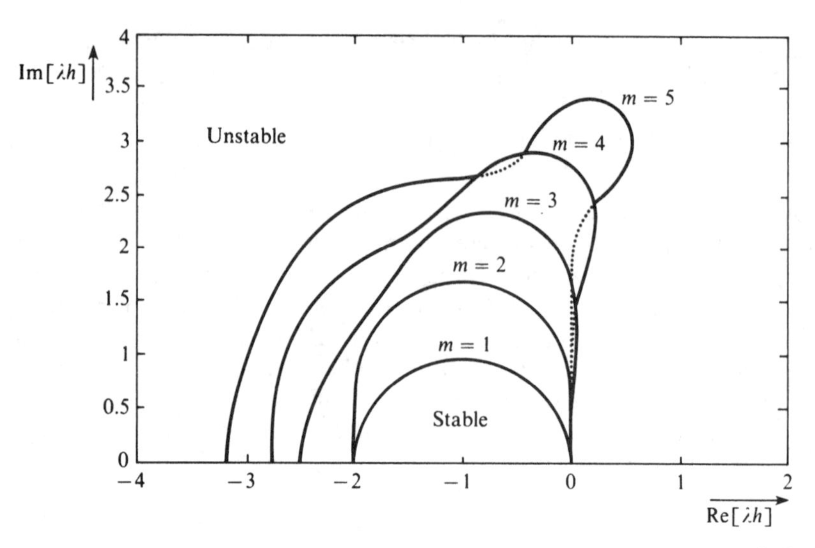


Figure 5.3 Stability regions for Euler ($m = 1$) and Runge–Kutta methods ($m = 2, 3, 4, 5$).

The same restriction holds for nonlinear systems of first order differential equations if $\lambda_i$ is one of the Jacobian matrix eigenvalues. But in the case of nonlinear systems the Jacobian matrix and the eigenvalues are time varying and so the algorithm can be stable at some working point but unstable in some other.

In the same way, the stability region for other higher order explicit Runge–Kutta algorithms can be determined in the $\lambda h$ plane. Fig. 5.3 shows the stability region for Euler and various Runge–Kutta methods (order 2–5). As for all other methods, only the regions in the upper part of the $\lambda h$ plane are shown. The methods are stable inside the closed curves.

Although the high order Runge–Kutta methods improve accuracy, the stability regions of various orders do not differ greatly. We can conclude that Runge–Kutta methods are stable if the selected calculation interval $h$ approximately satisfies the equation

$$|\lambda_{\max}|h < 3 \tag{5.24}$$

where $\lambda_{\max}$ is the maximal eigenvalue of the Jacobian matrix. As $\lambda_{\max}$ is inversely proportional to the minimal time constant $T_{\min}$ of the system, we can write the relation

$$h < 3T_{\min} \tag{5.25}$$

But this is only the stability condition. It is well known that the calculation interval must be smaller than the minimum time constant to minimize the error (see Fig. 5.1).

Single-step implicit methods have larger stability regions. The procedure for determining them is the same as for the explicit methods. The Gauss methods of the first, second (trapezoidal rule) and third order are stable in the whole of the left part of the $\lambda h$ plane. The Radau and Lobatto methods have smaller stability regions but, nevertheless, these are greater than those of the explicit methods ($|\lambda_{\max}|h < 5$).

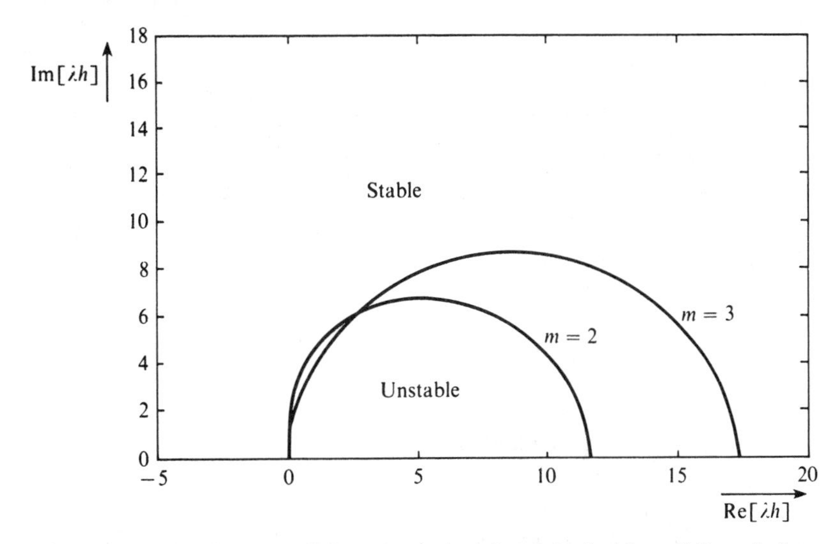


Figure 5.4 Stability regions for Rosenbrock methods (2. and 3. order).

Good stability properties also characterize the semiimplicit methods. Fig. 5.4 represents the stability regions for the Rosenbrock semiimplicit methods of second and third order. The methods are unstable inside the circles.

Implicit and semiimplicit methods enable relatively large calculation intervals from the stability point of view. But the computer time saved due to the large calculation interval is partly spent on the iterative solution of implicit equations in the case of implicit methods, or for Jacobian matrix evaluation and inveısion in the case of semiimplicit methods. These methods are particularly important for the simulation of systems with very different eigenvalues or time constants (stiff systems).

### 5.1.4 Multistep Integration Methods

Single-step methods, especially when they are of the same complexity as the fourth order Runge–Kutta method, seem to be computationally uneconomical because they demand several derivative evaluations in one calculation interval. Schemes which take into account the results of several preceding steps are less expensive in terms of the required number of derivatives evaluations per calculation interval and are thus much faster.

The multistep integration formula is defined by the equation

$$\begin{aligned}\mathbf{x}_{k+1} &= a_0\mathbf{x}_k + a_1\mathbf{x}_{k-1} + \cdots + a_p\mathbf{x}_{k-p} \\ &\quad + h[b_{-1}\mathbf{f}(t_{k+1}, \mathbf{x}_{k+1}) + \cdots + b_p\mathbf{f}(t_{k-p}, \mathbf{x}_{k-p})] \\ &= \sum_{i=0}^{p} a_i\mathbf{x}_{k-i} + h\sum_{i=-1}^{p} b_i\mathbf{f}(t_{k-i}, \mathbf{x}_{k-i}) \qquad (5.26)\end{aligned}$$

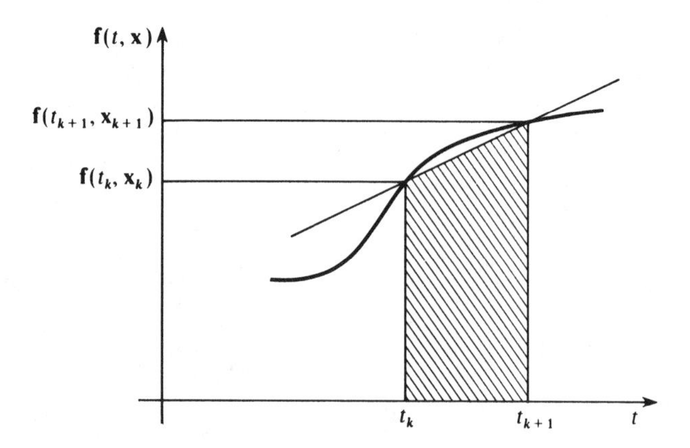


Figure 5.5 Integration using the trapezoidal rule.

and requires the state vector and its derivative in $p$ past values ($p + 1$ step method). If $b_{-1} = 0$, then the rule is called an explicit or a predictor, since the future value $\mathbf{f}(t_{k+1}, \mathbf{x}_{k+1})$ is not used. For $b_{-1} \neq 0$ the rule is an implicit (closed rule) or a corrector rule and an iterative procedure must be performed in each calculation interval to solve Eqn (5.26). As past values of $\mathbf{x}_k$ and $\mathbf{f}(t_k, \mathbf{x}_k)$ are stored, multistep methods require only one derivative evalution per calculation interval.

The coefficients $a_i$ and $b_i$ are selected so as to make the Eqn (5.26) exact for a polynomial of degree $m$ ($m + 1$ coefficients) with which the state $\mathbf{x}$ can be expressed. In this way $m + 1$ coefficients from $2p + 3$ coefficients in Eqn (5.26) are determined with the method of undetermined coefficients. Other coefficients are determined to minimize errors and to maximize the stability region.

Multistep rules present problems at the beginning of a solution, where past values of $\mathbf{x}_k$ and $\mathbf{f}(t_k, \mathbf{x}_k)$ are not available. This usually means that multistep rules have to be started by some form of single-step, self-starting algorithm of consistent accuracy. Problems also appear when derivatives show discontinuities during the simulation study. In this case the approximation Eqn (5.26) is inaccurate as the state $\mathbf{x}$ cannot be expressed accurately by the $m$th order polynomial. The only possibility to use these methods for discontinuous systems is to restart the integration procedure at the discontinuity points.

The local truncation error depends upon the order $m$ of the polynomial which expresses the state of the system and is proportional to $h^{m+1}$.

Using $p = 0, a_0 = 1, b_{-1} = \frac{1}{2}, b_0 = \frac{1}{2}$ the very popular trapezoidal rule is obtained:

$$\mathbf{x}_{k+1} = \mathbf{x}_k + \frac{h}{2}(\mathbf{f}(t_k, \mathbf{x}_k) + \mathbf{f}(t_{k+1}, \mathbf{x}_{k+1})) \tag{5.27}$$

The principle of this integration is shown in Fig. 5.5.

The integral of the derivative function on interval $t_k$, $t_{k+1}$ is evaluated by the trapezoidal area.

Table 5.2 Coefficients for Adams methods ($a_0 = 1$, all other $a_i = 0$)

| $m$ | $b_{-1}$ | $b_0$ | $b_1$ | $b_2$ | $b_3$ | *Name* |
|---|---|---|---|---|---|---|
| 1 | 0 | 1 | 0 | 0 | 0 | explicit Euler |
| 2 | 0 | $\frac{3}{2}$ | $\frac{-1}{2}$ | 0 | 0 | explicit trapezoidal |
| 3 | 0 | $\frac{23}{12}$ | $\frac{-16}{12}$ | $\frac{5}{12}$ | 0 | Adams–Bashforth 3rd order |
| 4 | 0 | $\frac{55}{24}$ | $\frac{-59}{24}$ | $\frac{37}{24}$ | $\frac{-9}{24}$ | Adams–Bashforth 4th order |
| 1 | 1 | 0 | 0 | 0 | 0 | implicit Euler |
| 2 | $\frac{1}{2}$ | $\frac{1}{2}$ | 0 | 0 | 0 | implicit trapezoidal |
| 3 | $\frac{5}{12}$ | $\frac{8}{12}$ | $\frac{-1}{12}$ | 0 | 0 | Adams–Moulton 3rd order |
| 4 | $\frac{9}{24}$ | $\frac{19}{24}$ | $\frac{-5}{24}$ | $\frac{1}{24}$ | 0 | Adams–Moulton 4th order |

As the method needs one previous derivative evaluation $\mathbf{f}(t_k, \mathbf{x}_k)$ $(p = 0)$, it is in fact the implicit single-step method and so is a special example of the multistep method. It can be used efficiently for simpler problems due to its calculation efficiency and good numerical stability properties.

Adams methods are the most popular multistep methods. They can be explicit (Adams–Bashforth) or implicit (Adams–Moulton). Adams–Bashforth methods are obtained by

$$p = m - 1 \qquad a_0 = 1 \qquad a_1 = a_2 = a_3 = \cdots = a_{m-1} = 0 \qquad b_{-1} = 0 \tag{5.28}$$

To obtain the $m$th order method, $p+1 = m$ previous values must be used. This is the $m$ step method.

Adams–Moulton methods are obtained by

$$p = m - 2 \qquad a_0 = 1 \qquad a_1 = a_2 = a_3 = \cdots = a_{m-2} = 0 \tag{5.29}$$

To obtain the $m$th order method, only $p+1 = m - 1$ previous values must be used. Such a method is the $m - 1$ step. Unfortunately, the iterative procedure for nonlinear equations must be performed in each calculation interval as the method is implicit. To ensure unity gain for integration of a constant, the sum of coefficients $b_i$ must be equal to 1:

$$\sum_{i=-1}^{p} b_i = 1 \tag{5.30}$$

Table 5.2 represents the coefficients of the commonly used Adams methods. Some single-step formulae can also be interpreted as special examples of multistep methods.

**Predictor–corrector methods**

When using implicit single or multistep formulae the main problem arises from the iterative algorithm (e.g. Newton–Raphson) to solve Eqns (5.15), (5.16) and (5.26). The effective solution to this problem is to use an explicit single or multistep rule to estimate $\mathbf{x}_{k+1}$. This estimation $\hat{\mathbf{x}}_{k+1}$ is then used to evaluate the derivative $\mathbf{f}(t_{k+1}, \hat{\mathbf{x}}_{k+1})$. Using this derivative the implicit method can be used as a corrector to obtain the value $\mathbf{x}_{k+1}$. Such methods are well known as predictor–corrector methods (PC). The order of the corrector method must always be greater than or the same as the order of the predictor method.

A simple predictor–corrector method is to use the explicit Nystrom method (midpoint rule) as the predictor and the implicit trapezoidal rule as the corrector:

$$\hat{\mathbf{x}}_{k+1} = \mathbf{x}_{k-1} + 2h\mathbf{f}(t_k, \mathbf{x}_k) \tag{5.31}$$

$$\mathbf{x}_{k+1} = \mathbf{x}_k + \frac{h}{2}(\mathbf{f}(t_{k+1}, \hat{\mathbf{x}}_{k+1}) + \mathbf{f}(t_k, \mathbf{x}_k)) \tag{5.32}$$

We can see that two derivative evaluations are used per calculation interval. The advantages of these methods are the good stability properties and the reduced local truncation error. The local error can be evaluated by the difference between the predicted value $\hat{\mathbf{x}}_{k+1}$ and the corrected value $\mathbf{x}_{k+1}$. If the difference between $\hat{\mathbf{x}}_{k+1}$ and $\mathbf{x}_{k+1}$ is not in the tolerance region, the corrector formula can be used iteratively. Each iteration demands one additional derivative and one additional corrector evaluation. The procedure converges for small calculation intervals. Where it does not converge, the calculation interval should be reduced. It is usually enough to perform a few iterations only to obtain satisfactory results.

The Adams–Bashforth and Adams–Moulton methods of fourth order are commonly used in predictor–corrector methods.

### Estimation of the local truncation error

The procedure for the evaluation of the local truncation error for multistep methods is the same as that for single-step methods. For the fourth order Adams–Bashforth and Adams–Moulton methods the approximation of the local error and a possible improvement is given by the formulae

$$\mathbf{e}_{k,h} \approx \frac{1}{15}(\mathbf{x}_{k,h} - \mathbf{x}_{k,2h}) \tag{5.33}$$

$$\mathbf{x}^*_{k,h} = \frac{1}{15}(16\mathbf{x}_{k,h} - \mathbf{x}_{k,2h}) \tag{5.34}$$

$\mathbf{x}_{k,h}$ is the solution obtained with calculation interval $h$, and $\mathbf{x}_{k,2h}$ that obtained with the interval $2h$, while $\mathbf{x}^*_{k,h}$ is the improved solution.

In predictor–corrector methods the procedure of local truncation error evaluation requires much less CPU time. In this case the local truncation error can be evaluated directly from the difference between the predictor and corrector solutions. If Adams–Bashforth of the fourth order as the predictor, and Adams–Moulton of the fourth order as the corrector are used, the local error and possible improvement is

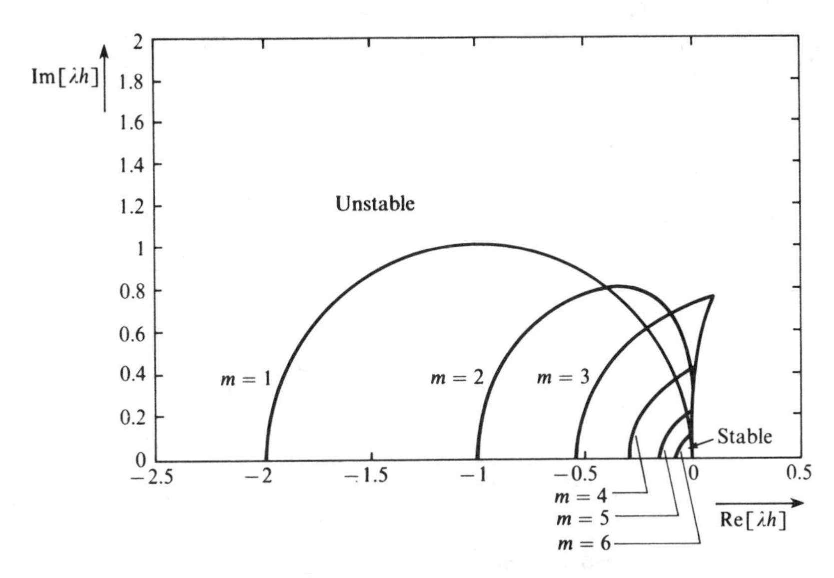


Figure 5.6 Stability regions for the Adams–Bashforth method.

given by the equations

$$\mathbf{e}_k \approx -\frac{1}{14}(\mathbf{x}_{k,C} - \mathbf{x}_{k,P}) \tag{5.35}$$

$$\mathbf{x}_k^* = \frac{1}{14}(13\mathbf{x}_{k,C} + \mathbf{x}_{k,P}) \tag{5.36}$$

**Stability of the multistep methods**

Explicit multistep methods have small stability regions in the $\lambda h$ plane. These regions become even smaller with increasing order of a method. Fig. 5.6 illustrates these regions for Adams–Bashforth methods, which are stable inside closed regions.

Implicit multistep methods have much larger stability regions and so enable larger calculation intervals. The implicit Euler method and the implicit trapezoidal rule are stable in the whole left part of the $\lambda h$ plane. Fig. 5.7 represents these regions for Adams–Moulton methods, which are stable inside the closed regions.

The stability properties of predictor–corrector methods depend in general on the predictor, the corrector and the number of corrector iterations. The stability region is greater than the stability region of the predictor (explicit method) but smaller than the corrector region (implicit method).

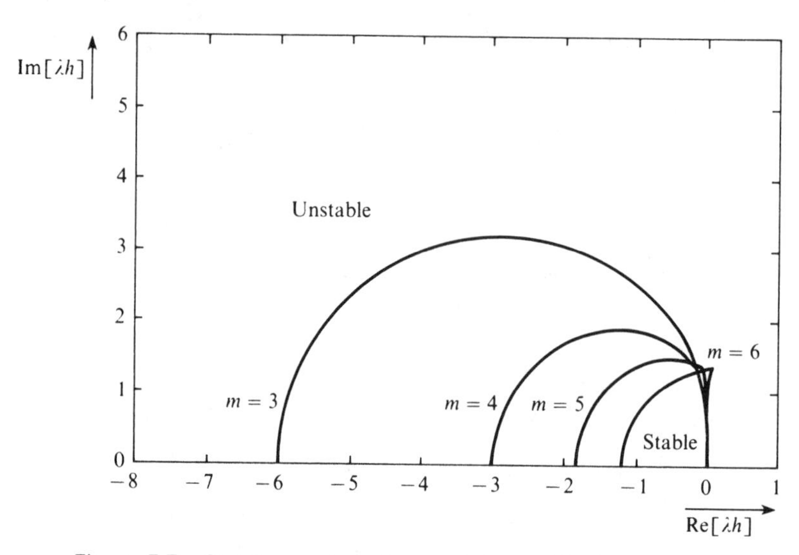


Figure 5.7 Stability regions for the Adams–Moulton method.

### 5.1.5 Extrapolation Methods

In comparison with the methods discussed previously, the extrapolation methods for solving ordinary differential equations are rather specific. The main idea of this approach is to integrate equations in the interval from $t_k$ to $t_{k+1}$ with different calculation subintervals, where every interval is smaller than the preceding one. Let us sign the solutions obtained $\mathbf{x}_{k+1,h_1}, \mathbf{x}_{k+1,h_2}, \mathbf{x}_{k+1,h_3}, \ldots, \mathbf{x}_{k+1,h_r}$. This principle is shown in Fig. 5.8.

The sequence of solutions $\mathbf{x}_{k+1,h_i}$ and the calculation subintervals $h_i$, $i = 1, 2, \ldots, r$ are used to obtain an analytical function $\mathbf{x}_{k+1}(h)$. Using this function the extrapolation procedure to obtain the solution $\mathbf{x}_{k+1,h_\infty}$ is performed, where $h_\infty$ is the zero calculation subinterval which would produce the accurate result.

Two types of extrapolation algorithms are normally used:

- Richardson extrapolation;
- rational extrapolation.

Richardson extrapolation is obtained when the analytical function $\mathbf{x}_{k+1}(h)$ is realized in polynomial form:

$$\mathbf{x}_{k+1}(h) = \mathbf{x}_{k+1,h_\infty} + \alpha_1 h + \alpha_2 h^2 + \cdots + \alpha_r h^r \tag{5.37}$$

For many applications, however, the analytical function $\mathbf{x}_{k+1}(h)$ in rational form enables more accurate results. Such an extrapolation procedure is called the 'rational function extrapolation' and is, for one component of the vector $\mathbf{x}_{k+1}(h)$, given by

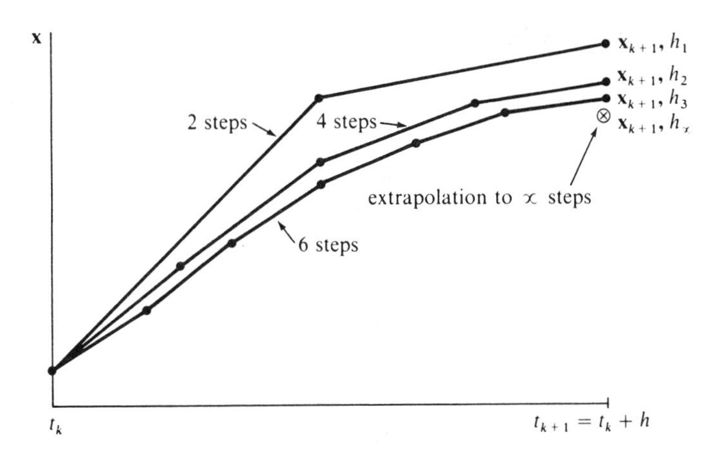


Figure 5.8 The principle of extrapolation integration method.

the equation

$$\left.\begin{aligned} x_{k+1}(h) &= \frac{p_0 + p_1 h + p_2 h^2 + \cdots + p_r h^r}{q_0 + q_1 h + q_2 h^2 + \cdots + q_r h^r} \\ x_{k+1,h_\infty} &= \frac{p_0}{q_0} \end{aligned}\right\} \tag{5.38}$$

Several sequences of calculation subintervals are used. The best results are probably obtained with the sequence $\frac{h}{2}, \frac{h}{3}, \frac{h}{4}, \frac{h}{6}, \frac{h}{8}, \frac{h}{12}, \ldots$, where $h$ is the calculation interval.

After each integration, the extrapolation is applied. It returns both the extrapolated value and the local error estimate. If the error is not satisfactory another integration with a smaller calculation subinterval is performed. If it gives satisfactory results the integration proceeds with the next calculation interval.

One approach to elementary extrapolation methods is represented by the Euler–Romberg method. The integration on different calculation subintervals is performed by the Euler method. The trapezoidal rule can also be used to obtain better stability and accuracy. Both approaches use polynomial extrapolation. But the principal drawback of the trapezoidal rule is that implicit equations must be solved in each calculation interval.

So the best extrapolation methods are probably based on the Bulirsch–Stoer–Gragg approach. Various methods use the explicit midpoint integration method and rational extrapolation.

The stability and error analysis of extrapolation depend upon the integration and extrapolation algorithms used. Extrapolation methods usually demand more derivative evaluations than the Runge–Kutta methods for a moderate level of accuracy. However, they are very efficient for higher levels of accuracy. Some authors also recommend them for systems with discontinuities.

### 5.1.6 Integration Methods for Stiff Systems

Many physical systems are described with models giving very different magnitudes of eigenvalues. Such situations arise in the study of flow dynamics, process dynamics and control, in electrical circuits, in chemical reactions, etc. Systems with very different eigenvalues or time constants (this term is preferred by engineers and physicists) are called 'stiff systems'. Good digital simulation systems must include at least one stiff integration method.

The system is usually defined to be stiff if

$$\frac{\max \text{Re}[\lambda_i]}{\min \text{Re}[\lambda_i]} > 100 \tag{5.39}$$

where $\max \text{Re}[\lambda_i]$ is the maximum value and $\min \text{Re}[\lambda_i]$ the minimum value of the Jacobian matrix real part of the eigenvalues. The problem associated with stiff systems is twofold: stability and accuracy. If a method with a finite absolute stability boundary is used (e.g. Runge–Kutta methods), large negative real parts of some $\lambda_i$ will result in the calculation interval length being excessively small. So CPU costs increase considerably when using classical Runge–Kutta methods, and due to the large number of calculation intervals in one simulation run the accumulated roundoff error can also increase. On the other hand, if a method that is stable in the whole left part of the $\lambda h$ plane is used (e.g. the trapezoidal rule) the stability problem is avoided, but for a reasonable calculation interval length $h$ the solution component corresponding to the largest eigenvalue will be approximated very inaccurately.

In addition, the convergence requirements for the iterative solution of the nonlinear algebraic equations arising at each step in an implicit method place restrictions on the largest value of $h$ that can be used. These restrictions vary considerably depending on the particular iterative technique used.

Because of its good stability properties, implicit and semiimplicit methods are generally used for stiff integration procedures. The substantial computing effort per calculation interval is often balanced by the possibility of using much larger calculation steps. Unfortunately, only low order single or multistep methods are stable in the whole left part of the $\lambda h$ plane. Gear also developed his famous stiff stable methods as multistep methods for orders greater than two. His approach is based on the definition of the stiffly stable method. This definition provides high accuracy and stability for major system eigenvalues of small magnitude and stability for only relatively unimportant short time constants associated with eigenvalues of large magnitudes. Using this definition he developed methods of orders one to six. The Newton–Raphson iteration is used to solve implicit equations. The Jacobian is recomputed only when tests indicate that the current approximation is no longer suitable. The local error is controlled using variable order, variable calculation interval strategies. Fig. 5.9 represents the stability regions for Gear stiff integration algorithms for orders one to six. The methods are stable outside the closed regions.

There are many other methods which are more or less suitable for stiff systems. One of them is the semiimplicit, embedded Rosenbrock–Wanner method. Another

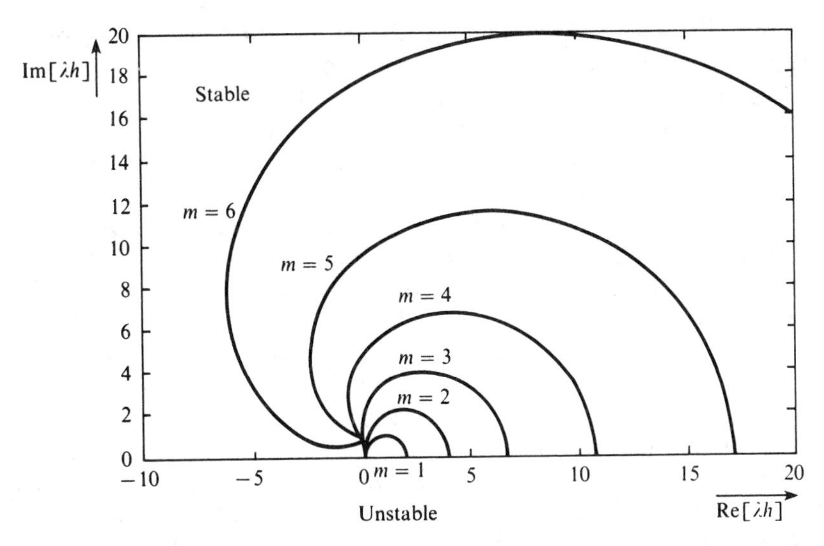


Figure 5.9 Stability regions for Gear stiffly stable methods.

promising approach represents the extrapolation methods (e.g. extrapolation methods using the implicit midpoint rule).

### 5.1.7 Choice of the Calculation Interval and the Strategies of its Control

#### Choice of the calculation interval

When we discussed various typical errors and integration methods, we saw that the use of smaller calculation intervals has the following effects:

- It reduces local (and so also global) truncation error.
- It increases roundoff error.
- It improves numerical stability properties.
- It improves the convergence of iterative procedures used in implicit methods.
- It increases CPU costs.

Fortunately, all modern digital simulation systems nowadays have algorithms for automatic adjustment of the calculation interval during simulation. So the interval is relatively large if the derivatives only change moderately slowly, and becomes smaller whenever the change is more rapid. This procedure confines error to within the prescribed tolerance band and is extremely effective when a system's or input signal's dynamics vary during simulation (nonlinear systems, time varying eigenvalues, discontinuities in the state and input variables and their derivatives).

So from the user's point of view the problem of the calculation interval determination is transformed into the problem of the appropriate prescribed error choice. But to choose this error correctly the user must have some basic knowledge of the model or, better, of how accurately the model is to be simulated. Model accuracy depends upon the accuracy of the data used in the modelling phase, but also upon the purpose of the model.

As well as the prescribed errors, the user can also define the initial calculation interval, which is usually selected to be the same as the interval in which the user wishes to observe the simulation results (communication interval). But it is sometimes possible to decrease CPU costs with a more suitable magnitude of initial calculation interval. For this purpose the user needs some knowledge about the model's dynamics (eigenvalues, time constants) and about input signals (frequency spectra) as well. Choice of a calculation interval such that the condition for numerically stable behaviour is fulfilled is recommended (e.g. $|\lambda| h_{max} < 3$ for Runge–Kutta methods). Also, we should choose a calculation interval such that at least some calculation points are used within the interval of one time constant (minimum time constant for nonstiff and maximum time constant for stiff methods). Taking into account the input signals, the user can consider the Shannon theorem (theoretically, at least two calculation points in one period of maximal frequency; in practice, five to ten).

**Calculation interval control strategies**

As stated earlier, all modern simulation systems have algorithms for the automatic adjustment of the calculation interval during simulation, usually in a band between a maximum value (communication interval) and a prescribed minimum value. All strategies are based on the local per calculation interval truncation error estimate, which is usually for an $m$th order method estimated by using the equation

$$\mathbf{e} = \mathbf{\Phi} \cdot h^{m+1} \tag{5.40}$$

where $\mathbf{\Phi}$ is a vector that depends upon the solution of differential equations. The estimation vector $\mathbf{e}$ is calculated at each calculation interval and is specific for the integration algorithm used. However, in order to use the calculated error vector in the calculation interval control strategy, a scalar error estimate (the norm of the error) must be calculated and compared with the user defined prescribed error or tolerance. Different norms, usually including mixed relative and absolute errors, are used. A typical example is

$$\|\mathbf{e}\| = \max_i \left| \frac{e_i}{|x_i| + \eta_i} \right| \tag{5.41}$$

where $e_i$ is the $i$th component of $\mathbf{e}$ and $\eta_i$ is a scaling factor for the $i$th component (usually set to one), which means that for $|x_i| \gg \eta_i$ the norm represents the relative error and for $|x_i| \ll \eta_i$ the norm represents the absolute error. In this case, the prescribed tolerance supplied by the user also has absolute–relative character.

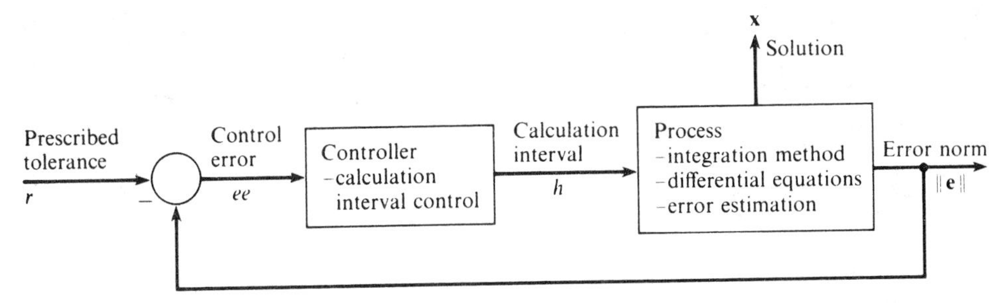


Figure 5.10 The system for calculation interval control.

The classical strategy for the control of the calculation interval is based on halving or doubling the calculation interval. If the error norm exceeds the prescribed tolerance, the interval is halved and integration is repeated on the same interval. However, if the norm multiplied by an integer (e.g. 5) is smaller than the prescribed tolerance the calculation interval is doubled. An integer greater than one realizes the dead zone for the algorithm, which suppresses possible oscillations in $h$. Errors in the dead zone result in an unchanged calculation interval.

The approach described can be regarded as a kind of three position nonlinear control. But it is well known from control theory that better control performances can be obtained with continuous controllers. The principle is shown by the scheme in Fig. 5.10. The norm of the local error $\|\mathbf{e}\|$, which is the output of the process, is compared with the prescribed tolerance $r$ and the result is fed back and used to determine the new calculation interval $h$. The controller should keep $\|\mathbf{e}\|$ close to $r$. A good control algorithm enables a smooth time response, i.e. a smooth calculation interval $h$.

Unfortunately, the results of modern control theory could not be applied to this problem in the past. Many modern integration methods (e.g. Gear stiff, Gear, 1971) use the integral control algorithm, which means that the calculation interval is proportional to the integral of the difference between the prescribed tolerance and the error norm. But the integral controller is probably not the best solution for a large class of problems. It is well known that such a controller has poor stabilizing capabilities. Oscillations of the calculation interval $h$ can clearly be seen when applying such algorithms for solving stiff problems using nonstiff methods (e.g. Runge–Kutta).

Gustaffson (Gustaffson, 1988; Gustaffson, 1990) tried to implement modern control theory for the control of the calculation interval. He implemented the discrete proportional-integral controller in the explicit Runge–Kutta method. He also derived the discrete dynamic model for the process (see Fig. 5.10) which was used to tune the controller parameters. The proportional-integral controller gave better overall performance for little extra calculation effort.

Finally, we must emphasize that the self-adjusting calculation interval procedure can easily be implemented in single-step methods. However, there are some problems using multistep methods because after a change of the calculation interval the previous values, which are used in the algorithm, must be obtained by an interpolation procedure; thus additional CPU time is required. It must also be pointed out that methods which include automatic adjustment of the calculation interval cannot be used in real time simulation applications.

### 5.1.8 Integration Method Selection

The basic goal when selecting an appropriate integration method is to achieve an acceptable limit of error with a near-minimum amount of computing effort. But it is very important to be aware of the fact that no algorithm exists which is equally suitable for all problems. In fact, it is possible to simulate elementary and simple problems without any additional knowledge of integration methods; but as soon as we deal with real problems a great deal of computing time can be saved, if an appropriate method is used. We shall therefore try to help an unskilled user to select the appropriate integration method (Cellier, 1979).

When solving a problem by simulation, several features which are important for integration method selection must be extracted from the problem. The first is closely associated with *accuracy* requirements. The dependence between CPU time and relative accuracy for various orders of integration is shown in Fig. 5.11. An integration method of low order is used if a problem does not demand high accuracy (e.g. significant simplifications in modelling phase, inaccurate measurements, etc.) If high accuracy is requested, a higher order method is used as it enables, in comparison with a low order method, more accurate results for the same CPU time. The commonly used rule states that for relative accuracy of $10^{-m}$, the $m$th order algorithm would be close to optimal.

Fig. 5.12 shows the simplified relationship of CPU time versus length of simulation run using single-step and multistep methods. Multistep methods require more CPU time during their initial phase but they are more economical for longer

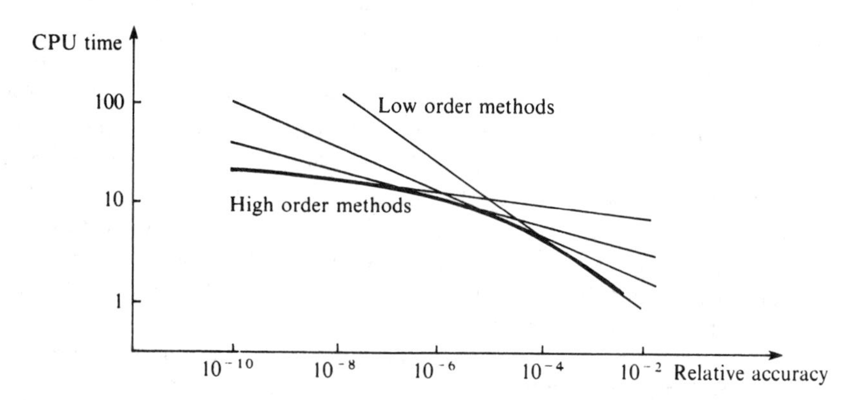


Figure 5.11 CPU time versus relative accuracy.

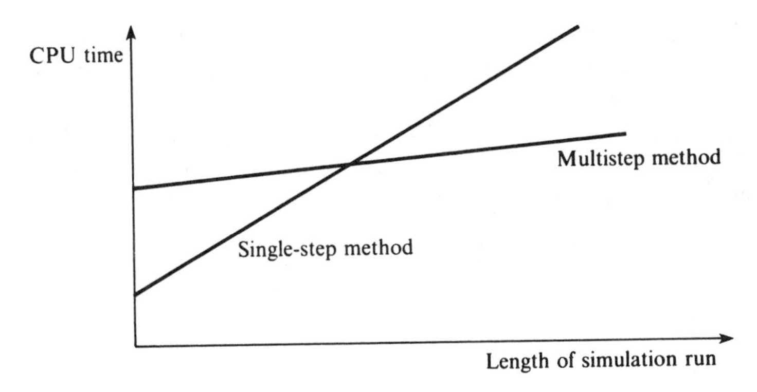


Figure 5.12 CPU time versus the length of the simulation run.

simulation intervals. This can be explained by the fact, that single-step methods (e.g. Runge–Kutta) are self-starting because they do not require the preceding values. Multistep methods (e.g. Adams–Moulton), on the other hand, demand more complicated initialization using single-step methods for the calculation of preceding values. This fact is extremely important where discontinuities in state variables or their derivatives appear frequently in the system. So the second feature that must be extracted from the problem shows whether the signals are smooth or nonsmooth during the simulation, i.e. whether *frequent discontinuities* are present. To integrate accurately over a discontinuity, modern algorithms identify the time of its appearance. Multistep integration algorithms must be restarted in each discontinuity point, so they are not suitable if discontinuities appear frequently in the system (e.g. bang-bang control, function generators, etc.).

The third feature that significantly influences the integration method selection is the *stiffness* of the system. For this reason the eigenvalue distribution of the Jacobian must be determined. If the ratio between the minimum and the maximum real parts of the eigenvalues is greater than 100, then a stiff method (e.g. Gear stiff) is recommended.

Finally, highly *oscillatory systems* (with dominant poles close to the imaginary axis) also demand special treatment. It is important to use integration methods that are numerically stable in the whole left part of the $\lambda h$ plane (low order implicit and semiimplicit methods).

The information provided by these four features (accuracy, smoothness, stiffness and oscillation) determines the most suitable integration algorithm for the majority of application problems. Let us end with some suggestions for integration rule selection, which in one sense is always limited by the capabilities of the simulation language used.

- Use high order methods for high accuracy and low order methods for low accuracy.
- Use Runge–Kutta–Fehlberg (4, 5) if you are unfamiliar with numerical problematics in simulation or if your system is nonsmooth or nonstiff.

- Use Gear's stiff method, the extrapolation method for stiff systems, or some other implicit or semiimplicit method (trapezoidal rule, implicit Euler rule) for stiff systems.
- Use Adams–Moulton multistep methods for smooth nonstiff systems.
- Use the extrapolation method if extremely precise results must be obtained.
- Use low order implicit or semiimplicit methods for oscillatory systems.

Finally, it must be pointed out that the methods for calculating the Jacobian matrix during simulation become wasteful of calculation effort with increasing order of differential equations (the order of Jacobian), and then some of the above suggestions become questionable.

## 5.2 SOLVING THE ALGEBRAIC LOOP PROBLEMS

In well-modelled systems, all the model variables can usually be calculated from state variables (outputs of integrators or delay elements). In this case, the sorting algorithm can easily sort the statements (blocks) from the simulation source program to the procedural order in the derivatives evaluation subprogram (DERIV), as described in Chapter 3. Each problem that can be sorted appropriately has the feature that each loop includes at least one block with a delay attribute (integrator, delay, lag, etc.). The model should be written in so-called canonical form, which usually eliminates all problems connected with algebraic loops. However, due to programming mistakes (undefined constants, input signals, typing mistakes in variable names, etc.), bad modelling procedures, or due to some other exotic modelling reasons, the input of a block can become an algebraic function of its outputs. That means that the model has closed loops which include only static components as summers, multipliers, etc. Such loops are called *algebraic loops*. They can cause instability problems even on analog computers due to the nonideal dynamic characteristics of analog components, and can create severe problems when using sequentially oriented digital computers for solving parallel systems.

In most cases, algebraic loops can be eliminated with explicit procedures by preliminary algebraic manipulations on the original system of equations. As calculation costs can be reduced significantly and accuracy can be increased in this way, the explicit procedure should represent the first attempt to solve algebraic loop problems.

Fig. 5.13 shows the simulation scheme of a second order system. When this scheme is analysed, three feedback loops can be found. The first and the third loops contain the integrators, but the second loop, which is shown with bold connections, gives the algebraic equation

$$y = f(y) \tag{5.42}$$

In this case, the algebraic loop can easily be eliminated explicitly. As

$$y = v - \mathrm{d}z = wy - \mathrm{d}z \tag{5.43}$$

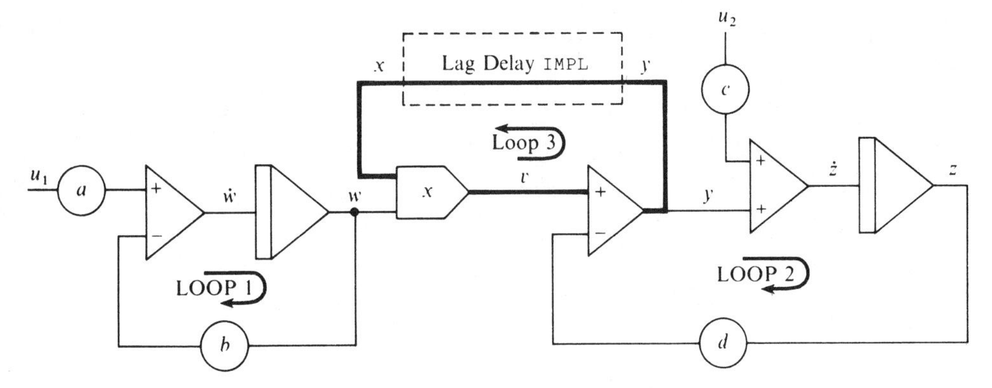


Figure 5.13 The simulation scheme of a second order system.

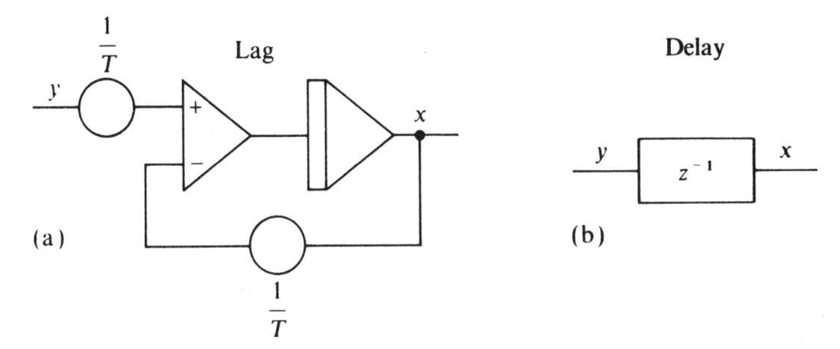


Figure 5.14 Systems for breaking an algebraic loop: (a) first order system; (b) discrete delay.

$y$ can be expressed by the equation

$$y = \frac{dz}{w - 1} \tag{5.44}$$

Sometimes, however, problems cannot be solved in such explicit ways. If the scheme shown in Fig. 5.13 is programmed, the compiler of a simulation language would send the message that all statements cannot be sorted. If the simulation language possesses no feature for the so-called implicit handling of algebraic loop problems, the loop can be artificially broken by inserting a lag (first order) system with unity gain and short time constant ($T$) or discrete delay. These possibilities are shown in Fig. 5.14. In the case of continuous systems the time constant $T$ must be much shorter than other time constants. The discrete delay is usually chosen to be one communication interval. Theoretically, the time constant and time delay must be as short as possible, but in practice this causes stiff problems in the simulation study.

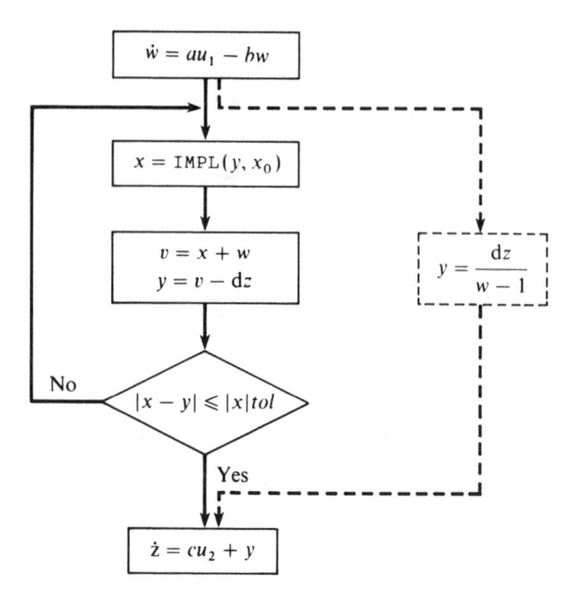


Figure 5.15 The calculation statements order for derivatives evaluations.

Many simulation languages can solve algebraic loop problems in an implicit way. In this case, the loop is broken with the so-called implicit block (operator `IMPL`), as shown in Fig. 5.13. This block is accepted by the compiler as a block with delay attribute. So the statements can be sorted with the `IMPL` block at the beginning of the algebraic loop statements. During derivatives evaluations the `IMPL` block puts the initial value of the variable $x$ on its output. From this value, the variable $y(x)$ is calculated as specified by the loop. Using this value the new estimation of $x$ is calculated inside the `IMPL` operator, usually with the aid of the Newton–Raphson iteration procedure. This procedure is repeated until the difference $|x - y(x)|$ is under the tolerance $|x|tol$ ($tol$ is the relative tolerance).

For efficient calculation, only those statements that are involved in the algebraic loop must be used in the iterative procedure. The order of calculations for derivatives evaluation with appropriate implicit loop is shown in Fig. 5.15. The simplification of the problem using the explicit approach is indicated by the dashed lines. The statements which can also be sorted without the `IMPL` operator are executed first. Then the statements which form the algebraic loop are included in the implicit loop and solved iteratively. The statements which can only be calculated using the results of implicit calculations but are not in the algebraic loop are placed (executed) at the end.

As stated, the Newton–Raphson iterative procedure is generally used. This procedure can be started by the initial values

$$x_0 \ldots \text{defined by the user} \qquad x_1 = x_0 + 0.0001x_0$$

and is defined by the algorithm

$$\left.\begin{aligned} c_n &= \frac{y(x_n) - y(x_{n-1})}{x_n - x_{n-1}} \\ x_{n+1} &= \begin{cases} [y(x_n) - c_n x_n]/(1 - c_n) & c_n \neq 1 \\ y(x_n) & c_n = 1 \end{cases} \end{aligned}\right\} \qquad (5.45)$$

for $n = 1, 2, 3, \ldots$. The initially defined value $x_0$ is used only for calculations at the beginning of a simulation run. In further calculation intervals, each initial value is chosen to be equal to the solution at the previous calculation interval ($x_0(t_k) = x(t_{k-1})$). Besides the initial value $x_0$ the user must usually also define the relative tolerance ($tol$) as the condition for iterative procedure termination, the maximum number of iterations (this is important if the procedure does not converge) and the value used to increment the variable $x$ to estimate the derivative (0.0001 in our case).

It is important that the iterative algorithms in the `IMPL` operator are defined in such a way that the convergence of the iterations is as efficient as possible. But it is not guaranteed in all cases that the algorithm used leads automatically to converging iterations.

The implicit iterative procedure drastically increases the simulation execution time. Even if the iteration procedure takes only a few steps, it is repeated several times in each calculation interval (depending on the integration algorithm used). Therefore, it is worth spending some time trying to avoid implicit loops by correct modelling procedure whenever possible.

## BIBLIOGRAPHY


BUTCHER, J.C. (1964), 'Implicit Runge–Kutta processes', *Math. Comp.*, **18**, 50.

CELLIER, F.E. (1979), Combined Continuous/Discrete System Simulation by Use of Digital Computers: techniques and Tools, PhD Thesis, Swiss Federal Institute of Technology, Zürrich.

GEAR, C.W. (1971), *Numerical Initial Value Problems in Ordinary Differential Equations*, Prentice Hall, NJ.

GEAR, C.W. (1984), 'Efficient step size control for output and discontinuities', *Trans. Soc. Comput. Sim.*, **1**, pp. 27–31.

GEAR, C.W. (1986), 'The potential for parallelism in ordinary differential equations', *Proc. Int. Conf. Numerical Mathematics*, pp. 33–48.

GUSTAFFSON, K. (1988), Stepsize Control in ODE-Solvers-Analysis and Synthesis, Licentiate Thesis, TFRT-3199, Department of Automatic Control, Lund Institute of Technology, Lund, Sweden.

GUSTAFFSON, K. (1990), 'Using control theory to improve stepsize selection in numerical integration of ODE', *Preprints of 11th IFAC World Congress*, Tallin, Estonia, USSR, vol. 10, pp. 139–44.

HAIRER, E. and C. LUBICH (1988), 'Extrapolation at stiff differential equations', *Numer. Math.*, **52**, pp. 377–400.

HAIRER, E. and G. WANNER (1990), *Solving Ordinary Differential Equations II. Stiff and Differential-Algebraic Problems*, Springer Verlag, Berlin.

HALL, G. and J.M. WATT (1976), *Modern Numerical Methods for Ordinary Differential Equations*, Clarendon Press, Oxford.

KORN, G. and J.V. WAIT (1978), *Digital Continuous System Simulation*, Prentice Hall, NJ.

LAPIDUS, L. and J.H. SEINFELD (1971), *Numerical Solution of Ordinary Differential Equations*, Academic Press, NY and London.

PRESS, W.H., B.P. FLANNERY, S.A. TEUKOLSKY and W.T. VETTERLING (1986), *Numerical Recipes – The Art of Scientific Computing*, Cambridge University Press, Cambridge, UK.

# 6

# Influence of Software Engineering and Modern Technology on Simulation

Software engineering concepts drastically influenced the development of new software structures of simulation languages in the 1980s. These new structures enable very powerful modular modelling and simulation, as well as other more complex procedures. Simultaneously, modern technology has enabled dramatic development in computer hardware. Using special purpose simulation computers and workstations, supercomputers and modern hybrid systems, simulation has become a methodology that can be used almost without limit; it is often also used in applications that are extremely wasteful of calculation time, as well as real time and hardware-in-the-loop applications.

## 6.1 INFLUENCE OF SOFTWARE ENGINEERING ON MODERN SIMULATION LANGUAGES

It is not enough that a simulation language offers the possibility of obtaining accurate solutions. It must also ensure minimization of cost in the phases of model development and implementation, model modification and bug elimination. It is well known that approximately 75% of costs for the development of a software product goes to maintaining and modifying existing software. The field that evolved to improve the development of software is *software engineering*. Since models can be very large and complex, simulation is an important area in which software engineering techniques should be applied. It is true to say that software engineering considerations can be taken into account in all project phases: in the development and management of the software system design, in the cost estimation, in the requirements analysis and documentation, in data base definition, in analysis and design of algorithms, in program structure definitions, in program coding, in testing, in software maintenance, etc.

However, the most popular and commercially attractive languages for the simulation of continuous dynamic systems were developed in the period when software engineering concepts had not yet been developed.

### 6.1.1 Need for New Standards and Concepts in Simulation Languages

During the last decade the new technologies, the software engineering proposals and, in this connection, the development of hardware and software equipment, as well as numerical algorithms, have resulted in a divergence of the existing languages from the CSSL'67 standard. Experience showed that the CSSL'67 languages were very restrictive in various applications despite their never-ending evolution. It was obvious that the new trends demanded a new standard which should have included not only the functional but, above all, structural characteristics of simulation languages. Several efforts to establish a new language standard were made in 1981 by two international groups (Technical Committee TC3 on simulation software of the IMACS – International federation for MAthematics and Computer Simulation; and the CSSL Committee of the SCS – Society for Computer Simulation) which considered suggestions from within the simulation user community. Their aim was to define a CSSL standard for the 1980s and 1990s. Unfortunately, the situation in 1981 was radically different from that in 1967. The members of both groups were also developers of simulation software and they tried to promote their own product to become the accepted standard. So the groups were not able to define the new standard but, rather, only to propose facilities to be included in new languages. These proposals are sometimes considered as the *CSSL'81 standard* (Hay and Crosbie, 1981; Crosbie and Cellier, 1982).

One of the languages that took serious account of the recommendations was the simulation language ESL (Crosbie *et al.*, 1985; Crosbie, 1986; Hay and Crosbie, 1986). The development of this language began in 1983 at the University of Salford as a project supported by the European Space Agency (ESA). Other new generation simulation languages, e.g. SYSMOD (Baker and Smart, 1983) and COSMOS (Kettenis, 1986) differ considerably in some aspects from the CSSL'81 proposals. They were developed under the hard influence of software engineering concepts.

It must also be mentioned that discrete event simulation languages (e.g. SIMCRIPT, SIMULA), having more complicated structures, also forced developers to consider software engineering proposals in the past. So the structures of these languages also influenced the development of continuous simulation languages.

Analysing the CSSL'81 proposals, software engineering proposals and the new generation simulation languages, we summarize the most important features of the new generation simulation languages as follows:

- model definition modularity;
- flexible program and data structures;
- characteristics of combined simulation;
- numerical robustness;
- complex experimentation with simulation models; and
- user friendly user interface.

### 6.1.2 Model Definition Modularity

The concept of *modularity* is perhaps the most important concept of each structured program. Using this principle, large simulation models are not developed as single, monolithic units but are subdivided into modules which operate as independent functional components of the overall system. Such hierarchical decomposition of a model gives the following advantages:

- It is possible to focus attention on each component as a small problem. If a component is still too complicated it can be further divided into smaller components until each module is of manageable size. Modules which are too complicated cause details to be forgotten and interactions to be confused, which results in many mistakes.
- Several modellers can work simultaneously on a modelling and simulation project, since they work on different modules whose interactions must be carefully defined in advance.
- It is easy to implement changes and corrections in one or a few simple modules.
- The modular structure enables the partial testing of a model, i.e. testing component by component.
- The modular developed model is extremely convenient for documentation purposes.
- The modular structure enables the application of different algorithms to different subsystems (e.g. the model is divided into stiff and nonstiff submodels using stiff and nonstiff integration algorithms during simulation).
- The modular concept is very important when the model is simulated on parallel processing systems where submodels are targeted for different processors.
- Finally, the modular model can be very realistic, reflecting the hierarchical structure inherent in the system. It can even be said that investigations of natural but somewhat hidden hierarchical structures of real systems influence the development of modern modelling and simulation tools (Marquardt, 1991).

Having the modular hierarchical structure in mind we shall discuss the model as a module at the top of the hierarchical structure, and the submodels at lower levels.

Thus, strongly influenced by software engineering proposals, the modern simulation languages introduce very flexible possibilities for the simulation of hierarchical models. They enable the use of many preprogrammed submodels from the particular language library and the creation of user defined submodels which can first be tested separately and then included in the library. Using such a library, a general purpose simulation tool can become a kind of application oriented tool and it is sometimes difficult, from the simulation model listing, to recognize which language is being used.

We shall describe the structure of model and submodel definitions in modern simulation languages, as well as some types of submodel concepts.

### The structure of the model or submodel definition

The structures by which models and submodels can be described in modern simulation languages are very different. Trying to unify them we shall use the classical terminology of the CSSL languages, although the terms used in different languages are quite different.

In modern simulation languages a model or a submodel is described by INITIAL, DYNAMIC and TERMINAL structural sections. Some languages also have a special declarative section where data, variables, procedures, types of processes, etc., are declared, and a special communication section which contains input and output specifications and possible operations which do not need to be executed in each calculation interval.

Figure 6.1 shows the possible structure of a model or a submodel.

The INITIAL section is used for the definition of constants, initialization of variables, statements, etc., i.e. for all operations that must be executed before integration starts. It is usually defined by a procedural code, which means that the operations are executed in the defined statements order. In some languages the INITIAL section can be executed at an arbitrary time rather than only at the beginning.

The DYNAMIC section usually consists of DYNAMIC continuous and DYNAMIC discrete sections. The continuous section describes the continuous part of a model or submodel and consists of DERIVATIVE and PROCEDURAL sections. The DERIVATIVE section contains representation statements (equations), which must be simulated in 'parallel' and are defined by a nonprocedural code, i.e. by arbitrary statements order. The PROCEDURAL section includes the code, which need not be executed in every calculation interval but perhaps only in each communication interval. It is written with procedural code. The DYNAMIC discrete (sometimes

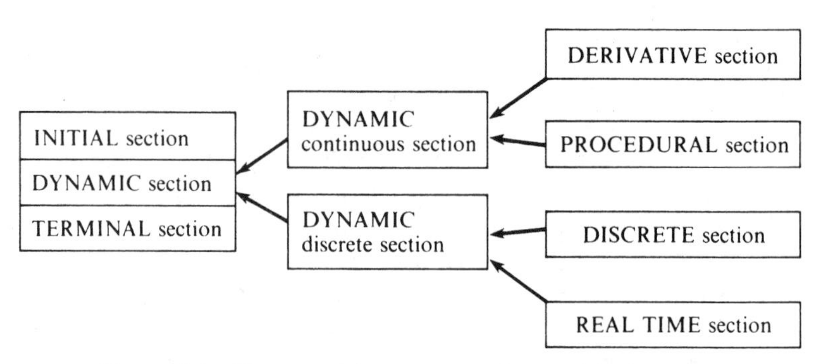


Figure 6.1 The possible structure of a model of a submodel.

also called DYNAMIC sampled) section is characterized by the fact that it communicates with the other DYNAMIC sections only at fixed time instants. This section can consist of the DISCRETE section, which contains a discrete algorithm written in a procedural code (e.g. a discrete controller) and sometimes also of the REAL TIME section, which is in fact an interface with the physical system. The realization of the latter section is specific to a particular hardware and demands simulation in real time. The DYNAMIC discrete section can be executed periodically, but it can also be invoked from the DYNAMIC continuous section (e.g. using operator `WHEN`).

The programming possibilities of the DYNAMIC section are of course extremely important for efficient modelling and simulation. Besides standard simulation oriented operations, modern languages enable the inclusion of transfer functions (univariable, multivariable), state space descriptions (univariable, multivariable), user predefined hierarchical submodels, higher order differential equations, etc. Control statements (loops, branches) can be included directly.

The TERMINAL section is used for simulation results postprocessing. There is no need to use it for experimentation control, as in classical CSSL languages, as experimentation is realized by means of the experimental environment.

## The use of segments

The realization of so-called *segments* (e.g. ESL (Crosbie *et al.*, 1985; Crosbie and Hay, 1985; Crosbie, 1986; Hay and Crosbie, 1986)) is a feature of some modern simulation languages and is an additional contribution to the model definition modularity. Segments are in fact those parts of a model which are simulated separately. They communicate with each other only at defined prescribed time instants. The final goal of segments implementation is to simulate a model in a multiprocessor environment. However, they can be also used on sequential machines where the multiprocessor environment is emulated.

Alongside the benefits, such a segmental approach introduces many problems into the simulation study. Signals, which are transferred between segments, are treated as constant between communication intervals, which results in amplitude and phase distortion. Due to the staircase shape of these signals, the frequency region is expanded, resulting in the possible appearance of stiff charactersistics. So it can be the case that an integration method which successfully simulates the unsegmented system needs a much shorter calculation interval for the appropriate simulation of the segmented system. These problems can be partly eliminated by appropriate filtering and interpolation techniques.

The use of segments is very efficient in simulations where the model can be partitioned into high and low frequency parts. In this case, the low frequency segments can be simulated with much larger calculation intervals, which results in an efficient and cost-effective simulation. There are some questions as to how to transfer high frequency signals into low frequency segments without disturbing the integration algorithm, but these can usually be solved by appropriate filtering.

Hay and Crosbie (Hay and Crosbie, 1986) recommended the emulation of segments in simulation languages on sequential processors. Using this feature, the user can obtain the answer as to whether true parallel simulation in a multiprocessor environment has any benefits at all. These authors also recommend that the unsegmented model be simulated first to obtain a reference set of results which can be used to validate the results using the segmentation approach.

**Submodel realizations**

There are different ways in which submodels are realized in simulation languages.

The MACRO feature has been the most important facility that enabled modular modelling and simulation for a long period. It is still implemented in many modern simulation languages. MACRO is a submodel which is defined using language syntax and some special macro statements. During processing, all MACRO calls in the source simulation model are replaced by appropriate MACRO definition bodies and their formal arguments are exchanged for the real ones. In further processing the sorting algorithm can arbitrarily change the order of all statements, so the MACRO feature does not cause any problems with the sorting algorithm. The main disadvantage of this approach is that at each structural change all model components must be reprocessed. So a MACRO feature cannot be treated as a modern hierarchical approach.

A MODUL can be treated as a generalization of the MACRO feature. Whereas MACRO implies that its inputs and outputs are known in advance, this is not always necessarily the case. For example the same model resistor can be described by the equation $I = U/R$ or $U = IR$. So the MACRO feature demands two submodels for the same physical law. Using the MODUL submodel, both equations can be realized with one submodel, because the formula manipulation routine sorts the equation in the 'horizontal direction' depending on the context in which the equation is used. So it is immaterial whether the resistor is coded as $U = IR$ or $I = U/R$ or $U - IR = 0$. MODUL is rarely used in simulation languages (COSY; Cellier, 1979) and it was mainly used as a point of theoretical investigation.

*Separately compiled submodels* represent much more flexible submodel features in the new generation simulation languages. They are defined in a manner similar to subprograms in a general purpose language. These submodels are separately compiled and enable true hierarchical modelling and simulation structures, as they can be nested to any depth.

The production of a simulation tool that supports separately compiled hierarchical structures is a very exacting and complicated task. Central manipulation of state variables and their derivatives or predictions and self-configuration, which means that each submodel reserves its storage requirements and establishes necessary pointers, must be performed. But the most difficult task is to define the hierarchical data base properly and to implement the sorting algorithm correctly. The calculations which realize a submodel containing another submodel and/or blocks with delay attributes, must be due to the sorting algorithm performed in several stages. The realization is particularly complicated as submodels can in general have several inputs and outputs.

SYSMOD and COSMOS have very powerful hierarchical model structures. Highly influenced by software engineering proposals, they enable dynamical creation of submodels during simulation (e.g. at the beginning the simplified model is simulated; when a condition is fulfilled, a complex model is taken into account).

### 6.1.3 Flexible Program and Data Structures

Taking into account the development of structured programming techniques, it becomes apparent that one of the quality measures of a simulation language is its ability to support *flexible, structured control statements.* Although modern programming languages such as PASCAL, MODULA, ADA, etc., provide these statements, they are rarely available in simulation languages. However, a good simulation language should provide at least a well-structured `IF-THEN-ELSE` statement and at least one form of `DO-WHILE` statement for repetitions.

The *structure of data* used in a simulation program is important, too. The simulation language must provide all the basic types of data: integer, logical, real array with user defined upper and lower bounds, character string, and also the extended data structures which consist of the basic data structures. Every data item should be declared as in modern procedural languages.

The *matrix data structures* are of extreme importance in current and future simulation languages (Rimvall and Cellier, 1985). Matrix operations are perfected in some mathematical libraries and computer aided design packages (LINPACK, MATLAB, CTRL-C, MATRIXx, IMPACT, etc.). But this is unfortunately not the case for simulation languages.

### 6.1.4 Characteristics of Combined Simulation

In recent years, the interest in systems with continuous and discrete characteristics, i.e. *combined systems,* has grown significantly. To explain this approach let us consider the system shown in Fig. 6.2. In this process the metal ingot is placed into a furnace. Then it is heated to a prescribed temperature and after that it is removed from the furnace. The process can be described by discrete events when the ingot enters and when it leaves the furnace, However, the furnace containing the ingot is best represented by a continuous system, since the temperature behaviour can be described by differential equations. Further, there are interactions between the discrete and continuous subsystems which affect both components. There are not many simulation

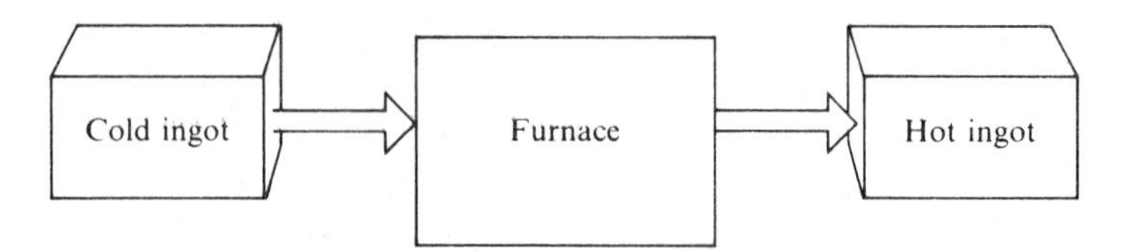


Figure 6.2 Ingot processing system.

languages that can be used efficiently for such purposes (GASP IV, COSY, SIMAN, COSMOS, SYSMOD).

The structures that enable an efficient simulation of continuous or discrete event systems are well known. What makes the combined simulation more difficult, and of particular interest for the designers of modern simulation languages, are the interactions that may occur between the discrete and continuous components of the simulated system. These interactions can appear in the following cases:

- A discrete event makes a change in a parameter or in a structure of a continuous system (e.g. a discrete event closes a valve and changes the flow rate to zero).
- A change in a continuous system's variable can activate a discrete event (e.g. the ingots in our example are removed when they reach the prescribed temperature).

These interactions imply the most advantageous characteristic of the combined language: *system components can be activated and deactivated dynamically* during the simulation of a model.

In his prototype simulation language PROSIM, Golden (Golden, 1985) defined the *process* to be the basic module for building a model. This is a kind of subroutine that can have either discrete or continuous attributes. A process may be thought of as a pattern or template but does not actually occupy any data storage or cause any activity during simulation. During simulation, one or several *process instances*, i.e. copies of the process, are created. They can also be created and destroyed, scheduled for execution and suspended. Of course, each process instance has its own data storage area as it has specific inputs, outputs, states and parameters. The concept of creating multiple copies of a system component is well known in discrete event oriented languages but is rarely met with in continuous simulation. So this feature indicates one of the promising directions for the future development of continuous simulation languages.

Although the current commercially attractive CSSLs enable the inclusion of discrete events in a continuous model, this concept is far from that mentioned earlier. For this purpose ACSL has a so-called DISCRETE section which is executed when an event is detected. This detection is performed in the DERIVATIVE (continuous) section with the aid of special operators. The SKEDTE operator is used for the time condition and the SKEDSE operator for the state condition. It is also possible to implement a periodic triggering of the DISCRETE part of a model by 'sampling time' specification.

To explain the difference between combined prototype simulation language and current CSSLs, let us take the example of a continuous discrete model given by Cellier (Cellier, 1979). The system consists of identical pieces of a domino game which are placed into an array in which all elements are equally distanced. If the first piece is pushed, a chain reaction is started and after some time, all pieces fall flat. Using a current CSSL language the user should simulate all the pieces. But the combined prototype simulation language offers the possibility to model the system in such a way that only the moving pieces will be presented in the model. Each time a piece touches the next one, that next piece is created, and from that moment on it will be part of the model. Each time a piece touches the ground, it will be terminated and will no longer be a part of the model.

### 6.1.5 Numerical Robustness

The numerical robustness of the built-in numerical algorithms influences the accuracy and reliability of solutions, as well as the simulation time. Numerical problems appear with the following:

- discontinuities in variables;
- very different time constants (stiff systems);
- algebraic loops;
- tabular defined signals;
- time delays;
- steady state calculations; and
- partial derivatives evaluations for Jacobian or Hessian matrices.

There are only a few languages that can successfully handle all the above problems. In this sense CSSL IV (Nilsen, 1985) is perhaps the best language as it uses well-tested libraries EISPACK, LINPACK and IMSL. It also has efficient numerical diagnostics.

Some of the above-mentioned numerical problems were described in Chapter 5. Here we shall discuss only the problems of *discontinuities* in variables, as this is a numerical problem that is properly treated only in modern simulation languages.

In the preceding subsection we mentioned that elements of combined simulation represent important features of new generation continuous simulation languages. Besides the very complicated structural elements that enable interactions between continuous and discrete sections, these languages must also possess the correct numerical treatment of discontinuities in derivatives which frequently appear in this type of simulation.

#### The numerical treatment of discontinuities

Real physical systems are continuous by nature but during the process of modelling, complicated relations are often represented by discontinuous characteristics. Standard

CSSL'67 defined the use of typical discontinuous model components. So simulation languages have these components in their libraries but, until recently, only a few CSSLs implemented these nonlinear functions in a manner which guaranteed correct numerical treatment. The problem is particularly emphasized where multistep integration algorithms are used.

Discontinuities can be divided into the following two groups:

1. discontinuities that appear in precisely defined time instants;
2. discontinuities that are caused by a state of the system and it is not possible to know the time of appearance in advance.

The first problem is typical for control systems, for example where the control variables are modified by a control law at explicitly stated instants (periodically). Numerical solution of this problem is not complicated. Because the time of the appearance of the discontinuities is known in advance, the last calculation interval should be reduced so that integration ends just before the discontinuity appears. The moment the discontinuity appears, the integration procedure must be restarted.

The second type of discontinuity is more complicated. Regular numerical treatment is particularly complicated when conditions for discontinuity appearance contain nonlinear relations, e.g. hysteresis, backlash, coulomb friction, etc. In this case the so-called switching function must be included in the integration procedure. This function detects that discontinuity has taken place during a calculation interval. After detection, the time of occurrence of the discontinuity is exactly located by some iterative algorithm. These algorithms must have two properties: fast convergence and wide convergence area (Cellier, 1979). Once the time of occurrence is identified, further procedure is the same as for the first discontinuity. The procedure entails additional CPU costs which are repaid in the reliability of the solution.

### 6.1.6 Complex Experimentation with the Simulation Model

The CSSL'67 standard gave, along with the structures for model definition, hints for the features of an interactive run time interpreter. These were actually the first elements of experimentation capability in simulation languages. Unfortunately, the hints were based on the notion that a simulation run is the only experiment that can be done with a model. So the interactive interpreter was used for very simple operations during simulation runs, e.g. changing the model constants, simulation run control parameters, etc. But it was soon realized that other, more complex, experiments such as optimization, linearization, parameter study, sensitivity study, etc., are desirable features of modern simulation languages.

The only way to implement experiments in traditional CSSL languages is to use the INITIAL, DYNAMIC and TERMINAL model descriptions. Input signals for simulation studies, criterion function and constraints for optimization are directly included in the model and cannot easily be changed. This means that complex experiments must be programmed together with the model description using sophisticated loops and tricky flags. This all leads to 'grown' implementations which do not obey software engineering proposals at all.

In accordance with software engineering and modelling principles Ziegler and Ören (Ziegler, 1976; Ören and Ziegler, 1979) proposed the concept of *experimental frames*. This concept is based on the separation of model, experimental frame and execution control. It is probably the first hierarchical concept because it defines an experiment as a function of a model and an experimental frame. Model specifications and experimental frame specifications are on the same hierarchical level above the level of execution control.

Breitenecker (Breitenecker and Solar, 1986) proposed the terms *model*, *method* and *experiment* for the new hierarchical concept. Their meanings are as follows:

- Model is the description of a system using appropriate mathematical formulations and a certain, usually nonprocedural, language.
- Method is an operation or an algorithm which defines an action using a model (whatever can be done with a model).
- Experiment is the performance of a certain method with a certain model where all aspects of execution control are included.

Using this concept, very simple methods (e.g. the change of a model constant), the simulation run as a basic method (i.e. integration of differential equations) and complex methods (e.g. optimization, which uses the methods simulation run, parameter change, readout variables) are treated uniquely. But to make the concept more clear the formation of the following groups of methods is proposed:

- Simulation run of one model or of several models.
- Linearization of a model. The method gives the state space description (**A**, **B**, **C**, **D** matrices).
- Linear analysis using a linearized model. The calculation of eigenvalues, eigenvectors, observability, controllability, steady state, Bode analysis, Nyquist analysis, some design procedures, etc., can be included in this group.
- Parameter changing methods. The methods of single model constant change to repetitive simulation studies with different constants (parameter study) and optimization as a complex method are included.
- Generation and changing a model. The group includes the model definition or changing, possible compilation, model data base generation, etc.
- Generation and changing a method. This is a very interesting group that illustrates the efficiency of the proposed concept. Using this feature, a simulation language can be seen as an open system for any application as it is possible to create a 'shell' for some special purpose application so that no knowledge of the language is required.
- Experiment results presentation. This group enables the documentation of results with prints and plots. The results of several runs of one or of several models can be presented and analysed (zooming, tracing, etc.).

The new concept allows the inclusion of dynamic features in a method, so that the method and the model communicate during simulation. For example, input signals for the simulation run or criterion function and constraints for optimization can be automatically included and easily changed. In the case of optimization, the model that calculates the criterion function is included in the method, and can easily be changed by the method–model change.

An experiment defines the formal parameters of a method, which also enables appropriate linking with some model parameters and variables which are included in the method (e.g. model parameters in optimization). If we take, for example, the parameter study method, the experiment must define, besides the choice of a concrete model and concrete method, also the selection of model constant to be swept, the initial, final and step values of the constant and the parameters of the basic method which is included in the parameter study, i.e. the simulation run (initial time, final time, calculation interval, integration algorithm, etc.).

Models and methods are totally separated only in some new generation simulation languages: ESL (Crosbie and Hay, 1985; Crosbie, 1986; Hay and Crosbie, 1986; COSMOS (Kettenis, 1986); SYSMOD (Baker and Smart, 1983; Baker and Smart, 1985)). SYSMOD, which is perhaps the most comprehensive simulation language, allows models and experiments to be compiled separately and included in libraries of models and experiments. Besides the experiment section, where the parameters of methods and models are initialized, the section which describes the user's interactions during the experiment, as well as some other sections, can be specified.

### 6.1.7 User Interface

The *user interface* is that part of a simulation language which enables efficient work with the tool. It communicates through the supervisor program with all parts of the simulation language and has a substantial effect on the user friendly environment.

Despite the importance of the user interface to the acceptance and success of simulation languages, user interface considerations have often played a less important role in their design. Command driven interfaces, particularly for experimental possibilities, have been very popular for a long period (SIMNON, ACSL). Some languages also enable the extension of functionality of the command language by adding interactively defined macros. So this interaction principle is very efficient for skilled users but the price is quite high for the novice user or for a person who is not expected to be a regular user of the package. Almost the same is valid for the model definition phase where the source program in the corresponding input language is usually specified by means of a universal editor.

The concept of the user interface drastically changed with the revolutionary advent of computer graphics on workstations as well as on personal computers. Graphically based systems represent an ideal environment for new generation user interfaces using 'pull-down' menus, 'fill-in forms' and 'point-and-click' techniques.

A primary feature of a modern graphics user interface is to exploit the full power of modern graphic software and hardware possibilities. It is characterized by its natural nature, obtained by the use of proven techniques of interaction (Shneiderman, 1987). The environment caters for novice users and for the experts as well. At no stage does the user need to understand anything of the computer hardware and software, but can concentrate only on the modelling and simulation problematics. The environment must support him with appropriate information on every stage of the creative process commencing with the earliest model definition stage and ending with the experimentation and results postprocessing stages (Barker, 1988; Barker *et al.*, 1988; Barker *et al.*, 1989; Barker *et al.*, 1990; Taylor *et al.*, 1990; Elmqvist and Matson, 1986).

Thus, the most important requirements for the user interface are considered to be (Britt *et al.*, 1991) the following:

- a natural environment which is similar to paper and pencil problem solving;
- user guidance instead of enforcing a certain sequence of actions;
- fast response;
- meaningful and consistent display;
- equal support for novice and expert users.

In simulation languages the user interface consists of three important parts, each of them possessing some specifics. These are as follows:

- model definition;
- experimenting with the simulation model; and
- presentation of experiment results.

**Model definition**

Graphic presentation of simulation schemes has been found to be a natural and popular way of expressing models. People think in pictures, so the structure and flow of information often appears most clearly and simply in this form. Engineers always try to solve each problem with the aid of a block diagram as their favoured tool from their early education days.

Nowadays, practically every comprehensive simulation tool (traditional ones, too) possesses in its user interface a graphics editor which is usually implemented as a preprocessor for model definition. The graphic simulation scheme is defined in the form of the block diagram but sometimes also in the form of signal flow diagrams. The programmer defines a model using block icons from the library and places them on the screen. The blocks can be added, deleted and moved. After that they are connected appropriately. When the topology is described, the attributes of the blocks must be entered. It is very important that the graphics editor detects as many errors as possible during model definition so that the user can make appropriate corrections interactively.

Graphics editors must exploit all features of the simulation language: the inclusion of complex blocks (e.g. transfer functions, state space descriptions), hierarchical submodels, blocks written in high level language (with subroutine calls), blocks where complex equations can be included directly, as well as function generators, signals and nonlinearities. Besides these features, graphics editors should enable some functions which extend the modelling capabilities of the simulation language. So simulation schemes can be manipulated using block algebra rules, the transformation from block diagram to signal flow diagram and vice versa can be undertaken, etc. Only with such a comprehensive tool would an expert user who has perhaps been using the textual input language for years, be willing to use the graphics input.

There is of course no doubt that a graphics editor is much more convenient for beginners. In this sense it seems true that before graphics editor implementation, users were divided into two groups: expert users and unhappy or nonusers (Taylor *et al.*, 1990). By incorporationg graphics editors the number of effective users is increasing significantly as they are implemented in such a way as not to penalize the experts.

Finally, it must be emphasized that graphic presentation of models serves not only as an ideal method for model definition but also for model documentation purposes.

## Experimenting with the simulation model

This part of the user interface is closely linked with the simulation language supervisor. It must correspond well with the concept of the simulation language experiment. Thus it represents the user friendly environment for all possible experiments using predefined models and methods. Graphics menus with fill-in forms (for parameter specifications) sometimes have some drawbacks when compared to command language based user interfaces. If some operations are to be executed several times, it is frustrating to pull down the same menus and select the same operations time after time. In command language based systems, such loops of operation are easily implemented as a macro, a sequence of possibly parametrized commands. To avoid this disadvantage the best solution is for the user interface to act on two levels. At the lower level a specific command language can be used to obtain the macro facility. Such a facility can be a built-in menu driven graphic user interface.

## Presentation of experiment results

This is the area in which graphics play an important role. Dynamic simulation produces a vast amount of results which are comprehendable only by the graphic postprocessing as well as graphic run time presentation. The graphic postprocessing feature is mainly used for time history plots and phase portraits. However, modern graphic possibilities are exploited in particular for runtime graphic presentation. One of the modern features is to display the simulation results in the graphic simulation scheme which is defined by the aid of the graphics editor. Before the simulation, or even during simulation, the user can define 'monitors' by choosing appropriate signal lines. Time history plots, as well as digital or analog 'voltmeters' and bar graphs, can be used. If the modelling procedure is performed by the technological scheme the results can be presented efficiently with animation. If such a scheme works with real time simulation in the background it can be used for educational purposes as well as for operator training as it provides a real life environment for process monitoring and interactions.

Much effort has been directed towards the development of standards and tools for user interface improvement. Some interfaces were implemented with graphics standards GKS and PHIGS, but the future will see the full support of windowing standards (e.g. X-windows), together with such methods as object oriented programming (Barker *et al.*, 1990). Using the windowing concept, experimental and model definition actions, as well as the presentation of results, can be accessed in different windows which are present simultaneously on the screen.

## 6.2 INFLUENCE OF MODERN TECHNOLOGY ON THE DEVELOPMENT OF SIMULATION SYSTEMS USING SPECIAL PURPOSE COMPUTERS

There is no doubt that, nowadays, computers work correctly and reliably. But are they fast enough? This question is a permanent challenge to the programmers (to invent faster algorithms) and to the hardware designers (to develop faster circuits). As the dynamic simulation method is one of the most wasteful of CPU time it represents the area which, along with that of signal processing, has influenced the development of modern computer technology most drastically. Mini and microcomputers have become so powerful that complex simulation languages have been transferred from large mainframe computers without limitation. So simulation as a modern methodology has become accessible for almost everybody. Although simulation systems on such general purpose computers are appropriate for most simulation applications, there are still problems that demand special purpose computers. In this sense the introduction of the techniques of multiprocessing and pipelining in digital computer systems design has been of particular importance. It is therefore essential and very appealing to explore the speed up by orders of magnitude by taking advantage of the techniques mentioned (Chen, 1989; Worlton, 1989; Duncan, 1990).

So the purpose of this section is to describe the potential impact of major innovations in digital computer design upon the development of simulation systems. As hybrid computation was 'newly' born in 1983 with the hybrid mutiprocessor SIMSTAR, where the advantages of real parallel (analog) simulation, multiprocessing techniques and high level simulation software were combined, our review will finish with a short description of new trends in hybrid computer development.

### 6.2.1 Parallelism Concepts

A well-recognized limitation of conventional general purpose digital computers is the so-called von Neumann bottleneck. In this type of computer there can be just a single command (in the command register) and this command can affect an arithmetic or logical operation only upon a single data word (in the accumulator). Such machine organization is termed *single instruction–single data* (SISD).

An important class of digital computer systems possesses a number of processing elements, each consisting of an arithmetic unit and a memory unit. These two units are interconnected to form an array. One control unit can activate any or all of the arithmetic units. Each arithmetic unit performs an operation on different data. For this reason, this type of structure is termed *single instruction–multiple data* (SIMD). This structure is implemented in vector processing computers (supercomputers, superminicomputers) and in peripheral array processors (Spriet and Vansteenkiste, 1982).

With decreasing costs of digital hardware in general, replication of the control unit has become more and more feasible from an economic point of view. As a result, a wide variety of multiprocessor systems have been constructed or proposed. Each element of the array of processors now includes a control as well as an arithmetic and memory unit, i.e. the complete computer. Each processing element can carry out different arithmetic or logic operations. For this reason, systems of this type are termed *multiple instruction–multiple data* (MIMD).

Another type of parallelism which is also often included in special purpose digital computers is termed *pipelining*, which is a procedure where a sequential computational operation (e.g. floating point addition) is broken down into a series of subtasks (in the case of floating point addition these are, for instance, the modification of one exponent to become the same as the other, addition of the mantissae, normalization, round off), and separate hardware units are provided for each task. These units operate concurrently so that several pairs of numbers can be in various stages simultaneously. So when the pipeline is full, each hardware unit is engaged in processing different data, and after completion of the task, the results of its computation are passed to the next unit. Pipelining techniques are particularly effective in enhancing execution speeds when processing loops. They are implemented in supercomputers, in array processors and in multiprocessor systems.

### 6.2.2 Supercomputers

Supercomputers represent the most power digital computers of the present day (CRAY 1, CRAY X-MP, CRAY 2, CYBER 205, HITACHI, FUJITSU, NAS, etc.).

But, unfortunately, extreme computational efficiency must be payed for in high cost. Their development originated at the end of the 1960s when demands for the simulation of complex systems in meteorology, seismology, fluid dynamics, plasma physics, etc., arose. The extraordinary calculation efficiency of supercomputers was achieved by introducing single instruction multiple data (SIMD) hardware structures using parallelism and pipeline concepts.

As the work of supercomputers is based on SIMD concepts using so-called vector processing, the high efficiencies can be attained only if a considerable part of the code can be vectorized to take advantage of parallelism and pipeline concepts. Up to now, only modest speed up is achieved if available codes are ported without any hand coded optimization. There are many studies in the field of code optimization (Zitney, 1990) but there is unfortunately not much reported work on optimization of integration algorithms for large systems of OD or PD equations. As the vectorization of the derivative part represents an even bigger problem because many scalar operations appear in the model definition part we can conclude that dynamic systems simulation is not an application which easily obtains the advantages of supercomputers.

Nevertheless, rather good general purpose software equipment can be used nowadays for supercomputer programming (FORTRAN, PASCAL). Even some installations of commercially attractive general purpose simulation languages are known (CSSL IV, ACSL on CRAY, CYBER 205 – Colijn and Ariel, 1986). However, in general, supercomputers are only used for very special simulation applications with specially developed simulation software for solving problems in the above-mentioned fields.

Fortunately, there are two other possibilities for special simulation applications. These are as follows:

- array processors; and
- parallel multiprocessor systems.

As far as calculation efficiency is concerned, both possibilities approach the efficiency of supercomputers but at much lower cost. Unfortunately, the level of software equipment does not equal the level of hardware capability.

### 6.2.3 Array Processors

Array processors are special purpose digital computing devices that are designed to serve as peripherals for conventional digital computers (acting as a host), so as to enhance the performance of the host in specific numerical computing tasks (Karplus, 1984a, b; Pearce *et al.*, 1985). They achieve high calculation efficiency through extensive use of parallelism and/or pipelining, using an arithmetic section containing at least one adder and one multiplier. Some array processors possess several adders and several multipliers; some also have dividers and square roots. Often, these arithmetic elements are internally pipelined so that a new arithmetic computation can begin while the previous operation is still in progress. Interfaces are available for particular array processors for various host computers.

The software equipment and the programming possibilities of array processor systems are far below the level of general purpose computers. Often, the peripheral array processor systems are completely closed and application oriented systems are programmed by the users in an application oriented fashion (e.g. dynamic simulation workstation XANALOG-1000 (Schrage and McArdle, 1986; Schmidt and Scheider, 1988) which is completely graphically programmed and real time simulation system SYSTEM 100 (Grierson, 1986), which can be programmed with CSSL type simulation language).

Peripheral array processors were initially introduced in the mid-1970s, in signal processing applications (particularly those arising in medical tomography, seismic data handling, image and acoustic signal processing, etc.). Such applications typically involve long arrays of data and relatively short computations. An array processor must therefore have a high I/O rate, small program memory and fast arithmetic units which mainly consist of multipliers and adders. The AP-120B array processor is an important representative of the devices which were optimized to handle vectors of high dimension efficiently.

The simulation of dynamical systems became an increasingly significant application for peripheral array processors at the end of the 1970s and represented 30% of array processor applications (Karplus, 1984b) by the mid-1980s. Special simulation applications which demand special purpose computers can be roughly divided into two categories: those which demand solution of ordinary differential equations in real time and with hardware-in-the-loop, and those which demand solutions of partial differential equations. Although both applications require relatively long and laborious computations, large memories for complex programs and function generator breakpoints, their overall computational requirements are sufficiently different from each other, which led to the development of distinct and specialized types of simulation oriented peripheral array processors.

Two modern simulation systems that are based on the simulation specialized array processors will be briefly presented: SYSTEM 100 and XANALOG-1000.

## SYSTEM 100

In the mid-1970s researchers at Applied Dynamics International concluded that analog–hybrid systems were no longer adequate to meet the demands for greater simulation accuracy. The decision was to design an all-digital system to satisfy these growing requirements. The first system, which was named SYSTEM 10, was based on the AD 10 array processor. In 1985 Applied Dynamics introduced SYSTEM 100 which was based on the AD 100 processor (Grierson, 1986). Nowadays, this system represents one of the most efficient tools for real time simulation applications.

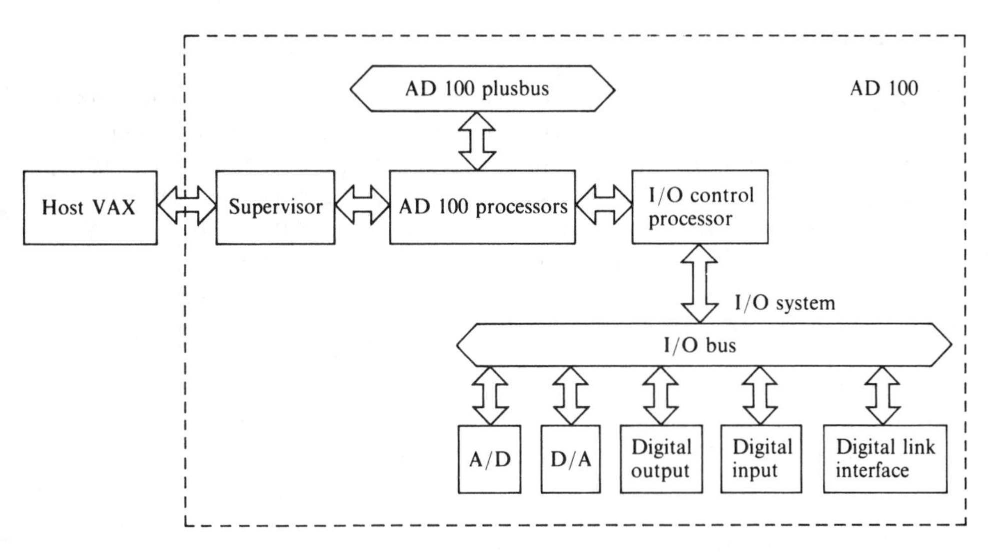


Figure 6.3 SYSTEM 100.

SYSTEM 100 consists of the Applied Dynamics AD 100 high speed multiprocessor, the VAX/VMS host computer and the ADSIM simulation language as the programming environment. The hardware structure is shown in Fig. 6.3.

The *AD 100* is a high resolution, high speed digital computer designed with gate arrays and emitter-coupled logic. Its special purpose architecture is based on pipelined parallel processing and a high speed synchronous bus, and consists of several different processors. All AD 100 processors have their own memory. The input/output control processor enables the AD 100 to communicate with external hardware using a real time input/output system which contains analog to digital and digital to analog converters as well as digital inputs and outputs.

*The VAX host computer* is used for program development and loading, for interactive simulation experimentation and for simulation results postprocessing. The AD 100 and the host computer are connected through the supervisor interface.

*ADSIM* (ADvanced SIMulation language) represents the dynamic system simulation environment. It is a CSSL-like simulation language. Structural programming using INITIAL, DYNAMIC and TERMINAL sections is possible. The model is defined and compiled for the AD 100 processor with the ADSIM compiler on the host computer. The compiler also optimizes program code by searching for mathematical operations (addition and multiplication) that can be performed concurrently, and organizes the code to enable overlapping operations whenever possible. After the compilation procedure, a highly interactive command language INTERACT is used: to load and initialize the AD 100, to start the simulation run, to display or change any value during simulation, to select the numerical integration algorithm and simulation run specifications, to present the results of simulation graphically, etc.

## XANALOG-1000

XANALOG-1000 is a very powerful modern simulation workstation designed to take advantage of the high cost-effectiveness of floating point array processors optimized for simulation applications (short vectors, a great deal of scalar code) (Schrage and McArdle, 1986; Schmidt and Scheider, 1988). The extreme effort required to implement an application on an array processor was totally eliminated by the introduction of graphics editor programming. The program is produced by choosing icons from the library and connecting them into a simulation scheme. Implemented hierarchical structures enable user modular programming using hierarchical sub-models. After scheme definition, the compiler generates the code for the array processor, which then enables a very fast simulation. Using optional input/output subsystems (A/D and D/A converters, etc.) XANALOG-1000 represents an effective tool for real time and hardware-in-the-loop applications.

### 6.2.4 Parallel Multiprocessor Systems

The simulation time on SISD or SIMD computers increases significantly with increased model complexity. If we want to avoid this disadvantage, at least partly, *parallel multiprocessor systems* using MIMD concepts must be used (Duncan, 1990). This is in fact an approach that originates in the behaviour of natural real systems or in analog–hybrid computation. Unfortunately, due to the inefficient programming capability, simulation on true parallel systems still runs mainly in academic research departments or in very special applications with specially developed software equipment. Nevertheless, parallel multiprocessor systems with a large number of nodes represent the most promising method for many applications especially in simulation.

It is obvious that the calculation efficiency of a parallel system depends upon the calculation efficiency of the individual processors that are connected to the multiprocessor system. However, it is well known, too, that communication ability between the processors represents a potential bottleneck in any parallel processing system. So it is extremely important to have fast communication channels between processors. For peak efficiency, communication bandwidth must be increased by adding busses when large number of processors are used.

An efficient parallel hardware is a necessary, but not the sufficient, condition to obtain fast simulation solutions. Good performance requires exploitation of the hardware architecture together with appropriate programming. Unfortunately, only modest parallelization and thus inefficient computation can be achieved at present with automatic parallelizing compilers.

It is not our intention to review all the approaches which enable appropriate code parallelization on multiprocessor computers. We will explain briefly only some of the basics which must be known in order to understand the problems of code parallelization.

As the communication between processors represents an essential limitation in parallel processing systems, the following conditions must be fulfilled for efficient computation:

- Individual processors must take approximately the same calculation times in the phase, they do not need mutual communications.
- The intervals without communication must be as large as possible.
- During communication as small a number of data as possible should be exchanged.

Parallel operations in computer systems may be achieved at different levels. According to Gear (Gear, 1986), parallelization in the numerical solution of ODEs and PDEs can be classified into *parallelism across the method* (PAM) and *parallelism across the system* (PAS). In the former approach the sequential code of different parts of a method (including evaluations of model equations) is reorganized in order to obtain parallel execution on more processors. This is not a very efficient method for simulation applications. The second approach originates in the parallel structures that are inherently contained in the problem definition (model). Such parallel structures are characteristic for dynamic simulation and the schemes used for simulation on analog computers or with block oriented languages represent the starting point of the decomposition for the parallel computer (Boullard and Coen, 1985) in an algorithmic way. The basic elements of the algorithmic procedure are blocks (summer, integrator, transfer function, etc.), which are considered as indivisible tasks. The execution time of each block is fixed and known in advance. The algorithm demands that all model loops are broken at the outputs of the blocks with delay attributes (integrator, delay, zero order hold, lag, transfer function, etc.). Using an appropriate procedure the parallel paths originate in the outputs and terminate in the inputs of the blocks with delay attributes. Each simulation cycle begins with known outputs of these blocks and finishes with the evaluation of these outputs for the next simulation cycle. The number of parallel paths defines the maximum number of processors required in the parallel computer. Each parallel path is characterized by its execution time and the longest time influences the time needed for simulation execution.

Using similar approaches it would be possible to automatize the entire procedure completely. The user would only have to define his problem with a corresponding simulation language as on sequential computers. Nowadays, it is possible to use simulation languages on some parallel computers but the user himself must distribute the blocks on several processors (e.g. DELFT parallel processor with simulation language PARSIM (Bruijn and Soppers, 1986)).

We shall briefly present the applicability of single chip computers – transputers for dynamic simulation applications.

## Transputers

*Transputers* are VLSI single chip computers which enable a simple parallel digital computer of extreme calculation efficiency to be constructed at a very low cost (Hamblen, 1987; Eckelmann, 1987). Transputers represent one of the most promising trends in the field of parallel computer development and also in the field of digital simulation of dynamic systems.

Transputer T-800 (INMOS Corporation) contains a 10 MIPS (millions of instructions per second) 32-bit processor, 4K Byts of RAM memory, input/output interfaces and a timer. A bit serial interface is called a 'link'. Four links are used as communication channels for the connection of an arbitrary number of transputers in a parallel computer. The data is converted from parallel to serial, set out over the link and converted from serial back to parallel by the hardware. Communication between transputers is asynchronous and does not usually represent a bottleneck of the parallel computer calculations. A/D and D/A converters can be connected using link adapter chips which turn the link channels into parallel ports.

To support parallel processing the INMOS Corporation has developed OCCAM, a high level language. OCCAM is a simple block structured language that supports both sequential and parallel programming on one or more transputers. Statements that are to be executed in sequential order are placed in the `SEQ` block. Statements that can be executed in parallel are placed in the `PAR` block. An OCCAM program can be run on any number of transputers. If there are more parallel processes than transputers, multiprocessing is used to simulate parallel processing. The operations of data-sending to links and data-receiving from them must be defined in the program block for the appropriate communications between transputers.

Parallel decomposition of the simulation model can be performed in the same way as for the other parallel computers. However, transputer parallel computers usually consist of a large number of transputers so that a high level of model decomposition can be realized (e.g. $n$ transputers for an $n$th order differential equation). As with other parallel computers, computation on a transputer system is more effective in the case of large communication intervals. So the transputer system works efficiently in the case of more complex integration algorithms. Good programming can ensure that the simulation time is sufficiently short to be very close to a factor equal to the number of transputers, in comparison to the simulation time using a single transputer.

Transputers are ideal components for dynamic systems simulation. Nevertheless, a lot more work is needed on the development of software equipment development.

**Example 6.1 Simulation of the second order system in OCCAM language**

Let us simulate the differential equation

$$\ddot{y} + a\dot{y} + by = u \tag{6.1}$$

with zero initial conditions and with constants $a = 1.4$, $b = 1$ on a parallel transputer computer. For integration, we shall use the Euler algorithm with calculation interval $\Delta t = 0.01$ s.

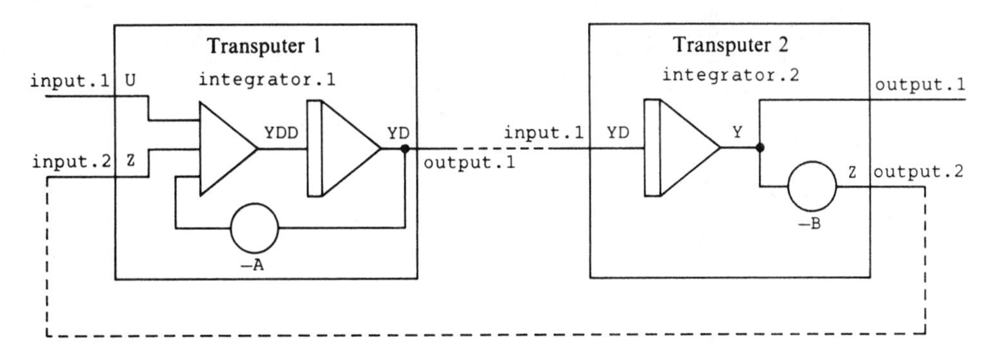


Figure 6.4 Transputer simulation scheme for the second order system.

Assuming that two transputers are used to simulate the problem, the first step is to determine which computations will be performed on which transputer. The simple method recommends implementation of the calculations for each state variable (output of a block with delay attribute) on a separate transputer. Fig. 6.4 represents the digital simulation scheme of a second order system with an appropriate distribution of blocks to both transputers. The computer variables `U`, `Z`, `YDD`, `YD`, `Y`, `A`, `B`, `DT` are used as a problem variables $u$, $z$, $\ddot{y}$, $\dot{y}$, $y$, $a$, $b$, $\Delta t$ respectively. The dotted lines represent hardware links between transputers. The first transputer needs three links (two inputs: `input.1`, `input.2` and one output: `output.1`) and the second one needs the same number of links (one input: `input.1` and two outputs: `output.1` and `output.2`).

The state variables (`Y`, `YD`) are known at the beginning of each calculation interval. So the transputers send outputs, which are related to the state variables, over links, receive their inputs over links, perform derivative evaluations (`YD`, `YDD`) and then compute the state variables (`Y`, `YD`) for the next calculation interval with the Euler integration method (see Section 5.1). This process of communicating and calculating is repeated in every calculation interval. Besides the data coming over links, the transputers still need the data from the local memory where also the processor programs reside. The arrival of input data from the links synchronize the transputers so that they work with the 'same calculation interval'.

Figure 6.5 shows the program in the OCCAM language. It consists of two processes: `PROC integrator.1` and `PROC integrator.2`. In `PROC` statements the physical link connections are assigned. Then the simulation constants and model variables declaration are specified. Both processes are then described with sequential (`SEQ`) and parallel (`PAR`) blocks. In the first sequential block the initial condition is specified and the simulation loop is realized with the `WHILE` statement. The next sequential block, which is embedded in the first one, realizes the appropriate simulation calculation for each calculation interval. The calculations that can be executed in parallel (in an arbitrary statements order) are specified in the parallel block (`PAR`). The lines with exclamation marks send data over a link and the lines

```
PROC integrator.1 (CHAN input.1, input.2, output.1)
  VAL A IS    1.4 (REAL32):
  VAL DT IS 0.01 (REAL32):

  REAL32 U,YDD,YD,Z,DT:

  SEQ
    YD:=0.0(REAL32)
    WHILE TRUE
      SEQ
        PAR
          output.1!YD
          input.1?U
          input.2?Z
        YDD:=(U+Z)-(A*YD)
        YD:=YD+(DT*YDD)

PROC integrator.2 (CHAN input.1, output.1, output.2)
  VAL B IS   1.0(REAL32):
  VAL DT IS  0.01(REAL32):

  REAL32 YD, Y, Z, DT

  SEQ
    Y:=0.0(REAL32)
    WHILE TRUE
      SEQ
        Z:=-B*Y
        output.2!Z
        PAR
          output.1!Y
          input.1?YD
        Y:=Y+(DT*YD)
```

Figure 6.5 The program in OCCAM language.

with question marks receive data (with appropriate wating) from a link. The last rows in both processes realize the Euler integration algorithm. The end of an OCCAM block is indicated by indenting only; no end statement is used.

In the example we did not deal with the problem of generation of the input signal $u$. It could be an external signal passing through an A/D converter to the appropriate transputer link. It is also possible to transfer the output signal $y$, using a D/A converter, to an external device. □

### 6.2.5 New Trends in Hybrid Computer Development

There is no doubt that digital computers are used and will be used in most applications. But there are, however, still application areas where analog computers have the advantage when used in hybrid configuration. They are used efficiently in real time simulation applications where high speed of computation is essential and where interfacing between the physical system (hardware-in-the-loop) and the computer represents an important part of the simulation.

The development of modern technology means that modern hybrid systems using parallel multiprocessor architecture organization enable automatic patching, automatic scaling and programming with the aid of a simulation language. This means that the work with the 'black box' hybrid computer is the same as work performed with digital simulation tool. The most important representative of these systems is the SIMSTAR computer.

## Hybrid system SIMSTAR

Using the above-mentioned concepts, the most modern and efficient hybrid system SIMSTAR came on to the market in 1983 (Landauer, 1988a, b).

SIMSTAR architecture is shown in Fig. 6.6. A minimum SIMSTAR configuration consists of the digital arithmetic processor (DAP) and one parallel simulation processor (PSP) which represents the analog part of the hybrid system and can be used efficiently to interface directly to external hardware with logic and analog signals. To this basic configuration can be added the function generation processor (FGP) for high speed function generation and the data conversion processor (DCP) for high speed analog to digital and digital to analog conversion. Data resolution is greater than four digits. Additional expansion can include a host digital computer to be used mainly for program preparation. A second parallel simulation processor can be added to this system. It is possible to connect three such configurations with one host computer in maximum configuration.

A common memory and shared data bus provide the communication path throughout the system. But the primary information flow for the simulation occurs inside the processors themselves.

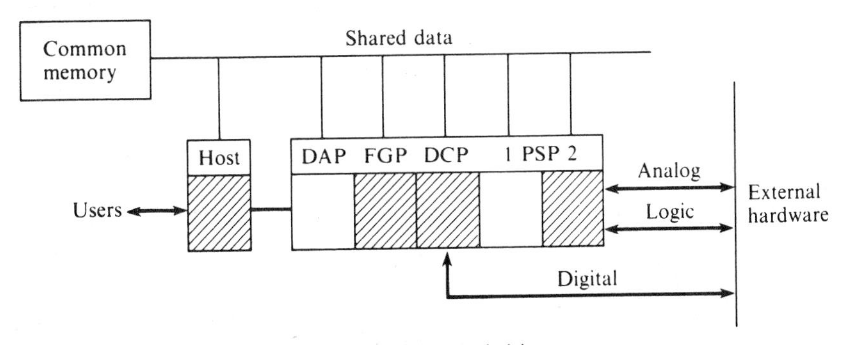


Figure 6.6 SIMSTAR hybrid system.

The *digital arithmetic processor* (DAP) is a computer from the GOULD family. During simulation it works in real time and is triggered by periodic interrupts. It can realize many simulation functions. Some typical operations which are executed at each interrupt are: reading the data of DCP, preparing the data for FGP, processing the data coming from PSP, detecting the error if DAP is not able to perform all the calculations in a frame time, etc.

The *data conversion processor* (DCP) enables very fast transfer of data between the DAP, PSP and the external hardware. A/D and D/A converters enable a sampling rate up to 300 KHz on each channel. A/D converters are automatically triggered by the frame rate. When the conversions are finished, the DCP interrupts the DAP, which receives and processes the data and sends them in the D/A conversion part of the DCP. So the processed data are transferred to the PSP or to the external hardware. The complete conversion procedure does not demand any DAP computation effort.

The *function generation processor* (FGP) is a multiprocessing system in itself. It possesses its own memory. Using appropriate tables in its memory it enables extremely fast generation of pseudocontinuous functions from one to four independent variables with linear interpolation between breakpoints. The function generation procedure is performed entirely in the FGP without any DAP calculation effort.

The *parallel simulation processor* (PSP) (the analog part of SIMSTAR) is designed and built for time critical simulation operations. Figure 6.7 shows the simplified diagram of the PSP. It consists of the parallel math unit (PMU) and the parallel logic unit (PLU). Both units are connected by threshold comparators. The PMU contains up to two hundred mathematical computing blocks (MCB) such as integrators, summers, multipliers, etc. The outputs of the MCBs are internally

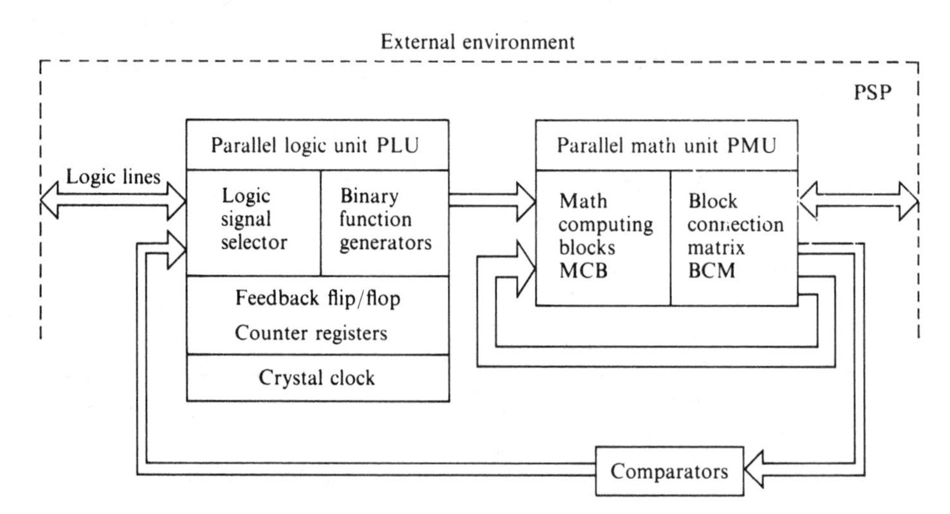


Figure 6.7 Parallel simulation processor.

connected to particular inputs of the block connection matrix (BCM). The BCM is set during simulation model compilation and enables the appropriate connection of the MCB outputs to the MCB inputs, comparators or to the external environment. In a similar way, the PLU is configured on the basis of the simulation program for the logical operations required.

The programming environment allows the use of the SIMSTAR hybrid computer with the high level simulation language STARTRAN, which is a CSSL-type simulation language. It is compiler oriented and compiles the source model into FORTRAN modules and an appropriate data base. The user defines a program through the definitions of the INITIAL, DYNAMIC and TERMINAL sections. The DYNAMIC section, which is used for model structure description, can consist of one to five DERIVATIVE sections. It is important to know that the first section has the highest priority and the last section has the lowest priority during simulation. Thus the first section must contain submodels with the highest frequency regions containing comparisons, switching and discontinuities.

When a model is defined the user must perform a 'targeting' procedure to define the processors which will execute the appropriate parts of the model. Most simulation functions can be executed either on the DAP or the PSP. This is an extremely important function which obtain efficient real time solutions. It is convenient to target the submodels with the highest natural frequencies, the shortest time constants, the most frequent discontinuities, algebraic loops and external analog/logic connections for the PSP. These features can be extracted from off-line solutions of the model, which can be obtained previously with digital simulation in ACSL (Mitchel & Gauthier, 1981).

The STARTRAN preprocessor handles the appropriate targeting statements and generates the sequential program and the parallel program. The sequential program is further processed by the DTRAN, a real time ACSL translator to produce FORTRAN code for running the submodels on the DAP. The parallel program is processed by the PTRAN, a special parallel translator which reduces the equations and ranges variables and coefficients to allocate available mathematical computing blocks (MCB). The result is an appropriate interconnection data file and an appropriate FORTRAN code for setup. Executable SIMSTAR code is generated from DTRAN and PTRAN outputs.

When the simulation model is processed it can be executed in an interactive environment. At any phase the model can easily be rerun in ACSL to compare particular SIMSTAR solutions.

## BIBLIOGRAPHY

ÅSTRÖM, K.J. (1985), 'Computer aided tools for control and system design', in *Computer-aided Control System Engineering* (eds M. Jamashidi and C.J. Herget), North Holland, Amsterdam, pp. 3–40.

BAKER, N.J.C. and P.J. SMART (1983), 'The SYSMOD simulation language', *Proceedings of the 1st European Simulation Congress*, Aachen, W. Germany, pp. 281–6.

BAKER, N.J.C. and P.J. SMART (1985), 'Elements of the SYSMOD simulation language', *Proceedings of the 11th IMACS World Congress*, Oslo, Norway, vol. 2.

BARKER, H.A. (1988), 'Graphical environments for computer aided control system design', *Preprints of the 4th IFAC Symposium on Computer Aided Design in Control Systems, CADCS'88*, Beijing, China, pp. 60–6.

BARKER, H.A., P. TOWNSEND, M. CHEN and J. HARVEY (1988), 'CES – a workstation environment for computer aided design in control systems', *Preprints of the 4th IFAC Symposium on Computer Aided Design in Control Systems, CADCS'88*, Beijing, China, pp. 248–51.

BARKER, H.A., T. HUYNH-QUOC and P. TOWNSEND (1989), 'A graphical pre-processor for continuous system simulators', *Proceedings of the 3rd European Simulation Congress*, Edinburgh, Scotland, pp. 349–55.

BARKER, H.A., M. CHEN, P.W. GRANT, I.T. HARVEY, C.P. JOBLIN and P. TOWNSEND (1990), 'The impact of recent developments in user-interface specification, design and management on the computer-aided design of control systems', *Preprints of 11th IFAC World Congress*, Tallin, USSR, vol. 10, pp. 110–15.

BOULLARD, L. and L.D. COEN (1985), 'A multi-microprocessor system for parallel computing in simulation', *Proceedings of the 11th IMACS World Congress*, Oslo, Norway, vol. 3, pp. 163–6.

BREITENECKER, F. and D. SOLAR (1986), 'Models, methods and experiments – modern aspects of simulation languages', *Proceedings of the 2nd European Simulation Congress*, Antwerp, Belgium, pp. 195–9.

BRITT, J.I., J.S. WARECK and J.A. SMITH (1991), 'A computer-aided process synthesis and analysis environment', in J.J. Siirola, I.E. Grossman and G. Stephanopoulos (eds), *Foundations of Computer-aided Process Design*, Elsevier, Amsterdam, pp. 281–307.

BRUIJN, M.A. DE and F.P.J. SOPPERS (1986), 'The Delft parallel processor and software for continuous system simulation', *Proceedings of the 2nd European Simulation Congress*, Antwerp, Belgium, pp. 777–83.

CELLIER, F.E. (1979), Combined Continuous/Discrete System Simulation by Use of Digital Computers: Techniques and Tools, PhD thesis, Swiss Federal Institute of Technology, Zürich.

CELLIER, F.E. (1983), 'Simulation software – today and tomorrow', *Proceedings of the IMACS Conference*, Nantes, France, pp. 3–19.

CHEN, S. (1989), 'Toward the future', in *Supercomputers: Directions and technology and applications*, National Academy Press, Washington, DC, pp. 51–65.

COLIJN, A.W. and P.D. ARIEL (1986), 'Continuous system simulation languages on supercomputers', *Proceedings of the 1986 Summer Computer Simulation Conference*, Reno, USA, pp. 3–7.

CROSBIE, R.E. (1984), 'Simulation on microcomputer – experiences with ISIM', *Proceedings of the 1984 Summer Computer Simulation Conference*, Boston, Massachusetts, USA, vol. 1, pp. 13–17.

CROSBIE, R. (1986), 'Developments in ESL and others for the 1990's', *Proceedings of the 1986 Summer Computer Simulation Conference*, USA, pp. 313–14.

CROSBIE, R.E. and F.E. CELLIER (1982), 'Progress in simulation language standards, an activity report for Technical Committee TC3 on simulation software', *Proceedings of the 10th IMACS World Congress*, Montreal, Canada, vol. 1, pp. 411–12.

CROSBIE, R.E. and J.L. HAY (1985), 'Applications of ESL', *Proceedings of the 11th IMACS World Congress*, Oslo, Norway, vol. 3, pp. 193–6.

CROSBIE, R.F., S. JAVEY, J.S. HAY and J.G. PEARCE (1985), 'ESL – a new continuous system simulation language', *Simulation* **44** (5), 242–6.

D'HOLLANDER, E.H. (1985), 'A multiprocessor for dynamic simulation', *Proceedings of the 1985 Summer Computer Simulation Conference*, USA, 112–7.

DUNCAN, R. (1990), 'A survey of parallel computer architectures', *IEEE Computer*, February, pp. 5–16.

ECKELMANN, P. (1987), 'The transputer – component for the 5 generation systems', *Proceedings VLSI and Computers. First International Conference on Computer Technology, Systems and Applications*, Hamburg, West Germany, p. 977.

ELMQVIST, H. and S.E. MATSON (1986), 'A simulator for dynamical systems using graphics and equations for modeling', *Proceedings of the IEEE Control System Society. Third Symposium on Computer Aided Control System Design*, Arlington, USA, pp. 134–9.

FADDEN, E.J. (1987), 'SYSTEM 100: Time-critical simulation of continuous systems', *Multiprocessor and Array Processor Conference*, publishing number 87 02mT15, San Diego, USA.

GEAR, C.W. and O. OSTERBY (1984), 'Solving ordinary differential equations with discontinuities', *ACM Trans. Math. Software*, **10**, 23–44.

GEAR, C.W. (1986), 'The potential for parallelism in ordinary differential equations', *Proc. Int. Conf. Numerical Mathematics*, pp. 33–48.

GOLDEN, D.G. (1985), 'Software engineering considerations for the design of simulation languages', *Simulation*, **45** (4), pp. 169–78.

GRIERSON, W.O. (1986), 'Perspectives in simulation hardware and software architecture', *Modeling, Identification and Control* (Norwegian Research Bulletin), **6** (4), 249–55.

GUSTAFSSON, K. (1988), 'Stepsize control in ODE solvers – analysis and synthesis', *Technical Report*, Department of Automatic Control, Lund Institute of Technology, Sweden.

HAMBLEN, J.O. (1987), 'Parallel continuous system simulation using transputer', *Simulation*, **49** (6), 249–53.

HAY, J.L. and R.E. CROSBIE (1981), *Outline Proposal for a New Standard for Continuous System Simulation Language (CSSL'81)*, Computer Simulation Centre, Department of Electrical Engineering, University of Salford.

HAY, J.L. AND R.E. CROSBIE (1986), 'Parallel processor simulation with ESL', *Proceedings of the 1986 Summer Computer Simulation Conference*, Reno, USA, pp. 959–64.

HUBER, R.M. and A. GUASCH (1985), 'Towards a specification of a structure for continuous system simulation languages', *Proceedings of the 11th IMACS World Congress*, Oslo, Norway, pp. 109–13.

KARPLUS, W.J. (1984a), 'Selection criteria and performance evaluation methods for peripheral array processors', *Simulation*, **43** (3), 125–31.

KARPLUS, W.J. (1984b), 'The changing role of peripheral array processors in continuous systems simulation', *Proceedings of the 1984 Summer Computer Simulation Conference*, Boston, Massachusetts, USA, vol. 1, pp. 1–13.

KETTENIS, D.L. (1986), 'COSMOS: a member of a new generation of simulation languages', *Proceedings of the 2nd European Simulation Congress*, Antwerp, Belgium, pp. 263–9.

KLEINERT, W., M. GRAFF, R. KARBA and B. ZUPANČIČ (1988), 'Simulation einer Destillationskolonne-Modellierung mit SIMCOS und Vergleich der Ergebnisse von ACSL und SIMSTAR Simulationen', *Proceedings of the 5th sympsoium Simulationstechnik*, Aachen, W. Germany, pp. 254–9.

LANDAUER, J.P. (1988a), *EAI STARTRAN Environment*, users' manual, Electronic Associates, Inc., West Long Branch, New Jersey, USA.

LANDAUER, J.P. (1988b), 'Real-time simulation of the Space Shuttle main engine on the SIMSTAR multiprocessor', *Proceedings of the SCS Multiconference on Aerospace Simulation III*, San Diego, USA.

MARQUARDT, W. (1991), 'Dynamic process simulation – recent progress and future challenges', *Preprints of CPC IV, Fourth International Conference on Chemical Process Control*, South Padre Island, Texas, USA.

MATKO, D., B. ZUPANČIČ and S. DIVJAK (1989), 'New concepts in control system simulation', *Proceedings of the 3rd European Simulation Congress*, Edinburgh, Scotland, pp. 444–6.

MITCHEL & GAUTHIER, ASSOC. (1981), *ACSL: Advanced Continuous Simulation Language* (user guide/reference manual).

NILSEN, N.R. (1985), 'Recent advances in CSSL IV', *Proceedings of the 11th IMACS World Congress*, Oslo, Norway, vol. 3, pp. 101–3.

ÖREN, T.I. and B.P. ZIEGLER (1979), 'Concepts for advanced simulation methodologies', *Simulation*, **322** (3), pp. 69–82.

PEARCE, J.G., P. HOLLIDAY and J.O. GRAY (1985), 'Survey of parallel processing in simulation', *Proceedings of the 1st European Workshop on Parallel Processing Techniques for Simulation*, Manchester, UK, pp. 183–202.

RIMVALL, M. and F.E. CELLIER (1985), 'The matrix environment as enhancement to modelling and simulation', *Proceedings of the 11th IMACS World Congress*, Oslo, Norway, vol. 3, pp. 93–96.

RIMVALL, M. and F.E. CELLIER (1986), 'Evaluation and perspectives of simulation languages following the CSSL standard', *Modeling, Identification and Control* (Norwegian Research Bulletin), **6** (4), 181–99.

SCHMIDT, A. and F. SCHEIDER (1988), 'Erfahrungen mit Hardware in-the-loop Simulation an der Workstation XANALOG XA-1000', *Proceedings of the 5th symposium Simulationstechnik*, Aachen, W. Germany, pp. 2–13.

SCHRAGE, M.H. and D.F. MCARDLE (1986), 'New array processor continuous simulation system uses tactile sensing icon programming', *Proceedings of the 1986 Summer Computer Simulation Conference*, Reno, USA, 138–42.

SHNEIDERMAN, B. (1987), *Designing the User Interface*, Addison-Wesley Publishing Company, USA.

SPRIET, J.A. AND G.C. VANSTEENKISTE (1982), *Computer-aided Modelling and Simulation*, Academic Press Inc., London.

STEPPARD, S. (1983), 'Applying software engineering to simulation', *Simulation*, January, pp. 13–19.

STRAUSS, J.C. (1967), 'The SCI continuous system simulation language', *Simulation*, (9), pp. 281–303.

TAYLOR, H., D.K. FRIEDRICK, C.M. RIMVALL and H.A. SUTHERLAND (1990), 'Computer-aided control engineering environments: architecture, user interface, data-base management, and expert aiding', *Preprints of 11th IFAC World Congress*, Tallinn, USSR, vol. 10, pp. 55–65.

TESLER, L.G. (1989), 'Achieving a pioneering outlook with supercomputing', *Supercomputers: Directions in technology and applications*, National Academy Press, Washington, DC, pp. 90–5.

WORLTON, J. (1989), 'Existing conditions', in *Supercomputers: Directions in technology and applications*, National Academy Press, Washington, DC, pp. 21–50.

ZIEGLER, B.P. (1976), *Theory of Modelling and Simulation*, John Wiley & Sons, Inc., NY.

ZIEGLER, B.P. (1987), *Simulation Objectives: Experimental frames and validity. System and Control Encyclopaedia*, Pergamon Press, Oxford, pp. 4388–92.

ZITNEY, S.E. (1990), 'A frontal code for ASPEN + on advanced architecture computers', *AIChE Annual Meeting*, Paper 74 d, Chicago.

ZUPANČIČ B. (1989), 'Digital simulation language synthesis for the computer aided control system design', PhD thesis, Faculty of Electrical and Computer Engineering, University of Ljubljana, Yugoslavia.

ZUPANČIČ, B., D. MATKO, R. KARBA, M. ATANASIJEVIĆ and Z. ŠEHIĆ (1991), 'Extensions of the simulation language SIMCOS towards continuous–discrete complex experimentation system', *Preprints of the IFAC Symposium, CADCS'91*, University of Wales, Swansea, UK, pp. 351–6.

# 7

# Case Studies

## 7.1 CASCADE CONTROL OF A HYDRAULIC PLANT

The purpose of the present case study is as follows:

- to illustrate the theoretical modelling procedure (in this case based on the mass balance principle);
- to stress certain aspects in connection with the modelling aims;
- to illustrate digital simulation programming in two different languages (the CSSL-like language SIMCOS and the digital simulation language SIMNON); and
- to comment on the capabilities of both languages in view of the cascade control problem (discrete controllers simulation, multirate sampling).

### Hydraulic plant modelling

The plant under investigation is shown in Fig. 7.1. As can be seen, the plant consists of three noninteracting liquid tanks, which means that the behaviour of each is completely independent of the behaviour of the others. Each tank has its hydraulic capacitance ($C_i$), while the output flow rate ($\Phi_i$) depends on the hydraulic resistance ($R_i$), which can be assured with the aid of the corresponding valves, and of course on the levels of liquid ($H_i$) in the tanks. In our case, the aim of the theoretical modelling is to obtain the mathematical form to be used in the evaluation and design of the cascade control scheme. Here, the control variable is the input flow rate ($\Phi_1$), the controlled variable is the liquid level in the third tank ($H_3$), while the auxiliary variable is the liquid level in the first tank ($H_1$). The resistances are unchangeable in our case. Due to the character of the problem it is obvious that the mass balance principle represents the basis of the theoretical modelling approach. Let us summarize briefly the procedure for modelling one tank in the example (Fig. 7.2). An unsteady-state mass balance (Weber, 1973) can be described as

*input − output = accumulation*

which can be expressed for our case in the form:

$$\Phi_I - \Phi_O = \frac{\mathrm{d}V}{\mathrm{d}t} = S\frac{\mathrm{d}H}{\mathrm{d}t} \tag{7.1}$$

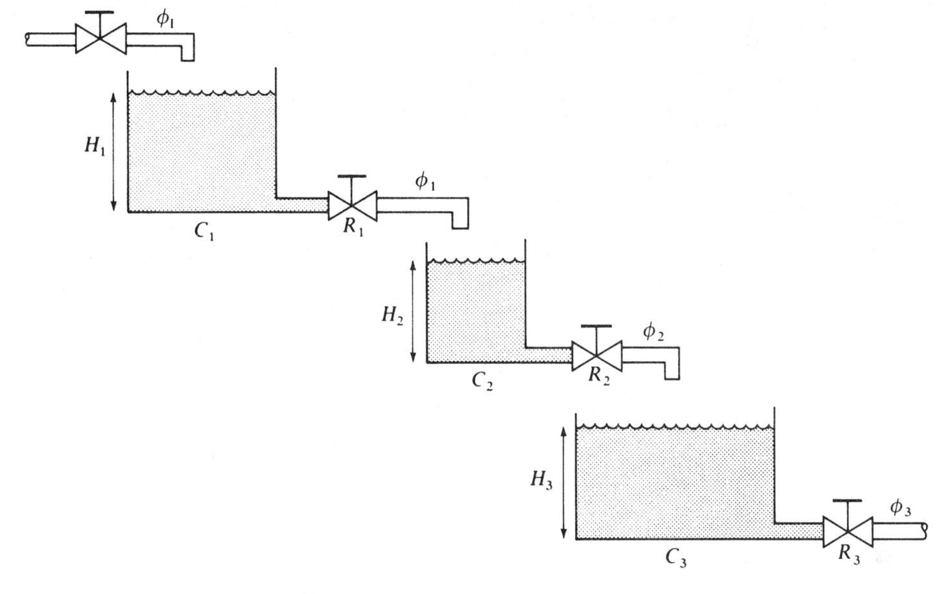


Figure 7.1 Hydraulic plant under investigation.

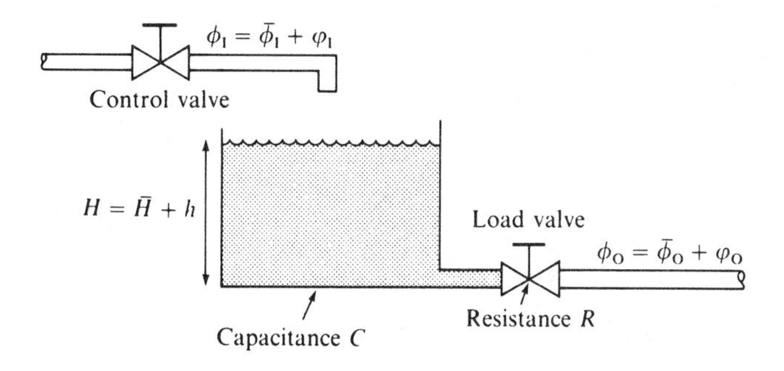


Figure 7.2 Liquid level system.

where $\Phi_I$ and $\Phi_O$ are input and output flow rates, respectively, while $V$ is the volume of liquid in the tank, $S$ is the cross-sectional area of the tank and $H$ is the liquid level or head.

As in the basic balance equations, the physical quantities in which one is interested for the particular situation are rarely present, the first step is to modify the basic equation in the sense of searching for the relations which would introduce the corresponding quantities in the mathematical model. In our case, such a relation should be the one which connects $H$ and $\Phi_O$ being defined as the input and output variables respectively. For the turbulent flow through the restriction (valve), the steady state flow rate is given by (Ogata, 1990)

$$\Phi_O = k\sqrt{H} \tag{7.2}$$

where $\Phi_O$ = steady state liquid flow rate ($m^3/s$); $k$ = coefficient ($m^{2.5}/s$); $H$ = steady state head (m).

Using Eqn (7.2) in the basic balance equation we obtain the nonlinear first order differential equation which connects the head with the input flow rate as required by the modelling aims. Note here that the level can also be controlled by the output valve. In such a situation the cross-sectional area of the valve must also be included in the mathematical model (in our case it is hidden in the coefficient $k$).

Let us now define the hydraulic resistance and capacitance. The former is expressed as the head change which causes the unit change in the flow rate. The resistance for turbulent flow $R$ can thus be obtained through the relation

$$R = \frac{dH}{d\Phi_O} \tag{7.3}$$

Since from Eqn (7.2) we obtain

$$d\Phi_O = \frac{k}{2\sqrt{H}} dH \tag{7.4}$$

we have

$$\frac{dH}{d\Phi_O} = \frac{2\sqrt{H}}{k} = \frac{2\sqrt{H}}{\Phi_O}\sqrt{H} = \frac{2H}{\Phi_O} \tag{7.5}$$

Thus

$$R = \frac{2H}{\Phi_O} \tag{7.6}$$

The value of the turbulent flow resistance $R$ depends on the head and flow rate and may be considered constant if changes in both quantities are small.

The hydraulic capacitance $C$ is defined as the change in the quantity of stored liquid necessary to cause a unit change in the potential (head). Consequently, the following relation is valid:

$$C = S \tag{7.7}$$

Note the difference between the capacitance ($m^2$) and the capacity ($m^3$) of the tank. If the cross-sectional area is constant, the capacitance is constant for any head.

At this moment the decision must be made if the model obtained is to be used in the nonlinear version, or if the linearization procedure is justified. In our case, the model will be used for the control design and evaluation where the controller has a relatively robust structure and where large deviations from the operating conditions are not foreseen. Therefore, the linear model is much more convenient.

To obtain a linear approximation of the nonlinear model (Ogata, 1990), only slight deviations of the variables from some operating condition are assumed. If the latter corresponds to $\bar{x}$ and $\bar{y}$, where $x$ is the input and $y$ is the output of the system, the nonlinear relation

$$y = f(x) \tag{7.8}$$

may be expanded into a Taylor series around the operating point. In the case of a small variation $x - \bar{x}$, the higher order terms may be neglected. Thus we obtain

$$y - \bar{y} = k(x - \bar{x}) \tag{7.9}$$

where

$$\bar{y} = f(\bar{x})$$

and

$$k = \left.\frac{\mathrm{d}f}{\mathrm{d}x}\right|_{x = \bar{x}}$$

Eqn (7.9) indicates that the deviation in $y$ is proportional to the deviation in $x$, giving the linear model of the nonlinear relation in Eqn (7.8) near the operating points $x = \bar{x}$ and $y = \bar{y}$. The procedure is also similar for functions with two or more variables. It is important to note that a certain mathematical model used for analysis and design may be accurate for one operating condition but may not work satisfactorily for others.

In applying the linearization approach in our case, let us consider the liquid system shown in Fig. 7.2. At the operating point, the input flow rate is $\Phi_I = \bar{\Phi}$ and the output flow rate is $\Phi_O = \bar{\Phi}$, while head is $H = \bar{H}$. For the turbulent flow, Eqn (7.2) is also valid for the operating point.

Assume that at $t = 0$ the input flow rate changes by $\phi_i(\Phi_I = \bar{\Phi} + \phi_i)$, which causes the head change $h(H = \bar{H} + h)$. In turn, this change causes the change of the output flow rate $\phi_o(\Phi_O = \bar{\Phi}_O + \phi_o)$. Combining these assumptions, Eqns (7.1), (7.2) and (7.7), we obtain

$$C\frac{\mathrm{d}H}{\mathrm{d}t} = \Phi_I - k\sqrt{H} \tag{7.10}$$

Let us define

$$\frac{\mathrm{d}H}{\mathrm{d}t} = f(H, \Phi_I) = \frac{1}{C}\Phi_I - \frac{k\sqrt{H}}{C} \tag{7.11}$$

Using the linearization technique the following equation can be written:

$$\frac{\mathrm{d}H}{\mathrm{d}t} - f(\bar{H}, \bar{\Phi}_I) = \frac{\partial f}{\partial H}(H - \bar{H}) + \frac{\partial f}{\partial \Phi_I}(\Phi_I - \bar{\Phi}_I) \tag{7.12}$$

Note that the operating condition is $(\bar{H}, \bar{\Phi})$ and since here $\mathrm{d}H/\mathrm{d}t = 0$ we have

$f(\bar{H}, \bar{\Phi}) = 0$. As

$$\left.\frac{\partial f}{\partial H}\right|_{H=\bar{H}, \Phi_1=\bar{\Phi}_1} = -\frac{k}{2C\sqrt{\bar{H}}} = -\frac{\bar{\Phi}}{\sqrt{\bar{H}}}\frac{1}{2C\sqrt{\bar{H}}} = -\frac{\bar{\Phi}}{2C\bar{H}} = -\frac{1}{RC}$$

where $R = 2\bar{H}/\bar{\Phi}$ and also

$$\left.\frac{\partial f}{\partial \Phi_1}\right|_{H=\bar{H}, \Phi_1=\bar{\Phi}_1} = \frac{1}{C}$$

then Eqn (7.12) can be rearranged into the form

$$\frac{\mathrm{d}H}{\mathrm{d}t} = -\frac{1}{RC}(H - \bar{H}) + \frac{1}{C}(\Phi_1 - \bar{\Phi}_1) \tag{7.13}$$

Since $H - \bar{H} = h$ and $\Phi_1 - \bar{\Phi}_1 = \phi_1$, Eqn (7.13) becomes

$$\frac{\mathrm{d}h}{\mathrm{d}t} = -\frac{1}{RC}h + \frac{1}{C}\phi_1 \tag{7.14}$$

or

$$RC\frac{\mathrm{d}h}{\mathrm{d}t} + h = R\phi_1 \tag{7.15}$$

which is the linearized version of Eqn (7.10) where $RC$ is the time constant of the system. Performing the Laplace transformation on both sides of Eqn (7.15), with zero initial conditions, we obtain

$$(RCs + 1)H(s) = R\Phi_1(s) \tag{7.16}$$

where $H(s) = \mathscr{L}[h(t)]$ and $\Phi_1(s) = \mathscr{L}[\phi_1(t)]$.

Since in our case $\phi_1$ is considered to be the input and $h$ the output transfer function of the system, which represents the last step of modelling, is

$$\frac{H(s)}{\Phi_1(s)} = \frac{R}{RCs + 1} \tag{7.17}$$

If, however, $\phi_O$ is taken as the output using the same input as previously, the transfer function is

$$\frac{\Phi_O(s)}{\Phi_1(s)} = \frac{1}{RCs + 1} \tag{7.18}$$

where the relationship

$$\frac{\Phi_O(s)}{H(s)} = \frac{1}{R} \tag{7.19}$$

was used.

In our case, as in many others, the representation of the single tank with the transfer function is convenient due to the fact that the development of the whole plant model is easier. Manipulation with transfer functions is much simpler than combining the differential equations. Due to the prescribed structure of the cascade control to be developed and studied, two transfer functions for the plant in Fig. 7.1 must be obtained, namely $H_1/\Phi_1$ and $H_3/H_1$. Due to the noninteracting connection of the tanks in the whole plant, by analogy with one tank the following transfer functions can be written:

$$\frac{H_1}{\Phi_1} = \frac{R_1}{C_1 R_1 s + 1} = G_{p_1}(s) \tag{7.20}$$

$$\frac{\Phi_1}{H_1} = \frac{1}{R_1} \tag{7.21}$$

$$\frac{H_2}{\Phi_1} = \frac{R_2}{C_2 R_2 s + 1} \tag{7.22}$$

$$\frac{\Phi_2}{H_2} = \frac{1}{R_2} \tag{7.23}$$

$$\frac{H_3}{\Phi_2} = \frac{R_3}{C_3 R_3 s + 1} \tag{7.24}$$

As can be seen, Eqn (7.20) directly represents one of the required transfer functions. The second one can be developed in the same way; $\Phi_1$ is expressed from Eqn (7.21) as

$$\Phi_1 = \frac{H_1}{R_1} \tag{7.25}$$

It is then substituted in Eqn (7.22) to obtain

$$\frac{H_2}{H_1} = \frac{R_2}{R_1(C_2 R_2 s + 1)} \tag{7.26}$$

Similarly, $\Phi_2$, obtained from Eqn (7.23) as

$$\Phi_2 = \frac{H_2}{R_2} \tag{7.27}$$

is substituted in Eqn (7.24) to obtain

$$\frac{H_3}{H_2} = \frac{R_3}{R_2(C_3 R_3 s + 1)} \tag{7.28}$$

Eqns (7.26) and (7.28) are then multiplied, which gives the second required transfer function

$$\frac{H_3}{H_1} = \frac{R_3}{R_1(C_2 R_2 s + 1)(C_3 R_3 s + 1)} = G_{p_2}(s) \tag{7.29}$$

If for the plant studied in Fig. 7.1 $R_1 = R_2 = R_3 = 1$ is considered, the required transfer functions are, according to the actual cross-sectional areas of the tanks, $C_1$, $C_2$ and $C_3$, given in the form

$$G_{p_1} = \frac{1}{7.5s + 1} \tag{7.30}$$

$$G_{p_2} = \frac{1}{(5s + 1)(10s + 1)} \tag{7.31}$$

**Cascade control of the hydraulic plant**

The cascade control scheme for the linearized model of the hydraulic plant, given by Eqns (7.30) and (7.31), is shown in Fig. 7.3. It consists of two loops (auxiliary and main) which are controlled by two controllers.

Since for the purposes of this case study it is almost irrelevant why the discrete cascade control was chosen and how the controller parameters were designed, let us give their discrete transfer functions only. Some analyses show that it is convenient to choose the PI controller from Eqn (7.32) to represent the auxiliary controller in our cascade scheme. It should make the closed loop system quicker and thus sampling time $T_{S_2} = 1s$ was chosen:

$$G_{R_2}(z) = \frac{2 - 1.85z^{-1}}{1 - z^{-1}} \tag{7.32}$$

On the other hand, the main PID controller with discrete transfer function

$$G_{R_1}(z) = \frac{2.6723 - 3.3452z^{-1} + 1.036z^{-2}}{1 - z^{-1}} \tag{7.33}$$

was designed for the sampling time $T_{s_1} = 4s$, which should ensure that the overshoot of the controlled variable is less than 5%.

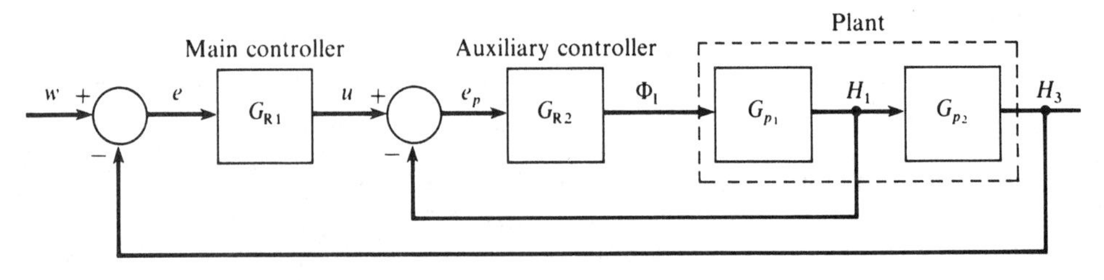


Figure 7.3 Cascade control scheme for the hydraulic plant.

**Comparison of programming two digital simulation languages in evaluating hydraulic plant cascade control**

In this subsection we want to show some properties and capabilities of two digital simulation languages with different characteristics, namely the CSSL-like language SIMCOS and the language SIMNON, with the aid of the hydraulic control scheme. As can be seen from the preceding subsections the plant is described by two continuous transfer functions and the main and auxiliary controller are given by two discrete transfer functions, which indicates that the plant is computer controlled. To simulate the whole control scheme (Fig. 7.3) the digital simulation language must have the capacity for discrete transfer function realization. For simulation in SIMCOS the control scheme is modified as in Fig. 7.4.

The corresponding program is given in Figs. 7.5 and 7.6. Inclusion of discrete transfer functions in SIMCOS is very simple due to the fact that some types of discrete functions are already programmed as SIMCOS blocks. For the realization of both controllers the block DPID is used. For each of the controllers the corresponding sampling time is chosen. The auxiliary variable is fed back through the zero order hold block, which causes the error in the auxiliary control loop to be constant between two consecutive samples. This block was inserted only due to the systematic reasons in the scheme and has no influence on the simulation results.

On the basis of the scheme shown in Fig. 7.4, the corresponding program in SIMNON was developed, which is shown in Figs 7.7, 7.8, 7.9 and 7.10.

The simulation language SIMNON does not include compiled discrete sub-models (transfer functions) in advance. It enables the specification of continuous and

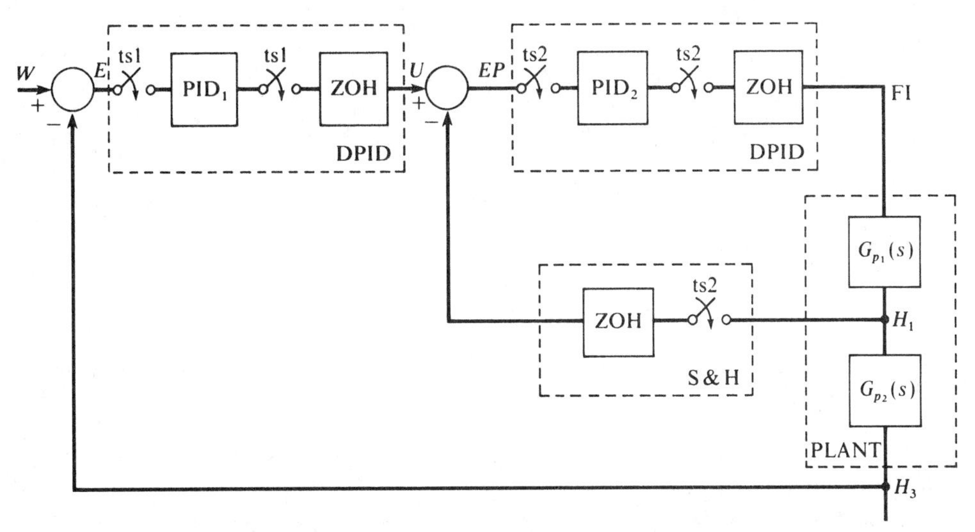


Figure 7.4 Cascade control scheme adapted for simulation in SIMCOS.

```
PROGRAM CASCADE CONTROL
''
''hydraulic system
''multirate sampling
''-----------------------------------------------------------------
''
''     W   ... reference
''     Q1  ... parameters of the main PID controller
''     TS1 ... sampling time in the main control loop
''     E   ... error in the main control loop
''     U   ... reference of the auxiliary control loop
''     Q2  ... parameters of the auxiliary PI controller
''     TS2 ... sampling time in the auxiliary control loop
''     EP  ... error in the auxiliary control loop
''     FI  ... controlling flow rate to the first tank
''     H1  ... head in the first tank
''     H2  ... head in the second tank
''     H3  ... controlled head in the third tank
''-----------------------------------------------------------------
''
''   duration of the simulation run
''
CONSTANT TFIN=30
''-----------------------------------------------------------------
''
''   main controller
''
ARRAY Q1(3),STATE1(3)
CONSTANT STATE1=3*0.
CONSTANT Q1=2.6723,-3.3452,1.036
CONSTANT W=1.,TS1=4.
E=W-H3
U=DPID(E,Q1,STATE1,TS1)
''-----------------------------------------------------------------
''
''   auxiliary controller
''
ARRAY Q2(3),STATE2(3)
CONSTANT STATE2=3*0.,STATE3=0.
CONSTANT Q2=2.,-1.85,0.
CONSTANT TS2=1.
H1V=SH(H1,STATE3,TS2)
EP=U-H1V
FI=DPID(EP,Q2,STATE2,TS2)
```

Figure 7.5 SIMCOS program for the simulation of the cascade control scheme (Part 1 of 2).

discrete submodels in the source program (CONTINUOUS SYSTEM, DISCRETE SYSTEM). The defined submodels are then connected by the compiler which uses the so-called CONNECTING SYSTEM to generate the simulation program. The built-in sorting algorithm then sorts the submodels. All the variables inside the

```
''------------------------------------------------------------
''
''  hydraulic process
''
H1=INTEG(FI/7.5-H1/7.5,0.)
H2=INTEG(H1/10.-H2/10.,0.)
H3=INTEG(H2/5.-H3/5.,0.)
''------------------------------------------------------------
''
''  simulation run control
''
NSTEPS NST=2
TERMT(T.GT.TFIN)
CINTERVAL CI=0.05
ERRTAG IERR
''
''  documentation
''
HDR CASCADE CONTROL
OUTPUT U,FI,H1,H2,H3
PREPAR U,FI,H1,H2,H3
END
```

Figure 7.6 SIMCOS program for the simulation of the cascade control scheme (Part 2 of 2).

```
continuous system HYDR
''
''  hydraulic system
''
''      FI ... controlling flow rate to the first tank
''      H1 ... head in the first tank
''      H2 ... head in the second tank
''      H3 ... controlled head in the third tank
''
''  input into the submodel
      input FI
''  outputs from the submodel
      output H1 H3
''
state HS1 HS2 HS3
der H1D, H2D, H3D
H1=HS1
H3=HS3
H1D=FI/7.5-HS1/7.5
H2D=HS1/10.-HS2/10.
H3D=HS2/5.-HS3/5.
HS1:0
HS2:0
HS3:O
END
```

Figure 7.7 SIMNON program for the simulation of the cascade control scheme (Part 1 of 4).

```
discrete system MAIN
''
''  main controller
''
''      W   ... reference
''      Q01,Q11,Q21 ... parameters of the main PID controller
''      TS1 ... sampling time in the auxiliary control loop
''      E   ... error in the main control loop
''      U   ... reference of the auxiliary control loop
''      H3  ... controlled head in the third tank
''
''  inputs into the submodel
     input W H3
''  output from the submodel
     output U
''
state W1 W2
new W1P W2P
time T
tsamp T1
E=W-H3
W1P=E+W1
U=Q01*W1P+Q11*W1+Q21*W2
W2P=W1
T1=T+TS1
TS1:  4.
Q01:  2.6723
Q11: -3.3452
Q21:  1.036
end
```

Figure 7.8 SIMNON program for the simulation of the cascade control scheme (Part 2 of 4).

submodels are of local type, while in the connecting program the submodel name must also be added to the variable name.

Comparing both languages, the following can be concluded:

- Simulation model generation in SIMNON requires more time than in SIMCOS, due to the fact that the programs for the main and auxiliary controller in SIMNON have to be completely written while in SIMCOS they are already in the library.
- If the same structure is used more times, several submodels must be generated in SIMNON whereas in SIMCOS the same discrete transfer function can be called several times with different inputs, outputs, parameters and states, of course.
- The simulation model generation for SIMNON corresponds to the macro capability in the language ACSL. But the macro can be called several times with different inputs, outputs, parameters and states as in SIMCOS. ACSL, however, does not allow direct definition of the discrete macro.

```
discrete system AUX
''
''  auxiliary controller
''
''      U   ... reference of the auxiliary control loop
''      Q02,Q12,Q22 ... parameters of the auxiliary PI controller
''      TS2 ... sampling time in the auxiliary loop
''      EP  ... error in the auxiliary control loop
''      FI  ... controlling flow rate to the first tank
''      H1  ... head in the first tank
''
''  inputs into the submodel
     input U H1
''  output from the submodel
     output FI
''
state W2
new W2P
time T
tsamp T2
EP=U-H1
W2P=EP+W2
FI=Q02*W2P+Q12*W2
T2=T+TS2
TS2:  1.
Q02:  2
Q12: -1.85
end
```

Figure 7.9 SIMNON program for the simulation of the cascade control scheme (Part 3 of 4).

```
connecting system CONNECT
''
''      W    ... reference
''      H3   ... controlled head in the third tank
''      U    ... reference of the auxiliary control loop
''      H1   ... head in the first tank
''      FI   ... controlling flow rate to the first tank
''      main ... main controller submodel
''      aux  ... auxiliary controller submodel
''      hydr ... hydraulic process
''
time T
W[main]=1
H3[main]=H3[hydr]
U[aux]=U[main]
H1[aux]=H1[hydr]
FI[hydr]=FI[aux]
end
```

Figure 7.10 SIMNON program for the simulation of the cascade control scheme (Part 4 of 4).

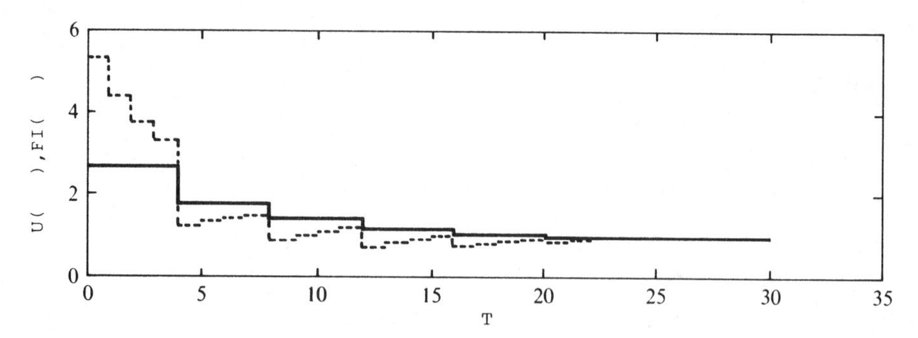


Figure 7.11 Outputs of the main ($T_{s1} = 4$ s) and auxiliary ($T_{s2} = 1$ s) controller.

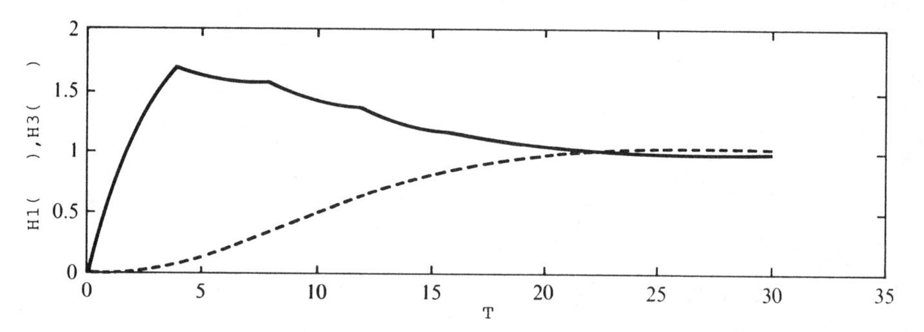


Figure 7.12 Time courses of the heads in the first (dashed line) and third tank for the unit reference signal.

- The simulation language SIMNON treats continuous and discrete systems in the same way. Therefore, it is more modular, particularly for hybrid models, than are ACSL and SIMCOS which are based on continuous simulation principles and, additionally, ACSL/SIMCOS extended with discrete capabilities.
- SIMCOS enables the simple use of the linear discrete transfer functions which were programmed in advance, while arbitrary nonlinear discrete submodels can be realized in SIMNON.

### Simulation results

The simulation results obtained with both simulation languages are the same. They are shown in Fig. 7.11 as well as in Fig. 7.12.

## 7.2 DISTILLATION COLUMN MODELLING

In the present case study we intend to:

- show the approach to the modelling of various rectification and distillation devices;
- show the procedure of theoretical modelling in the sense of model linearization and order reduction;
- stress the successive introduction of the assumptions, omissions and approximations which must be justified according to the modelling aims;
- show the use of the linear reduced order model obtained in the multivariable controller design procedure;
- show the testing of the designed controller with the aid of the nonlinear distillation column model.

### 7.2.1 An Approach to Model Development for Various Types of Distillation and Rectification Devices

Distillation is a typical energy-consuming process. Even for a nonproblematic distillation, successful control of the composition of both top and bottom products can yield substantial profit resulting from the potential saving in utility cost. Of course, no human supervision is needed in such cases. For the separation of components on the basis of volatility, different kinds of distillation are used in the chemical industry (Perry and Chilton, 1973; Shinskey, 1984; Stephanopoulos, 1984). The process under consideration, as shown in Fig. 7.13, consists of a reboiler where approximately constant vapour flow rate is ensured through corresponding heating. The vapour from the reboiler goes up through the column, gives up part of its energy at each plate and aids the vapourization of more volatile components. The role of the distillation column is to separate the mixture so that the distillate is of the prescribed concentration. In the condenser, sufficient heat exchange must occur as to ensure that all the vapour will condense. However, there is no need to cool the distillate, because part of the liquid is returned to the top of the column as the reflux flow with a temperature near to boiling point. This is done by the reflux distributor, which returns one part of the liquid from the condenser to the column, the other part being drawn off as a top product.

The quantities given in Fig. 7.13 are as follows: $L_i$ = liquid flow rate leaving plate (Kmol/s); $V_i$ = vapour flow rate leaving plate (Kmol/s); $x_i$ = liquid molar composition of more volatile component on plate $i$; $y_i$ = vapour molar composition of more volatile component on plate $i$; $M_i$ = liquid holdup on plate $i$ (Kmol); $m_i$ = vapour holdup above plate $i$ (Kmol).

The distillation devices which can in general separate two (binary distillation) or more components are in most cases used as *continuous distillation columns* where the mixture is fed continuously into the column at a prescribed position somewhere along the column. The index $f$ in Fig. 7.13 denotes the quantities associated with

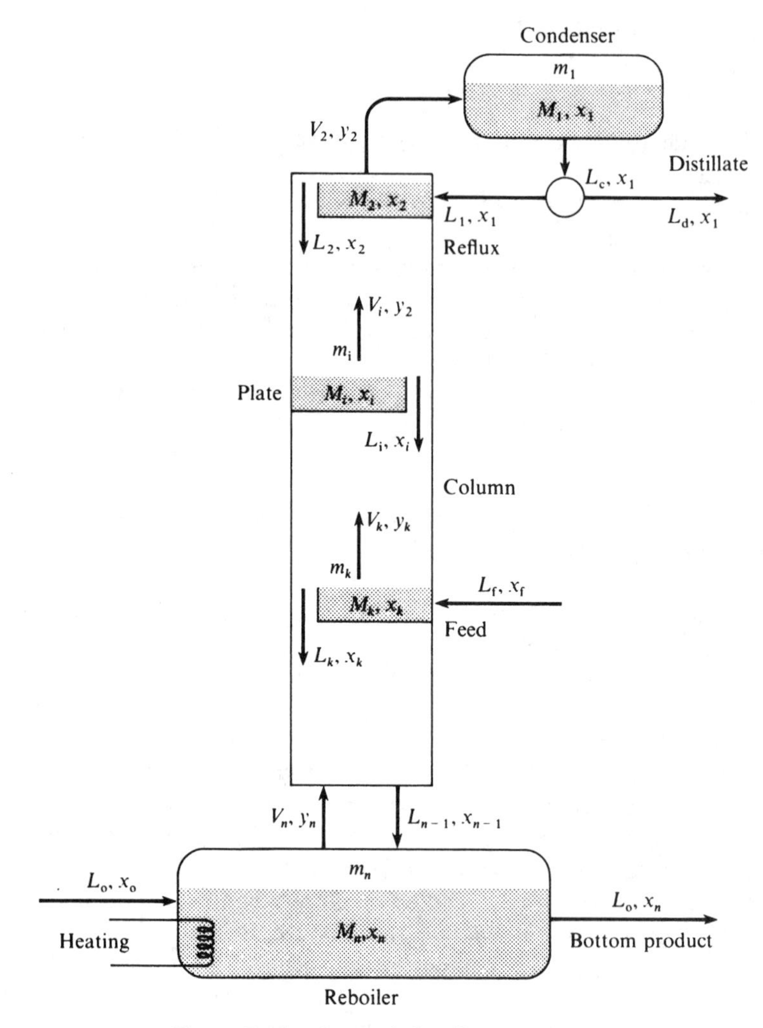


Figure 7.13 General distallation column.

this type of distillation. In the case of small production and frequently changing mixture, however, *batch rectification* (distillation is often called rectification for the batch-type régime of column operation) is suitable. Here, the mixture to be distillated is put into the reboiler and the separation procedure lasts till the concentration and/or mass of the mixture in the reboiler becomes so small that the prescribed quality of the distillate can no longer be attained. The fact that any impurities in the mixture remain in the reboiler and thus the column is not soiled, as in continuous operation, can be also regarded as an advantage. In some cases, this type of distillation can be improved by the use of *semibatch rectification*. As distinct from the batch operation, here the reboiler is continuously fed with the same flow of mixture as the flow of distillate which is fed into the accumulator. The advantages of this type of distillation are the enlarged mass of the final product and the constant level of the mixture in the reboiler, which, as a consequence, ensures approximately constant vapour flow rate (Atanasijević, Karba, Milanović *et al.*, 1986). Feed from such a régime of operation is denoted by the index $o$ in Fig. 7.13.

As suggested by the theoretic modelling procedure, the definition of the corresponding subsystems must be made initially. In column-type devices the basic subsystem is obviously the plate. So in the development of the nonlinear model which describes the time courses of the concentrations of the more volatile component along the column, the mass balance equations can be used as the basic relation. Taking into account the usual assumptions that the vapour holdup ($m_i$) is negligible and that the mixing of the liquid on the plate is complete and instantaneous, the total mass balance equation for the $i$th plate is

$$\frac{dM_i(t)}{dt} = L_{i-1}(t) - L_i(t) + V_{i+1}(t) - V_i(t) \tag{7.34}$$

while the mass balance equation for the more volatile component is given in the form

$$\frac{d(M_i(t)x_i(t))}{dt} = L_{i-1}(t)x_{i-1}(t) - L_i(t)x_i(t) + V_{i+1}(t)y_{i+1}(t) - V_i(t)y_i(t) \tag{7.35}$$

These two equations represent the basis for the mathematical model development of the three types of distillation. Another simplification can be made here: the relatively long distances and associated vapour or liquid transport times between condenser and column as well as between reboiler and column can be neglected and so the condenser can be observed as the first plate, while the reboiler can be looked upon as the $n$th plate.

## Nonlinear model for continuous distillation

Here the mixture is fed with flow rate $L_f$ and concentration $x_f$ to the $k$th plate, as shown in Fig. 7.13 (in this case $L_o = 0$). According to Eqns (7.34) and (7.35) the following relation can be written for the condenser (first plate):

$$\frac{dM_1(t)}{dt} = V_2(t) - L_c(t) \tag{7.36}$$

and

$$\frac{d(M_1(t)x_1(t))}{dt} = V_2(t)y_2(t) - L_c(t)x_1(t) \tag{7.37}$$

while the flow rate of the distillate to the accumulator is given by the relation

$$L_d(t) = L_c(t) - L_1(t) \tag{7.38}$$

and a similar relation can be written (multiplication of flow rate by the corresponding concentration $x_1(t)$) for the more volatile component.

For the arbitrary plate $i$ above the position of the mixture feed, Eqns (7.34) and (7.35) are valid. However, for the plate where the mixture is fed into the column, the following equations can be developed:

$$\frac{dM_k(t)}{dt} = L_f(t) + L_{k-1}(t) - L_k(t) + V_{k+1}(t) - V_k(t) \tag{7.39}$$

$$\frac{d(M_k(t)x_k(t))}{dt} = L_f(t)x_f(t) + L_{k-1}(t)x_{k-1}(t) - L_k(t)x_k(t) + V_{k+1}(t)y_{k+1}(t) - V_k(t)y_k(t) \tag{7.40}$$

For the arbitrary plate $j$ beneath the position of the mixture feed Eqns (7.34) and (7.35) are again valid for the index $j$. Finally, the equations for the reboiler ($n$th plate) are given in the form:

$$\frac{dM_n(t)}{dt} = L_{n-1}(t) - L_a(t) - V_n(t) \tag{7.41}$$

$$\frac{d(M_n(t)x_n(t))}{dt} = L_{n-1}(t)x_{n-1}(t) - L_a(t)x_n(t) - V_n(t)y_n(t) \tag{7.42}$$

**Nonlinear model for batch rectification**

For such operations the following equalities must be considered:

$$L_f(t) = L_o(t) = L_a(t) = 0$$

The dynamics of the condenser are described by Eqns (7.36), (7.37) and (7.38), while for the other plates along the column Eqns (7.34) and (7.35) are valid. The equations for the reboiler are given as

$$\frac{dM_n(t)}{dt} = L_{n-1}(t) - V_n(t) \tag{7.43}$$

$$\frac{d(M_n(t)x_n(t))}{dt} = L_{n-1}(t)x_{n-1}(t) - V_n(t)y_n(t) \tag{7.44}$$

**Nonlinear model for semibatch rectification**

This type of rectification prolongs the batch process by feeding the mixture in the reboiler, obeying the relation

$$L_d(t) = L_o(t) \tag{7.45}$$

where, of course, only the equality of the flow rates are considered. Therefore, no top product (distillate) is fed into the reboiler but, rather, only the same flow rate of mixture as already mentioned.

A similar relation (multiplication of the flow rates by the corresponding concentrations) is valid for the more volatile component:

$$L_d(t)x_1(t) = L_o(t)x_o(t) \tag{7.46}$$

The equations for the condenser are as Eqns (7.36), (7.37) and (7.38) and, for the plates along the column, as Eqns (7.34) and (7.35) respectively. The states in the reboiler are described by the following equations:

$$\frac{dM_n(t)}{dt} = L_o(t) + L_{n-1}(t) - V_n(t) \tag{7.47}$$

$$\frac{d(M_n(t)x_n(t))}{dt} = L_o(t)x_o(t) + L_{n-1}(t)x_{n-1}(t) - V_n(t)y_n(t) \tag{7.48}$$

### 7.2.2 Nonlinear Model Development for the Semibatch Distillation Column

In the process of dissolvents regeneration, the task is to separate methanol from water. These components are usually mixed in an impure lye and the semibatch distillation should enable the successful regeneration of methanol. Repeated use of the dissolvents means a substantial profit for the industry. In our case, the rectification device contains eight plates including the condenser and the reboiler ($n = 8$). As the modelling goal is to develop a model which can be used for the corresponding control design, the following additional assumptions are proposed:

- The liquid holdup is constant for particular plates. This assumption is justified, especially for describing the column behaviour in the steady state (not in the phase of putting the column in operation), which is usually the case. This assumption implies

$$M_i = const. \qquad i = 1, 2, \ldots, n \tag{7.49}$$

- The vapour flow rate along the column is approximately constant (it depends mainly on the heating in the reboiler if the device is thermally insulated). Therefore,

$$V_2(t) = V_3(t) = \cdots = V_n(t) = V(t) = f(Q) \tag{7.50}$$

- The liquid flow rate is a function of the reflux rate but is in general independent of location along the column. Thus the following equality can be written:

$$L_1(t) = L_2(t) = \cdots = L_{n-1}(t) = L(t) \tag{7.51}$$

The equations from the preceding subsection are simplified, taking into account the additional assumptions. For our case they have the form:

$$\frac{dx_1(t)}{dt} = \frac{1}{M_1}[V(t)y_2(t) - L_c x_1(t)] = \frac{1}{M_1}[y_2(t) - x_1(t)]V(t) \tag{7.52}$$

$$\frac{dx_i(t)}{dt} = \frac{1}{M_i}\{L(t)[x_{i-1}(t) - x_i(t)] + V(t)[y_{i+1}(t) - y_i(t)]\} \tag{7.53}$$

$$\frac{dx_n(t)}{dt} = \frac{1}{M_n}[L_o(t)x_o(t) + L(t)x_{n-1}(t) - V(t)y_n(t)]$$

$$= \frac{1}{M_n}\{[V(t) - L(t)]x_o(t) + L(t)x_{n-1}(t) - V(t)y_n(t)\} \tag{7.54}$$

The set of $n$ first order nonlinear differential equations obtained (Eqns (7.52)–(7.54) can be solved if the relation between the liquid and vapour phases on the plate is known, which can be expressed in the form

$$y_i = f_i(x_i) \tag{7.55}$$

The function connection in Eqn (7.55) represents one of the most important problems in distillation modelling. Here, a compromise between model simplicity and consideration of available data on the mixture properties and device characteristics must be found. Not to delve too deeply into the variety of possibilities which exist for the solution of the above problem let us mention only some of them (Perry and Chilton, 1973; Stephanopoulos, 1984; Buckley *et al.*, 1985):

- Finally, the relationships may in some cases be obtained from the *vapour–liquid equilibrium tables* which exist for certain mixtures and are obtained by complex and precise measurements.
- The *relative volatility*, being associated with the concept of plate efficiency, is also used as a kind of relation which provides the data on the $x$ and $y$ connections, especially when the relative volatility is supposed to be constant.
- The *distribution coefficients or k values*, which are defined as

$$k_i = \frac{y_i}{x_i}$$

  enable the successful description of the $x$ and $y$ relations. They are, of course, different for particular plates and are also time variable, which is in fact a consequence of changing conditions on the plates (temperatures and pressures). However, they can be taken as constant for the corresponding time intervals.
- Finally, the relationships may in some cases be obtained from the *vapour–liquid equilibrium tables* which exist for certain mixtures and are obtained by complex and precise measurements.

In our case the distribution coefficients approach is the most appropriate. Therefore, Eqns (7.52)–(7.54) are further simplified into the form

$$\frac{dx_1(t)}{dt} = \frac{1}{M_1}[k_2 x_2(t) - x_1(t)]V(t) \tag{7.56}$$

$$\frac{dx_i(t)}{dt} = \frac{1}{M_i}\{L(t)[x_{i-1}(t) - x_i(t)] + V(t)[k_{i+1}x_{i+1}(t) - k_i x_i(t)]\} \tag{7.57}$$

$$\frac{dx_n(t)}{dt} = \frac{1}{M_n}\{[V(t) - L(t)]x_o(t) + L(t)x_{n-1}(t) - V(t)k_n x_n(t)\} \tag{7.58}$$

The data base for the modelling is as follows:

- The concentration of the more volatile component in the mixture being fed into the reboiler is $x_o = 0.315$.
- The liquid holdups for the particular plates were estimated on the basis of data about the device (physical dimensions) as

  $M(1) = 105.99$; $M(2) = 5.35$; $M(3) = 5.33$; $M(4) = 5.29$

  $M(5) = 5.22$; $M(6) = 5.08$; $M(7) = 4.84$; $M(8) = 5441.13$
- The liquid and vapour flow rates, respectively, are given in Fig. 7.14 and represent the inputs to the model.
- The distribution coefficients were calculated for the particular plates and time intervals between the data for $x_i$, where they are supposed to be constant, from the corresponding vapour–liquid equilibrium table for the mixture of methanol and water (Gmehling and Onken, 1977). Typical results for the distribution coefficients are given in Fig. 7.15.
- The data for $x_i$ on the particular plates are shown, together with the model responses obtained ($x$-axis: time (s)) (Fig. 7.16).

As can be seen in Fig. 7.16, the nonlinear model responses (in this phase of modelling, all $x_i$ are interesting) are in relatively good agreement with the reference data, which indicates that the model is appropriate (this is also valid for the chosen assumptions).

### 7.2.3 Model Linearization

Nonlinear distillation column model simulation showed relatively slow changes of the more volatile component concentrations on the plates. This fact justifies the assumption that the system can be approximated by linearized models for the corresponding time intervals. Linearization by considering only first order terms in the Taylor series, which has already been mentioned in Section 7.1, transforms Eqns (7.56)–(7.58) into the following system of linear first order differential

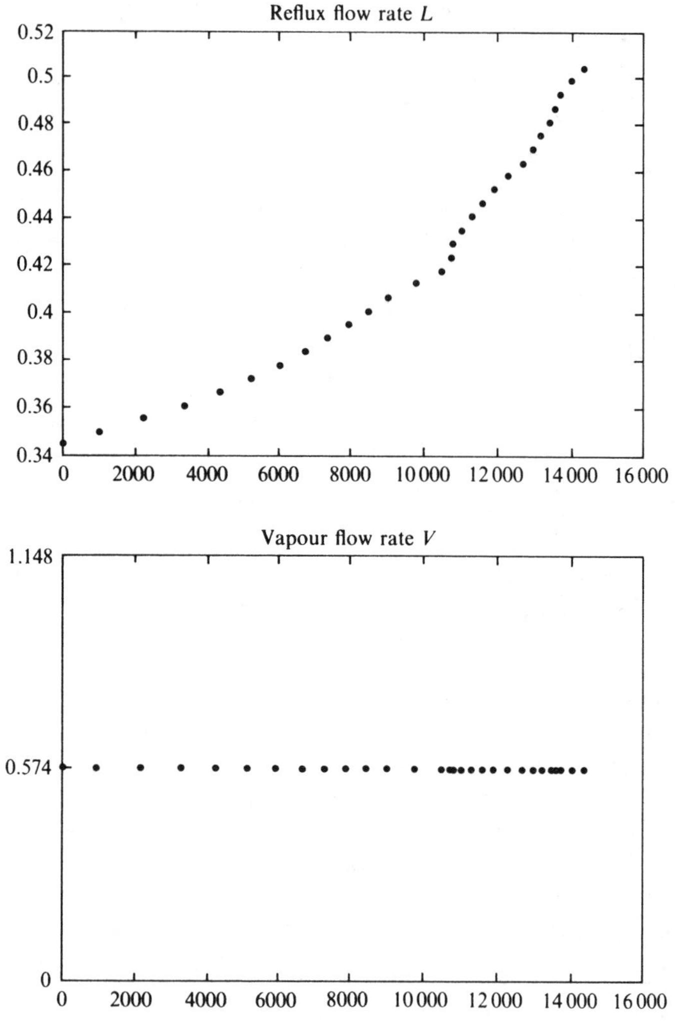


Figure 7.14 The data for $L$ and $V$.

equations:

$$\frac{d(\Delta x_1(t))}{dt} = \frac{1}{M_1}[-\bar{V}\,\Delta x_1(t) + (k_2\bar{V}\,\Delta x_2(t) + (k_2\bar{x}_2 - \bar{x}_1)\,\Delta V(t)] \tag{7.59}$$

$$\frac{d(\Delta x_i(t))}{dt} = \frac{1}{M_i}[\bar{L}\,\Delta x_{i-1}(t) + (-\bar{L} - k_i\bar{V})\,\Delta x_i(t) + k_{i+1}\bar{V}\,\Delta x_{i+1}(t) + (\bar{x}_{i-1} - \bar{x}_i)\,\Delta L + (k_{i+1}\bar{x}_{i+1} - k_i\bar{x}_i)\,\Delta V(t)] \tag{7.60}$$

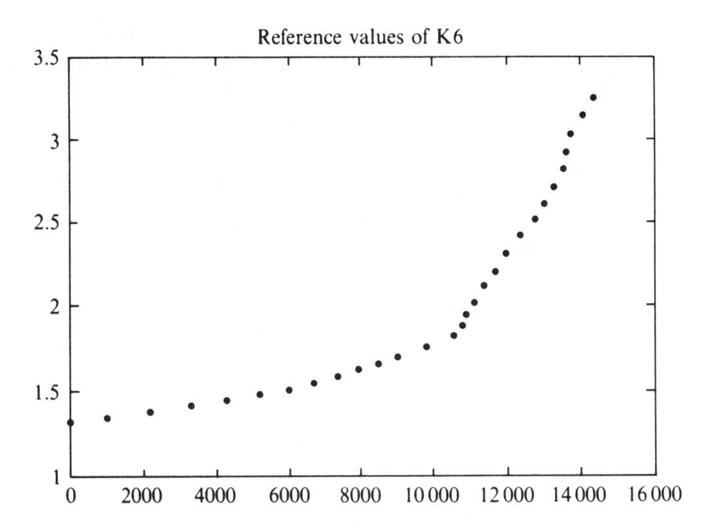


Figure 7.15 Typical data for the distribution coefficients (on the sixth plate $k_6$).

$$\frac{d(\Delta x_n(t))}{dt} = \frac{1}{M_n}[\bar{L}\,\Delta x_{n-1}(t) - \bar{V}k_n\,\Delta x_n(t) + (\bar{x}_{n-1} - \bar{x}_o)\,\Delta L(t) + (\bar{x}_o - k_n\bar{x}_n)\,\Delta V(t) + (\bar{V} - \bar{L})\,\Delta x_o(t)] \tag{7.61}$$

where $\bar{L}$ and $\bar{V}$ represent average values between two samples defining the particular working point.

These equations can also be expressed in the well-known state space formulation:

$$\dot{x} = \mathbf{A}x + \mathbf{B}u \tag{7.62}$$

$$y = \mathbf{C}x \tag{7.63}$$

Matrices **A**, **B** and **C** are constant. In the above description the state vector contains eight state variables $\Delta x_i$, $i = 1, \ldots, n$, while the input vector includes two controlling inputs $\Delta L(t)$ and $\Delta V(t)$, respectively, and the third one $(\Delta x_o(t))$, which represents the disturbance signal as the changes in the mixture concentration cannot usually be controlled. The first input, on the other hand, can be controlled by the reflux distributor, while the second depends on heating, which can of course also be controlled. It is obvious that the changes in mixture concentration have a greater influence on system behaviour at the beginning of the distillation process than later when the reflux rate is significantly increased. Among the concentrations of the more volatile component on particular plates, which are measured, the $\Delta x_2$ and $\Delta x_7$ were chosen to represent the outputs. It seems that a more logical choice would be the quantities $\Delta x_1$ and $\Delta x_8$ for which quality requirements are prescribed and which represent the products of the distillation. Our decision about the outputs which, in fact, are nearly of the same concentrations as in the condenser and reboiler, is based

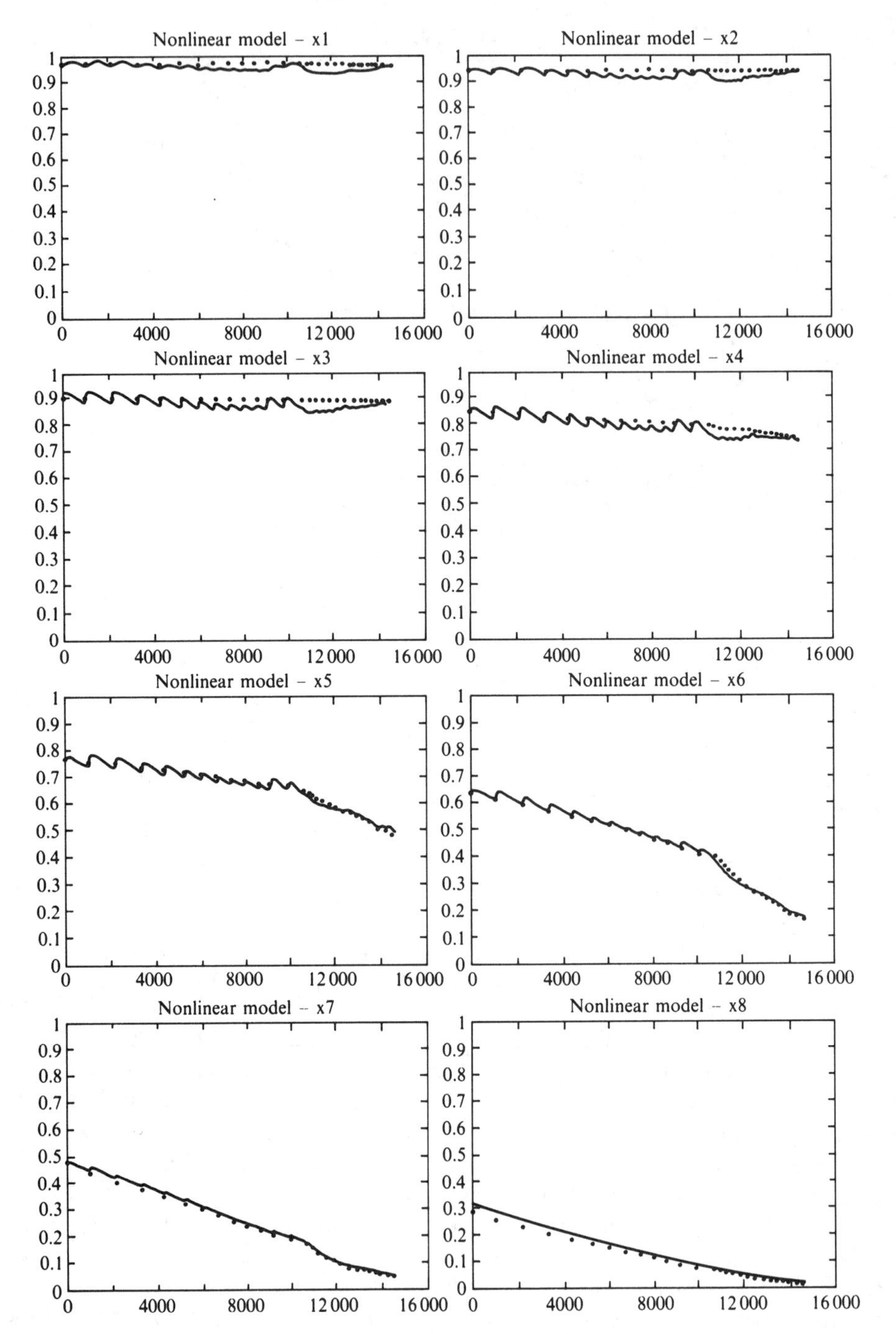


Figure 7.16 Reference data (circles) and nonlinear model responses (curves) for the particular plates of the modelled distillation column.

on the following facts:

- The time constants of the reboiler and condenser are significantly larger than those of the plates, which is caused by $M_1$ and $M_8$ as compared to $M_i$.
- The concentration of the more volatile component in the condenser is high.

Due to the facts itemized above, the eventual disturbances would need a relatively long time to be recognizable in the reboiler and condenser. Therefore, our choice of outputs is justified. From the stated model it can be seen that the semibatch distillation column is a multivariable system with two inputs and two outputs. The choice of the latter is reflected in the elements equal to 1 on the corresponding places in the $(2 \times 8)$ matrix **C** (Šega *et al.*, 1986).

Overlined quantities in Eqns (7.59)–(7.61) represent the values at working points around which the linearization is performed. It is clear that only one working point cannot cover all of the observation time. Therefore, a decision must be made as to how many working points are to be taken into account in the simulation. It is of course more advantageous to consider fewer working points but, on the other hand, this can cause a worse fit between the model response and the reference data. Some compromise must thus be found, which also depends on the modelling aims. Here we decided to suppose constant quantities between the particular reference data points, which means that twenty-eight working points were taken into account (the parameters are therefore the average quantities between two samples) and that the parameters were switched twenty-eight times to the correspondingly modified values during the simulation run. The results of the linearized model simulation can be seen in Fig. 7.17.

A very good fit between the model responses and the reference data in Fig. 7.17 shows that our choice of working points was suitable but, on the other hand, it is obvious that a more rational choice is possible from the point of view of the number of working points. This is especially true for the first half of the observation interval where the changes of the particular quantities are slow and where only a few working points would describe the system's behaviour just as well.

### 7.2.4 Model Order Reduction

As the goal in this case is the model development for control design purposes, the use of the model order reduction procedure is valid. Lower order models which still describe system behaviour satisfactorily are suitable for use in control design algorithms. However, after such simplification the model only provides information about the input–output relations for certain conditions of operation. In our case it would be preferable to obtain a low order model for the whole time of observation, instead of several linear models, as was the case in the preceding subsection. Therefore, we decided to use the black-box identification method in the discrete space included in the CACSD tool ANA (Šega *et al.*, 1985; Atanasijević *et al.*, 1985). As this method uses equidistant samples the data were taken from the responses of the linearized

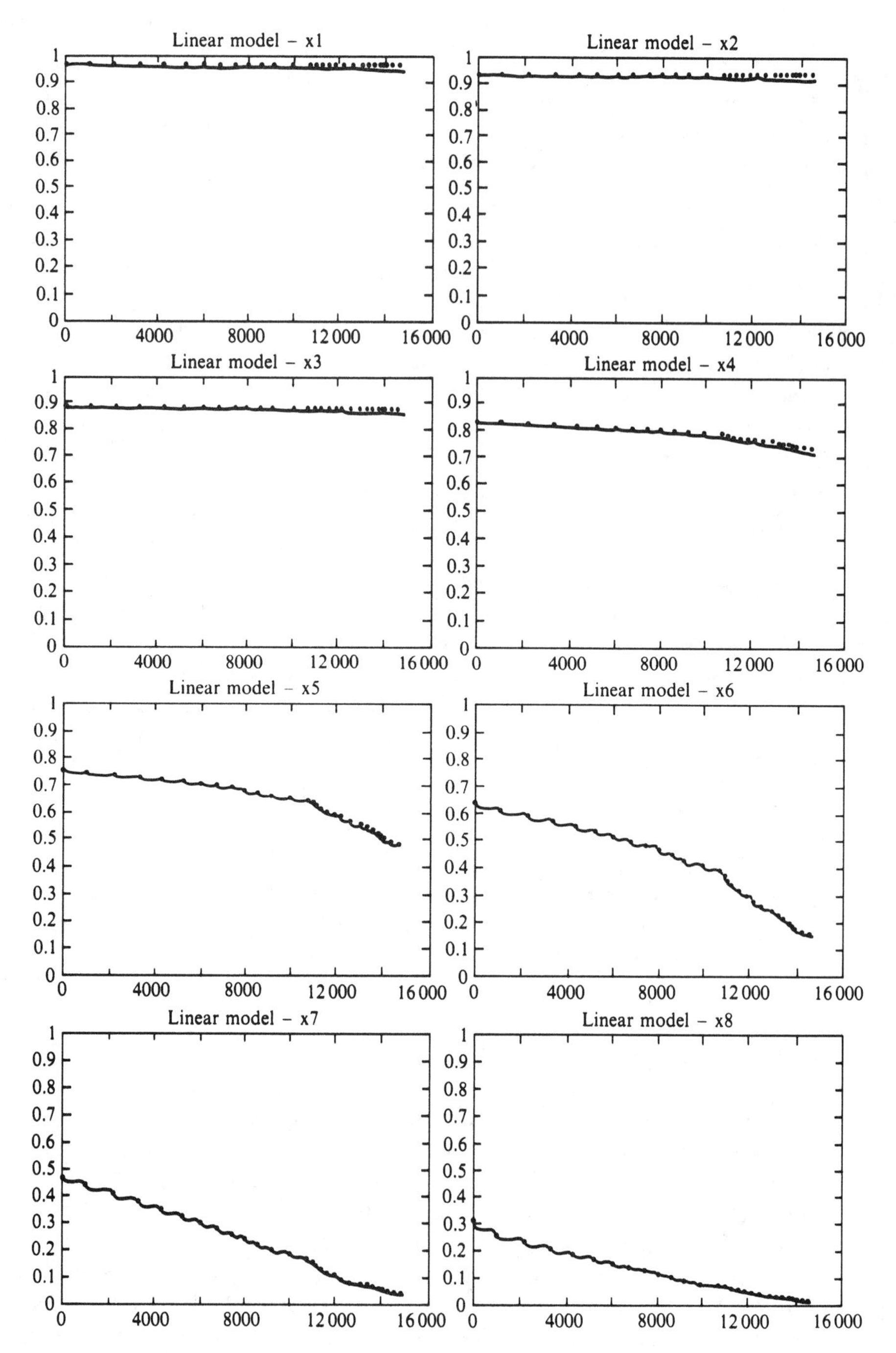


Figure 7.17 Reference data (circles) and linearized model responses (curves) for the particular plates of the modelled distillation column.

model and not directly from the reference data, which is justifiable because of the very good fit of the linearized model. This model was excited by the input disturbances represented by the deviations of $L$ and $V$ from the steady state values.

The result of identification was the third order model given by the following matrices:

$$\mathbf{A} = \begin{bmatrix} -0.0381 & 0.0394 & -0.0048 \\ -0.0200 & 0.0191 & 0.0035 \\ -0.0184 & 0.0204 & -0.0092 \end{bmatrix} \qquad \mathbf{B} = \begin{bmatrix} 0.0077 & -0.0043 \\ 0.0046 & -0.0022 \\ 0.0039 & -0.0019 \end{bmatrix}$$

$$\mathbf{C} = \begin{bmatrix} 1 & 0 & 0 \\ 0 & 0 & 1 \end{bmatrix} \qquad \mathbf{x}(0) = \begin{bmatrix} 0.042 \\ 0.041 \\ 0.010 \end{bmatrix} \qquad \mathbf{y}_{\text{off}} = \begin{bmatrix} -0.042 \\ -0.010 \end{bmatrix}$$

where vectors $\mathbf{x}(0)$ and $\mathbf{y}_{\text{off}}$ bring the system to the working point and the state space model describes its dynamics. The input signals and model responses, together with the responses of the linearized model representing the reference data, are shown in Fig. 7.18.

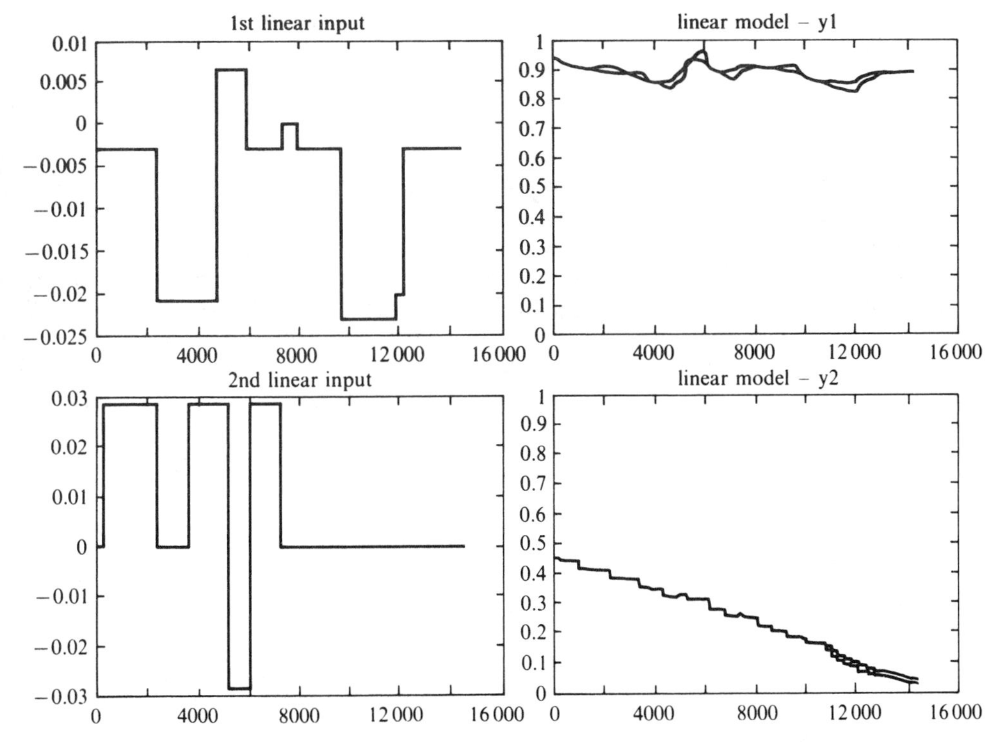


Figure 7.18 Input signals and reduced order model responses together with the results of the linearized model representing the reference data.

As the reduced order model responses satisfactorily fit the reference data regarding the use of the model, this can be the basis of the control design procedure.

### 7.2.5 Multivariable Controller Design for the Semibatch Distillation Column

In control theory, the methods for multivariable controller design represent a special area. Without going into detail, let us simply state that in our case we decided to use the so-called *pole assignment* method and that the multivariable control was designed with the aid of the corresponding program equipment (Atanasijević, Karba, Bremsǎk *et al.*, 1986; Atanasijević, Karba, Milanović *et al.*, 1986; Šega *et al.*, 1986). The block diagram of the control structure studied is shown in Fig. 7.19.

As can be seen in Fig. 7.19 the input disturbances ($DL$, $DV$) are included in the structure, enabling evaluation of the designed control quality. This was performed with the aid of the relationship between the uncontrolled nonlinear model responses and the reference data, which is compared with the similar relationship between the controlled nonlinear model responses and the reference data. First, the proportional multivariable controller was designed. The desired locations of the closed loop system poles were defined as the criteria at $[-0.01, -0.09, -0.08]$. The controller matrix $\mathbf{K}_P$ was calculated to be

$$\mathbf{K}_P = \begin{bmatrix} 0.552 & -3.185 \\ 0.029 & -3.654 \end{bmatrix}$$

The controller obtained was used in the closed loop system simulation, where the nonlinear model of the system was taken into account. The results are shown in Fig. 7.20, where the left side represents the behaviour of the uncontrolled system under the influence of disturbances and the right side represents the effects of the designed proportional controller, which tends to maintain the closed loop responses

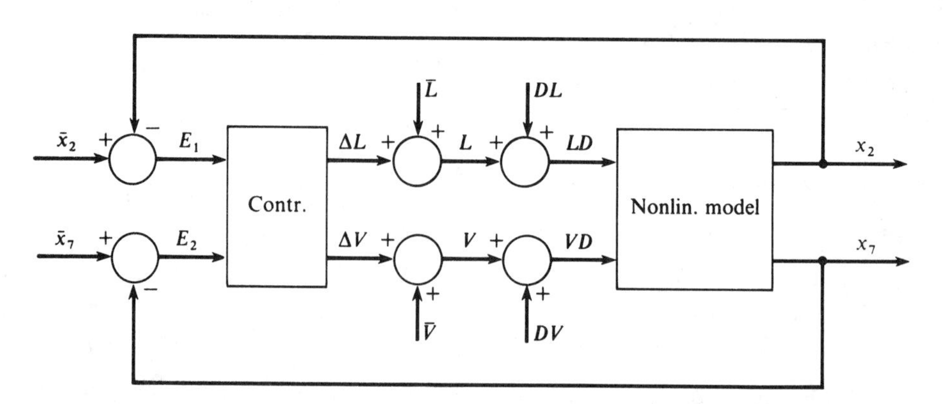


Figure 7.19 Structure of the semibatch distillation column control.

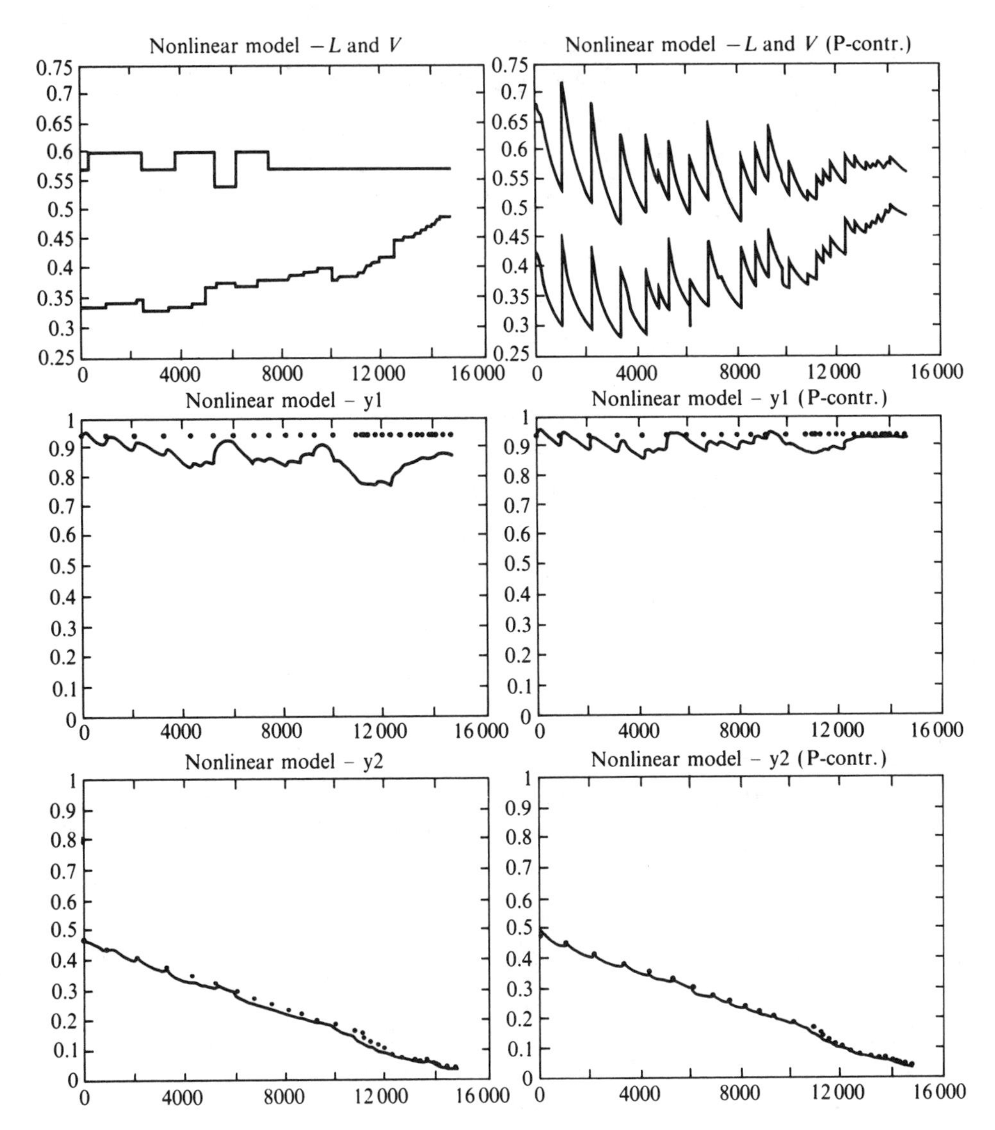


Figure 7.20 System inputs (*LD* and *VD*) and outputs ($x_2$ and $x_7$) time responses for uncontrolled (left side) and P-controlled system (right side) using a nonlinear model together with the reference date (circles).

as near to the reference data as possible. From Fig. 7.20 it can be seen that the use of the designed controller ensures better quality of column operation.

In the next step of controller design, the integral part was added to the proportional part. The controller matrix $\mathbf{K}_I$ was defined in the tuning procedure,

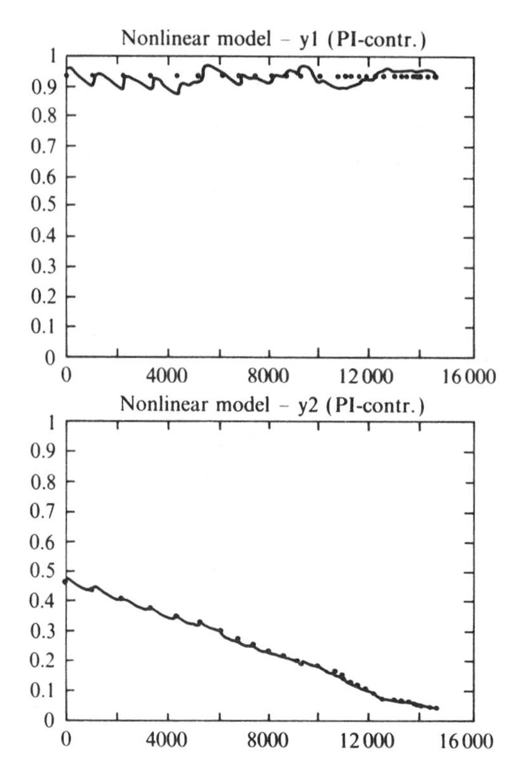


Figure 7.21 System outputs time responses using proportional-integral multivariable controller using a nonlinear model of the system together with the reference date (circles).

using the nonlinear model, as

$$\mathbf{K}_I = \begin{bmatrix} 0.0002 & 0.0002 \\ -0.0001 & 0.0003 \end{bmatrix}$$

The simulation results for such proportional-integral control are presented in Fig. 7.21. The results in Fig. 7.21 show that further improvement in distillation column behaviour is achieved by the designed PI controller. However, for more accurate information about column operation the controllers should be designed on the basis of low order models for the corresponding number of time subintervals in the observed time interval. Note that in the simulation study it transpired that the designed controller is robust against changes in the more volatile component concentrations in the mixture.

Due to the complexity of the models studied nothing was said in this case study about simulation and the problems associated with it. More about these aspects can be found in the literature (see Kleinert *et al.*, 1988).

## 7.3 CASE STUDY APPROACH IN DRUG PHARMACOKINETICS

In this example we wish to:

- illustrate the use of modelling and simulation in biomedicine;
- introduce the compartmental modelling approach briefly;
- stress the need for model validation; and
- comment on some properties and the role of analog and digital simulation in pharmacokinetic studies.

### Theoretical background and problem statement

When explaining some possible approaches to the modelling and simulation of drug pharmacokinetics under conditions of haemodialysis, some rudiments must first be given.

Selection of the most suitable drugs, as well as rational therapy design, are greatly dependent on the pharmacokinetical properties of the chosen dosage form. Pharmacokinetics deals with the rates of reactions of the drug in the human body and with their mechanisms. Passage of the drug from dosage form to the site of action is very complicated and is influenced by many biological and physicochemical factors. For the pharmacological action the rate of the drug appearance at the site of action is very important. This parameter is usually proportional to the rate of drug passage to plasma and so the drug concentration time responses in plasma are in general the most relevant data for the study of the commencement, efficacy and duration of the drug pharmacological action, taking into account known minimal effective concentrations (Karba *et al.*, 1979). Although pharmacokinetics is a relatively young science it has some interdisciplinary character and is becoming increasingly important in biomedicine. The speedy development of pharmacokinetics was initiated by the growing possibilities of chemical analysis of biological materials and new technologies and materials in the manufacture of dosage forms on the one side, as well as the system approach to the improvement of dosage forms and dosage régimen design, including modelling and simulation, on the other side (Mrhar *et al.*, 1987). The latter enable significant rationalizations in the number of necessary 'in vivo' studies, which are expensive, time consuming and sometimes ethically questionable.

When studying drug pharmacokinetics the corresponding pharmacokinetic model must first be defined. The latter is a simplified representation of the drug passage through the human body after application.

One of the most frequently used approaches to pharmacokinetic modelling is *compartmental analysis*, which is also used increasingly in all areas of biomedicine, ecology and chemical reaction kinetics, while in engineering it transpired that many models were of the compartmental type in spite of the fact that they were not explicitly recognized as such. Compartmental models consist of a finite number of homogeneous, well-mixed, lumped subsystems, called compartments, which exchange with each other and with the environment so that the quantity or concentration of material within each compartment may be described by a first order differential equation (Godfrey, 1983). The most general form representing mass balances is

$$\frac{dx_i}{dt} = f_{io} + \sum_{\substack{j=1 \\ j\neq i}}^{p} f_{ij} - \sum_{\substack{j=1 \\ j\neq i}}^{p} f_{ji} - f_{oi} \qquad i = 1, \ldots, p \tag{7.64}$$

where $x_i$ is the amount of material in the $i$th compartment of a $p$-compartment model, $f_{ij}$ is the flow rate from compartment $j$ to compartment $i$ and subscript o denotes the environment. If the flow rates from all compartments to the environment are zero ($f_{oi} = 0$, $i = 1, \ldots, p$) the system is closed, while in all other cases it is open. Compartmental models may be viewed as sets of constrained first order differential equations where the constraints draw their origin from the physical factor that flow rates are nonnegative. A source compartment is one which does not receive from any other compartment but may or may not excrete to the environment or other compartments. Source compartments thus behave as one-compartment systems (single exponential decays), independently of the compartments in the rest of the system. On the other hand, a sink (cumulative) compartment is one which does not excrete either to the environment or to any other compartment. It thus represents the pure integrator of the incoming flows. The main theory is concerned with linear time-invariant compartmental models for which the flow rates are directly proportional to the quantity in the donor compartment. So $f_{ij}, j \neq i$ are replaced by the products $k_{ij}x_j$, where $k_{ij}$ are the *rate constants* with the dimension of reciprocal time. Eqn (7.64) thus becomes

$$\frac{dx_i}{dt} = \sum_{\substack{j=1 \\ j\neq i}}^{p} k_{ij}x_j - \sum_{\substack{j=1 \\ j\neq i}}^{p} k_{ji}x_i - k_{oi}x_i + u_i(t) \qquad i = 1, \ldots, p \tag{7.65}$$

where the input flow rate $f_{io}$ from the environment is denoted by $u_i(t)$. Note that the flow rate from compartment $j$ to compartment $i$ ($k_{ij}x_j$) is independent of the quantity in the receptor compartment ($x_i$). Compartmental models are often given in the form of the special block diagram shown in Fig. 7.22. The notation of the particular inputs and outputs in Fig. 7.22 are general. In the areas mentioned, however, other symbols are used as well as some possible modifications of the graphic

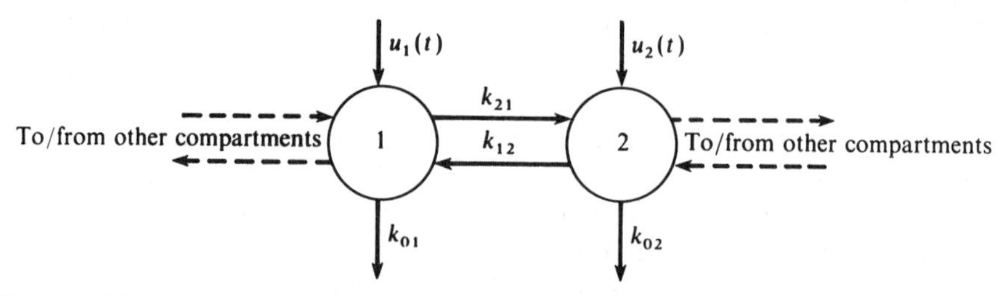


Figure 7.22 Two compartments of a linear, time invariant multicompartment model.

representation of the model and some variations in the mathematical model. In pharmacokinetics, for instance, it is sometimes convenient to define the excitation in the form of an initial dose or concentration which can be represented as the initial condition of the corresponding differential equation. One of the possible modifications of the mathematical model (Eqn (7.65)) is the so-called *null order* kinetics form, which is given by the following first order differential equation:

$$\frac{dx_i}{dt} = -k \qquad x(0) = x_0 \tag{7.66}$$

Unlike first order kinetics which gives results of exponential type, the result here is represented by a straight line from the value of the initial condition down to $x = 0$ (negative values of $x$ are, of course, impossible).

The transformation of the compartmental model in Fig. 7.22, which in pharmacokinetics is often called the pharmacokinetical model, to the mathematical one is, according to Eqn (7.65), very simple. For each compartment the first order differential equation is written where the products $kx$ for the inputs and outputs of the compartment have positive and negative signs respectively. So the mathematical representation of the model shown in Fig. 7.22 is

$$\frac{dx_1}{dt} = -k_{01}x_1 - k_{21}x_1 + k_{12}x_2 + u_1 \tag{7.67}$$

$$\frac{dx_2}{dt} = -k_{02}x_2 - k_{12}x_2 + k_{21}x_1 + u_2 \tag{7.68}$$

As mentioned previously in this section, the mathematical representations of the compartmental models are the sets of linear first order differential equations, which implies a close similarity with the well-known state space representation of the system (Godfrey, 1983). For a linear, time invariant $p$ compartment system, matrix **A** has dimensions $(p \times p)$ with the elements

$$a_{ij} = k_{ij} \qquad i \neq j \tag{7.69}$$

$$a_{jj} = -\sum_{\substack{i=1 \\ i \neq j}}^{p} k_{ij} - k_{0j} \tag{7.70}$$

Due to the constraints of the compartmental-type system, all elements on the main diagonal of **A** (Eqn (7.70)) must be nonpositive, while all other elements must be nonnegative (Eqn (7.69)). The following condition must be also fulfilled:

$$|a_{jj}| \geqslant \sum_{\substack{i=1 \\ i \neq j}}^{p} k_{ij} \tag{7.71}$$

where the equality stands only for $k_{0j} = 0$. The elements of the **B** matrix are all nonnegative and, due to the fact that in nearly all practical cases each input is introduced to one compartment only, **B** is the diagonal matrix with input gains which are often equal $b_1 = \cdots = b_p = 1$. Similar statements can be made for the **C** matrix. In pharmacokinetics, compartment models are often generated for quantities while the measured outputs are given as concentrations. In such case the elements of the **C** matrix are given as

$$y_i = \frac{1}{V_i} x_i \tag{7.72}$$

where $V_i$ is the *volume of distribution*.

Much less research was initially devoted to the nonlinear, time varying or even stochastic compartmental models, although it is obvious that they are unavoidable in real practical projects. Here the role of simulation is outstanding because it represents the only possibility for efficient model solving. The last statement must be viewed in the sense that nonlinearity, time variability and the random character of compartmental models do not represent serious problems in most cases and do not need a great amount of theory if some efficient simulation tool is available.

Haemodialysis is a clinical method which acts as an artificial eliminating organ for patients with impaired renal function. It utilizes the diffusion of different substances from the blood through an artificial membrane extracorporeally to the dialysate. The procedure is usually performed twice weekly and lasts for about four hours (Karba, Mrhar *et al.*, 1990). The main problem for patients on haemodialysis is whether therapy with the particular drug is discontinued after the procedure of haemodialysis or whether some modification of the dosage régimen must be made.

The latter question can be answered with the aid of modelling and simulation. For this example we chose a pharmacokinetic study of the modern quinolone antibiotic ciprofloxacin under conditions of haemodialysis. A corresponding 'in vivo' study was designed to evaluate ciprofloxacin plasma levels after single and multiple intravenous administrations in order to answer the question as to what dosage régimen for the period between two haemodialyses, during and immediately after haemodialysis, must be chosen (Drinovec *et al.*, 1987). Through the preliminary modelling and simulation study the optimal sampling protocol was provided which was used on eighteen long-term haemodialysed patients. The collected and analysed samples gave the individual and average plasma concentration profiles which served as a data base for model development and validation.

## Modelling

Ciprofloxacin was given in the form of 400 mg/h infusions lasting half-an-hour. So the compartmental pharmacokinetical model in Fig. 7.23 was proposed. The symbols used have the following meaning: $U$ = quantity of the drug; the subscripts I, P and D mean infusion, plasma and distribution compartments, respectively; $k_0$ = null order input constant, characterizing the rate of infusion; $k_{\mathrm{d}}$ and $k_{-\mathrm{d}}$ = first order

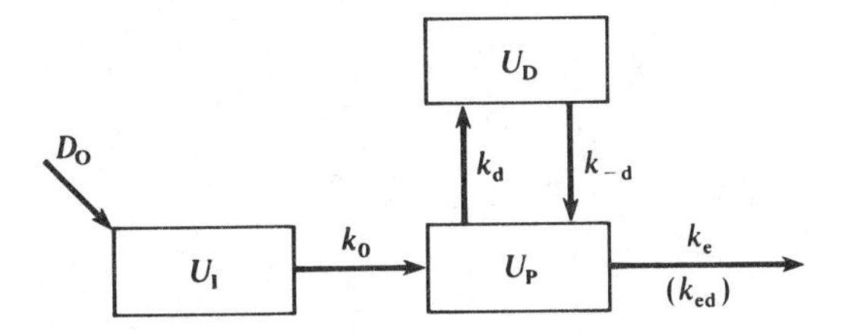


Figure 7.23 Pharmacokinetic model of ciprofloxacin under conditions of haemodialysis.

distribution rate constants; and $k_e$, $k_{ed}$ = first order elimination rate constants for the time between two haemodialyses and during haemodialysis, respectively, $D_O$ = dose.

The corresponding mathematical model can easily be derived from the model shown in Fig. 7.23. As mentioned, it consists of the set of linear first order differential equations which describe null or first order kinetics; in our case the $U_I$ compartment is given with null kinetics and the other two with first order kinetics. The following mathematical model is therefore valid for the problem

$$\frac{dU_I}{dt} = -k_O \qquad U_I(0) = D_O \qquad U_I \geqslant 0 \tag{7.73}$$

$$\frac{dU_P}{dt} = k_O + k_{-d}U_D - k_dU_P - k_xU_P \tag{7.74}$$

$$k_x = \begin{cases} k_e & 0 < t \leqslant 4\text{ h} \quad \text{and} \quad 8\text{ h} < t \leqslant T \\ k_{ed} & 4\text{ h} < t \leqslant 8\,h \end{cases} \tag{7.75}$$

$$\frac{dU_D}{dt} = -k_{-d}U_D + k_dU_P \tag{7.76}$$

where $T$ is the observation time, in our case given in hours.

We can see time variations of $k_x$ are introduced, simulating the elimination rate change due to haemodialysis, which is given in the form of a logic condition. Also, the time course of the drug quantity in $U_I$ must be arranged so that negative quantities cannot occur. In a similar way, time delays, nonlinearities and other parameter changes can be included in the model.

## Analog simulation

As a first approach, simulation on an analog–hybrid computer, EAI-580, is presented. The analog simulation scheme developed for this example is shown in Fig. 7.24. The scheme enables the simulation of single and multiple dosing of ciprofloxacin in the form of infusion, simulation of two haemodialysis procedures for arbitrary starting

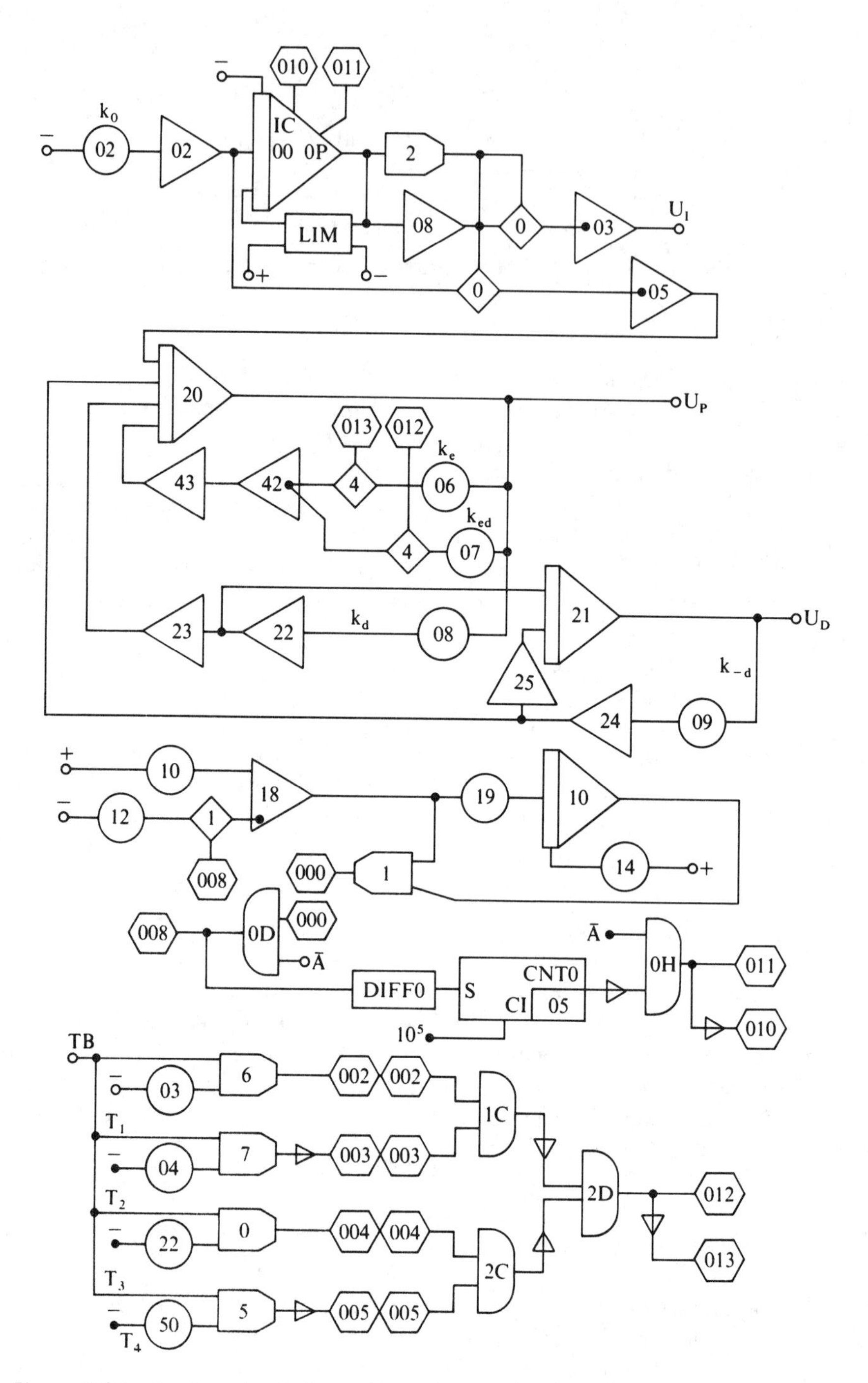


Figure 7.24 Analog simulation scheme for the developed compartmental model.

Table 7.1 Measured average concentrations of ciprofloxacin in plasma with corresponding quantities

| $T$ (h) | 0.5 | 1.5 | 3 | 4 | 4.5 | 6 | 7.5 | 8 | 10 | 12 |
|---|---|---|---|---|---|---|---|---|---|---|
| $C$ (mg/l) | 7.00 | 2.00 | 1.47 | 1.28 | 1.05 | 0.84 | 0.68 | 0.64 | 0.54 | 0.47 |
| $U$ (mg) | 124.8 | 35.7 | 26.2 | 22.8 | 18.7 | 15.0 | 12.1 | 11.4 | 9.6 | 8.0 |

$U_P = C_P V_D$ $V_D = 17.81$l.

and ending times, and also curve fitting of the model response to the measured data by manual changing of the potentiometers representing the model parameters. The upper part of the analog scheme simulates the time course of the infusion followed by the part representing the open two-compartment model for ciprofloxacin pharmacokinetic simulation. The logic circuit at the bottom of the scheme controls two switches for changing the elimination rate constant during haemodialysis. In the middle is the part for simulating multiple dosing with logic control of the infusion integrator.

The analog simulation scheme described realizes the model which was developed through the manual curve fitting procedure, which adjusts the model response to the measured data and is, due to the known properties of analog computers, very fast and illustrative. A man-in-the-loop procedure enables the potentiometers representing the system parameters to be changed manually, and very fast repetitive simulation runs enable on-line visual comparison between the model response and the measured data, which are reperesented by the function generator. When the model response is as close as possible to the measured values, the procedure is ended. At this point, some previously known properties, specifics, or weights of the particular measurements can easily be taken into account. Measured average concentrations, $C$, which are in our case converted to quantities, $U$, by multiplication by the volume of distribution, were introduced in the scheme by the function generator, as already mentioned. The conversion is summarized in Table 7.1.

The initial value for $k_e$ was obtained from the measured data using the stripping technique, while other rate constants had to be identified. The results of the curve fitting is shown in Fig. 7.25. The observation interval (problem time) is $T = 12$ h, while the computer time is $\tau = 12$ s. As the rate constants are given in $h^{-1}$, problem time in hours, computer time in seconds and time scaled rate constants ($k'$) in $s^{-1}$:

$$\frac{k'(s^{-1})}{k(h^{-1})} = \frac{T(h)}{\tau(s)}$$

and in our case

$$k' = k \tag{7.77}$$

Infusion lasts 0.5 h with a flow rate of 400 mg/h, which gives a dose of 200 mg. As the maximum values in all compartments are equal ($U_{\mathrm{Imax}} = U_{\mathrm{Pmax}} = U_{\mathrm{Dmax}} = D$), no amplitude scaling is needed in this case.

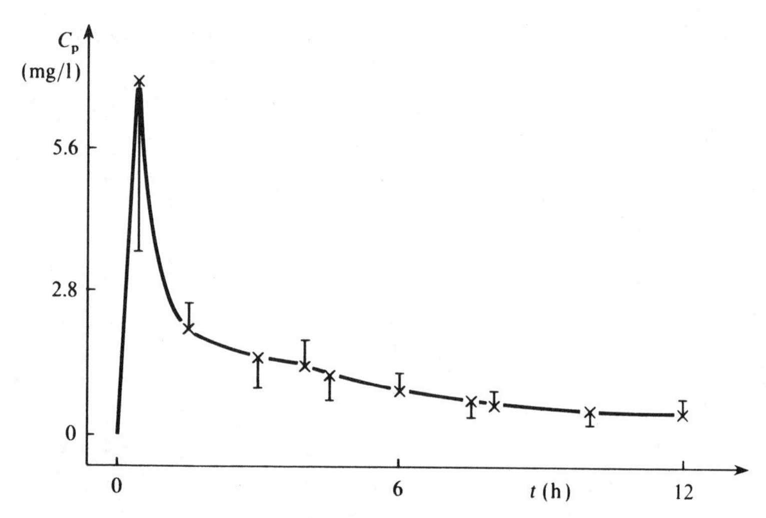


Figure 7.25 Ciprofloxacin plasma concentration profile after a single infusion lasting half an hour and haemodialysis from the fourth and eighth hours, where the crosses denote measured average concentrations with standard deviations and the curve represents the model response.

The null order constant $k_0$ is calculated from the relation

$$k_0 = \frac{1}{\tau}\frac{\tau}{t_{\text{inf}}\,(\text{h})} = 2$$

where $t_{\text{inf}} = 0.5$. For fitting the model response to the measured data in Fig. 7.25 the following values of constants were identified:

$$k_e = 0.62\ \text{h}^{-1} \qquad k_{ed} = 0.9\ \text{h}^{-1} \qquad k_d = 2.04\ \text{h}^{-1} \qquad k_{-d} = 0.73\ \text{h}^{-1}$$

Note here that, from the curve in Fig. 7.26 the end of the infusion can be clearly seen. The curve of plasma concentrations reach a peak at this moment. The model developed was also used for the prediction of the cirpofloxacin plasma concentration profile after multiple infusions (infusion every twelve hours) with first haemodialysis from the fourth to the eighth hour and the second one from the seventy-sixth to the eightieth hour (in the seventh dosing interval). The results are shown in Fig. 7.26. As can be seen from Fig. 7.26 the model was 'validated' (checked) with the aid of additional measurements (in the seventh dosing interval) which are summarized in Table 7.2. The results indicate that the model was chosen appropriately. In the case of ciprofloxacin, it is obvious that haemodialysis has relatively little influence on its plasma profiles and that no modification of dosage régimen is necessary.

## Digital simulation

A similar simulation, but only for a single dose, was then performed using the digital

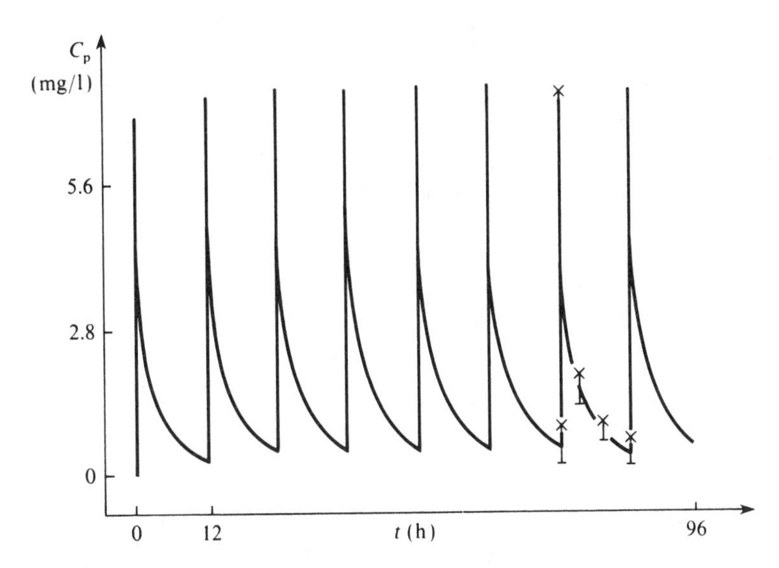


Figure 7.26 Ciprofloxacin plasma concentration profile after multiple infusions (every 12 h) and two haemodialyses in the first and seventh dosage interval, where the crosses denote measured average concentrations with standard deviations and the curve represents the model response.

Table 7.2 Measured average concentrations for ciprofloxacin in the seventh dosing interval with corresponding quantities

| $t$(h) | 72 | 72.5 | 76 | 80 | 84 |
|---|---|---|---|---|---|
| $C$ (mg/l) | 0.97 | 7.49 | 2.00 | 1.10 | 0.80 |
| $U$ (mg) | 17.3 | 133.5 | 35.7 | 19.6 | 14.3 |

simulation language SIMCOS. Due to its abilities in experimentation with models, the curve fitting procedure using the adaptive model was performed. The latter is shown in Fig. 7.27, which shows that the error between the measured data, obtained with the aid of the function generator, and the model response is used in the criterion function. In our case the integral of absolute error value was found to be acceptable. Through the optimization procedure, all model parameters are automatically changed in such a manner that the criterion function is minimized. Here the necessity for several simulation runs may represent a problem in the sense of the calculation time required. In this case the criterion function must be defined mathematically, which can again cause problems as the complexity of the criterion function increases the calculation time dramatically (limitations, weights, etc.). Additionally, the difference between the measured data and the model response for the curve with linear interpolations between measured values is taken into account, which may in

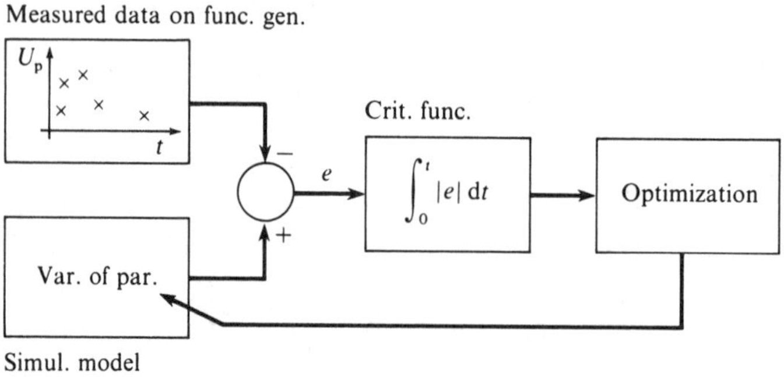

Figure 7.27 Curve fitting procedure using the adaptive model.

some cases lead to a prolonged or even unusable optimization procedure. The automatized curve fitting procedure, which is also known as 'identification' with the adaptive model, is not the only possibility for model parameter estimation; however, in the case of compartmental models it is in our opinion the most appropriate approach. The simulation program in SIMCOS is given in Fig. 7.28. As can be seen from this diagram, the measured values are given as the function generator (CIP) and the initial values of the parameters are included in the CONSTANT statement. Simulation of the model shown in Fig. 7.23 is performed by the corresponding first order differential equations and their integration. Due to the fact that we were not interested in the time course of $U_{\rm I}$ from Fig. 7.23, the infusion is simulated as the rectangular signal of amplitude 400 and duration 0.5 h. This signal is used as the input in the first order differential equation describing the time response of the plasma compartment. In our case we chose the realization with the aid of the built-in STEP function, while realization of the elimination rate constant change during the course of haemodialysis is obtained through the corresponding PROCEDURAL block. As in previous analog simulation, both phenomena can also be simulated with logical components but the approach used is more convenient in our opinion.

The results obtained after the optimization procedure in SIMCOS are shown in Fig. 7.29. After five iterations and an initial criterion function value of 59.5, its value was finally 25.0 and the optimal parameters obtained were

$$k_{\rm e}=0.55\ {\rm h}^{-1} \qquad k_{\rm ed}=0.79\ {\rm h}^{-1} \qquad k_{\rm d}=1.48\ {\rm h}^{-1} \qquad k_{-{\rm d}}=0.54\ {\rm h}^{-1}$$

Differences in the parameters values obtained after manual curve fitting on an analog computer, and automatic curve fitting with optimization in SIMCOS, originate from the characteristics of both procedures, which also reflect the properties of analog and digital simulation

## Concluding remarks

In our first case the curve fitting procedure was performed in a similar way to the man-in-the-loop simulation. The potentiometers simulating the system parameters

```
''
''Curve fitting with adaptive model
''Ciprofloxacin pharmacokinetics under the conditions of haemodialysis
''------------------------------------------------------------------
''
''Initial values of constants
CONSTANT KEE=0.5,KED=1.,KD=2.,KMD=1.
''
CONSTANT KO=400
''
''Model structure definition
UPP=KO+KMD*UD-KD*UP-KE*UP
UP=INTEG(UPP,0)
UDP=-KMD*UD+KD*UP
UD=INTEG(UDP,0)
''   Elimination rate constant change during haemodialysis
PROCEDURAL (KE=)
KE=KEE
IF((T.GT.4.).AND.(T.LT.8.))KE=KED
END
''   Simulation of infusion
KO=KO*(STEP(T,0.)-STEP(T,0.5))
''
''
''Criterion function generation
''  Ciprofloxacin measured data
TABLE CIP, 1,11,0.,0.5,1.5,3.,4.,4.5,6.,7.5,8.,10.,12.,...
               0.,124.8,35.7,26.2,22.8,18.7,15.0,12.1,11.4,9.6,8.4
MER=CIP(T)
ERROR=UP-MER
CRIT=INTEG(ABS(ERROR),0)
''
''Length of simulation run
TERMT (T.GT.12)
NSTEPS NST=2
CINTERVAL CI=0.1
''Integration algorithm
ALGORITHM IALGOR=1,JALGOR=3
''Output variables
OUTPUT UP,UD,MER,ERROR,CRIT
PREPAR UP,UD,MER,ERROR,CRIT
''
END
```

Figure 7.28 Source program in SIMCOS for the curve fitting procedure in ciprofloxacin pharmacokinetics under the condition of haemodialysis.

were changed manually, while the very fast repetitive simulation runs enabled on-line visual comparison between the model response and the measured data, which were given as the points of the function generator. When the model response was as close as possible to those points the procedure was finished. Here also some previously

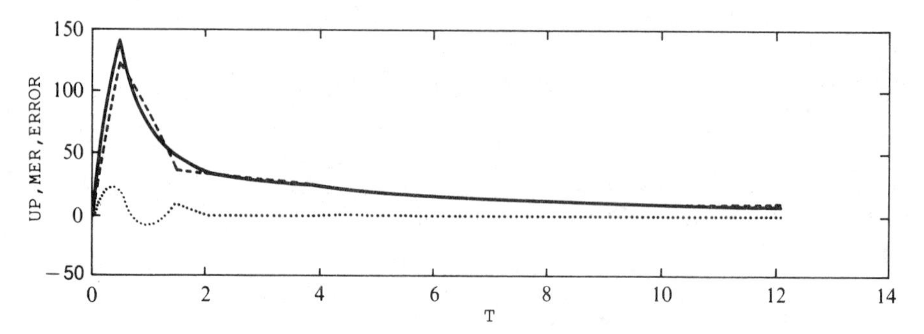


Figure 7.29 Ciprofloxacin concentration profile after a single influsion lasting half an hour and haemodialysis from the fourth and eighth hours, where the dashed line denotes measured data, the curve represents model response, while the error is given by the dotted line.

known properties or weights with which the particular measurements must be taken into account.

On the other hand, the second curve fitting procedure was performed automatically. Although the simulation runs were not of negligible time duration, the procedure was relatively short for our study of four parameters estimation. Here the comparison includes the whole course of the function generator output, which represents linear interpolation between measured points. Also, the choice of criterion function is of great importance. In our study we were interested in a good fit during the haemodialysis and so the integral of absolute error was suitable. Of course the optimization with some weight limitations in the criterion function is also possible but in such cases the procedure is more complicated and prolonged. All known properties of optimization must of course be taken into account. As the aim of the modelling was to develop a model which would enable the study of haemodialysis influence on drug pharmacokinetics, no identifiability question arose. We developed the model structure which, in our opinion, satisfactorily described the mechanisms in the observed process and there were no interesting connections between the model parameters and the physical factors which would indicate a need for the estimated model parameters to be uniquely determined.

Nevertheless, the possibility of experimentation with the model, where even complicated procedures such as curve fitting and optimization, as well as repeated simulation runs and changes of system parameters, can be performed in a user friendly manner, plays a significant role in the use of digital simulation in pharmacokinetics.

In our opinion this case study shows the wide possibilities of simulation in the biomedical sciences. Without going into detail in analysing the results of the example studied, especially from the pharmacokinetical point of view, it can be concluded that the modelling and simulation approach enables significant rationalization in all aspects of pharmacokinetical studies.

## 7.4 SOLUTION OF THE CANTILEVER BEAM EQUATION

The objective of this case study is to provide static and dynamic solutions of the cantilever beam equation. First the cantilever beam will be modelled and the corresponding partial differential equation and boundary condition will be developed. The static and dynamic solutions of this equation by simulation involve several specific approaches, such as the simulation of ordinary equations with boundary conditions, the simulation of partial differential equations and optimization.

### 7.4.1 Modelling the Cantilever Beam

Consider the thin (compared with its length $L$) beam shown in Fig. 7.30, where $x$ is measured in the direction of the beam and $y$ is its deflection perpendicular to $x$. The beam's equation of motion is obtained if the equilibrium of a small portion (with length $\mathrm{d}x$) of the beam is considered. It is supposed that an external force $F$ (which also includes gravity) per unit length is acting on the beam, so the equilibrium of the forces acting on the small portion of the beam in the $y$ direction can be written as

$$S + F\,\mathrm{d}x - \left(S + \frac{\partial S}{\partial x}\,\mathrm{d}x\right) = 0 \tag{7.78}$$

where $S$ is the transverse shear force as shown in Fig. 7.30. From this equation, the relation between the shearing force in the beam and the external force acting on it follows immediately:

$$\frac{\partial S}{\partial x} = F \tag{7.79}$$

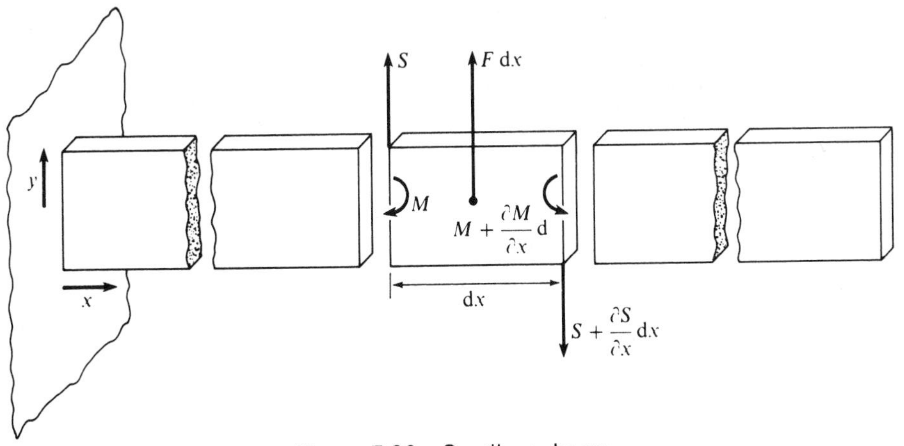


Figure 7.30 Cantilever beam.

Due to the external force, the beam is bent and at each point $x$ the bending moment of the external force is balanced by a resisting moment of the beam's material elasticity, denoted here by $M$. The equilibrium of moments acting on the small portion of the beam, as indicated in Fig. 7.30 yields

$$-M - S\frac{\mathrm{d}x}{2} - \left(S + \frac{\partial S}{\partial x}\mathrm{d}x\right)\frac{\mathrm{d}x}{2} + \left(M + \frac{\partial M}{\partial x}\mathrm{d}x\right) = 0 \tag{7.80}$$

and consequently, neglecting the second order term,

$$\frac{\partial M}{\partial x} = S \tag{7.81}$$

The resisting moment is due to the stretching of the fibres during bending. Due to Hook's law, the relative stretching of the fibre with cross-section $\mathrm{d}A$ is

$$\frac{\mathrm{d}l}{\mathrm{d}x} = \frac{\mathrm{d}F_1}{E\,\mathrm{d}A} \tag{7.82}$$

where $l$ is the length of the fibre, $F_1$ the stretching force in the fibre and $E$ is Young's modulus. If the displacement of the fibre from the longitudinal axis of the beam is denoted by $y$, the resisting moment can be expressed as

$$M = \iint yE\frac{\mathrm{d}l}{\mathrm{d}x}\,\mathrm{d}A \tag{7.83}$$

The stretching of the fibre $\mathrm{d}l$ is due to the bending of the beam and can, according to Fig. 7.31 be expressed as

$$\mathrm{d}l = y\,\mathrm{d}\alpha = y\frac{\partial^2 y}{\partial x^2}\mathrm{d}x \tag{7.84}$$

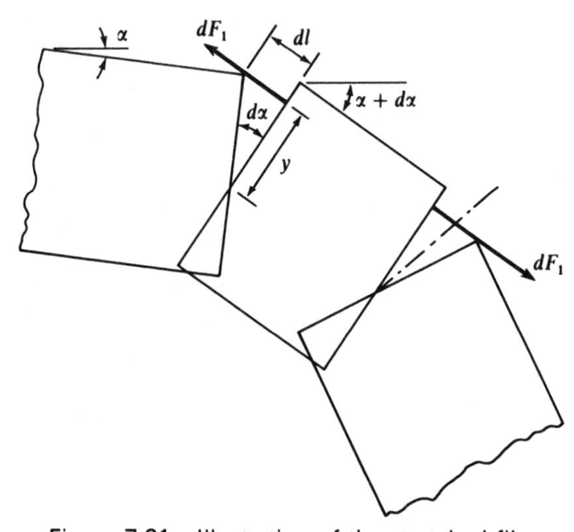


Figure 7.31 Illustration of the stretched fibre.

where the beam inclination change $d\alpha/dx$ is equal to the curvature $\partial^2 y/\partial x^2$. Substituting Eqn (7.84) into Eqn (7.83) and noting that $\int\int y^2 \, dA$ is the section's moment of inertia, denoted by $I$, the resisting moment can be expressed as

$$M = EI \frac{\partial^2 y}{\partial x^2} \tag{7.85}$$

Finally, the external force per unit length is supposed to consist of the inertia force and the gravity force

$$F = -\rho A \frac{\partial^2 y}{\partial t^2} - g\rho A \tag{7.86}$$

where $\rho A$ is the mass per unit length of the beam and $g$ is the gravity constant. The cantilever beam's equation of motion is thus obtained using Eqns (7.79), (7.81), (7.85) and (7.86) in the following form:

$$\frac{\partial^2}{\partial x^2}\left( EI \frac{\partial^2 y}{\partial x^2} \right) + \rho A \frac{\partial^2 y}{\partial t^2} = -g\rho A \tag{7.87}$$

The expression $EI$ is called the flexural rigidity of the beam and is in general a function of $x$, so it will be denoted in this book as $f_1(x)$. The mass per unit of the bar can also be a function of $x$ and will be denoted by $f_2(x)$:

$$\frac{\partial^2}{\partial x^2}\left( f_1(x) \frac{\partial^2 y}{\partial x^2} \right) + f_2(x) \frac{\partial^2 y}{\partial t^2} = -gf_2(x) \tag{7.88}$$

In the terminology of partial differential equations, a distinction is made between initial and boundary condition. As conditions with respect to time $t$ are usually defined at one point, they are called initial conditions, while conditions with respect to $x$ are called boundary conditions.

The boundary conditions are defined by clamping the beam at its ends, and the initial condition by its initial shape and its first derivative with respect to time. If the beam is built into the wall at $x = 0$, then the deflection and its first derivative at this point are zero all the time:

$$y(0, t) = 0 \qquad \left. \frac{\partial y(x, t)}{\partial x} \right|_{x=0} = 0 \tag{7.89}$$

If the beam is free at the other end, then the moment and its first derivative at $x = L$ are zero all the time and, consequently,

$$\left. \frac{\partial^2 y(x, t)}{\partial x^2} \right|_{x=L} = 0 \qquad \left. \frac{\partial^3 y(x, t)}{\partial x^3} \right|_{x=L} = 0 \tag{7.90}$$

If the beam is supposed to be straight (horizontal) and standing still at $t = 0$, then

$$y(x, 0) = 0 \qquad \left. \frac{\partial y(x, t)}{\partial t} \right|_{t=0} = 0 \tag{7.91}$$

Eqns (7.89) and (7.90) define the boundary conditions, while the initial conditions are given by Eqn (7.91). This completes the modelling of the cantilever beam.

### 7.4.2 Static Deflection of the Cantilever Beam

The static deflection of a cantilever beam of length $L$ under a transverse load will first be determined by simulation. The static deflection of the beam is obtained by setting the time-dependent term of Eqn (7.88) to zero, yielding the following ordinary differential equation:

$$\frac{\mathrm{d}^2}{\mathrm{d}x}\left(f_1(x)\frac{\mathrm{d}^2 y(x)}{\mathrm{d}x^2}\right) = gf_2(x) \tag{7.92}$$

where $y$ is the deflection, $x$ the distance along the beam from the point where the beam is fixed, $f_1(x)$ the flexural rigidity and $gf_2(x)$ the load intensity due to gravitation. The boundary conditions are given for the point where the beam is fixed:

$$y(0) = 0 \qquad \left.\frac{\mathrm{d}y(x)}{\mathrm{d}x}\right|_{x=0} = 0 \tag{7.93}$$

and for the free end,

$$\left.\frac{\mathrm{d}^2 y(x)}{\mathrm{d}x^2}\right|_{x=L} = 0 \qquad \left.\frac{\mathrm{d}^3 y(x)}{\mathrm{d}x^3}\right|_{x=L} = 0 \tag{7.94}$$

Recall that the indirect approach in solving an $n$th order differential equation also requires $n$ *initial conditions*, i.e. the $n$ values of the dependent variable and its first $n-1$ derivatives at a particular value of the independent variable (usually at $t = 0$).

In our case, two values of the dependent variable and its derivatives are specified at the beginning and two at the end of the independent variable, i.e. the *boundary conditions* are known. In general in such cases, the known initial conditions are set to their specified values and the unknown ones must be determined in such a way that the specified final conditions are met. Several methods exist to attain this and these will be reviewed briefly in the sequel:

- For low order equations (few unknown initial conditions) they can be determined by the *trial and error* procedure. This method is applicable on fast computers where the unknown initial conditions are adjusted manually until the specified final conditions are met.
- The procedure of adjusting the unknown initial conditions can be automated, i.e. an *optimization* algorithm is implemented and a criterion function such as the sum of squared final condition errors is minimized. This method requires an iterative simulation tool as well, but since no human feedback is implemented, the computational speed is not so important. However, the time for obtaining the optimal solution increases rapidly with the number of unknown (adjustable) initial conditions. The simulation tools with built-in

optimization are preferable. The optimization method is the only method which can be successfully implemented with *nonlinear* equations. It is not our aim here to go into the details of optimization; the reader is advised to consult the literature, where more about optimization, in conjunction with simulation, can be found in Amyot and van Blokland (1985); Amyot and van Blokland (1987); Birta (1977); Breitenecker *et al.*, (1988); and Cellier (1983).

- In the case of *linear systems* the principle of *superposition* can be used. According to this principle the general solution of a differential equation is the sum of a particular solution and the general solution of the homogeneous part of the equation. A question arises here as to what is the general solution of the homogeneous equation if it is solved by simulation. Actually, all solutions obtained by simulation are particular solutions. A general solution of the $n$th order differential equation is the weighted sum of $n$ particular solutions with orthogonal initial conditions. The weighting coefficients are the coefficients of the general solution. Orthogonal initial conditions can be obtained by setting $n-1$ initial conditions to zero and the remaining one to 1 (e.g. $1, 0, 0, \ldots, 0$; $0, 1, 0, \ldots, 0$; $0, 0, 0, \ldots, 1$). A system of $n$ linear equations is then obtained by writing the general solution of the differential (difference) equation for $n$ specified boundary conditions. The solution of this system is comprised of $n$ weighting coefficients which yield $n$ initial conditions. It must be mentioned that with $m$ known initial conditions, $m$ eqations in the system are trivial and so only $n-m$ equations remain to be solved.

As Eqn (7.92) is linear, the last method will be used here. According to the above discussion the general solution of the differential equation (7.92) is

$$y(x) = y_p(x) + K_1 y_1(x) + K_2 y_2(x) + K_3 y_3(x) + K_4 y_4(x) \tag{7.95}$$

where $y_p(x)$ is its particular solution, i.e. the solution for the given initial conditions

$$y_p(0) = \left.\frac{dy_p}{dx}\right|_{x=0} = \left.\frac{d^2 y_p}{dx^2}\right|_{x=0} = \left.\frac{d^3 y_p}{dx^3}\right|_{x=0} = 0 \tag{7.96}$$

and $y_1(x)$, $y_2(x)$, $y_3(x)$ and $y_4(x)$ are solutions of the homogeneous part of Eqn (7.92) for the initial conditions

$$\left.\begin{array}{llll}
y_1(0) = 1 & \left.\dfrac{dy_1}{dx}\right|_{x=0} = 0 & \left.\dfrac{d^2 y_1}{dx^2}\right|_{x=0} = 0 & \left.\dfrac{d^3 y_1}{dx^3}\right|_{x=0} = 0 \\
y_2(0) = 0 & \left.\dfrac{dy_2}{dx}\right|_{x=0} = 1 & \left.\dfrac{d^2 y_2}{dx^2}\right|_{x=0} = 0 & \left.\dfrac{d^3 y_2}{dx^3}\right|_{x=0} = 0 \\
y_3(0) = 0 & \left.\dfrac{dy_3}{dx}\right|_{x=0} = 0 & \left.\dfrac{d^2 y_3}{dx^2}\right|_{x=0} = 1 & \left.\dfrac{d^3 y_3}{dx^3}\right|_{x=0} = 0 \\
y_4(0) = 0 & \left.\dfrac{dy_4}{dx}\right|_{x=0} = 0 & \left.\dfrac{d^2 y_4}{dx^2}\right|_{x=0} = 0 & \left.\dfrac{d^3 y_4}{dx^4}\right|_{x=0} = 1
\end{array}\right\} \tag{7.97}$$

In the above equations the argument of the dependent variable was omitted for simplification. The general solution (7.95) is then written for all four boundary conditions:

$$\left.\begin{aligned}
y(0) &= 0 = y_p(0) + K_1 y_1(0) + K_2 y_2(0) + K_3 y_3(0) + K_4 y_4(0) \\
\left.\frac{dy}{dx}\right|_{x=0} &= 0 = \left.\frac{dy_p}{dx}\right|_{x=0} + K_1 \left.\frac{dy_1}{dx}\right|_{x=0} + K_2 \left.\frac{dy_2}{dx}\right|_{x=0} + K_3 \left.\frac{dy_3}{dx}\right|_{x=0} \\
&\quad + K_4 \left.\frac{dy_4}{dx}\right|_{x=0} \\
\left.\frac{d^2y}{dx^2}\right|_{x=L} &= 0 = \left.\frac{d^2y_p}{dx^2}\right|_{x=L} + K_1 \left.\frac{d^2y_1}{dx^2}\right|_{x=L} + K_2 \left.\frac{d^2y_2}{dx^2}\right|_{x=L} + K_3 \left.\frac{d^2y_3}{dx^2}\right|_{x=L} \\
&\quad + K_4 \left.\frac{d^2y_4}{dx^2}\right|_{x=L} \\
\left.\frac{d^3y}{dx^3}\right|_{x=L} &= 0 = \left.\frac{d^3y_p}{dx^3}\right|_{x=L} + K_1 \left.\frac{d^3y_1}{dx^3}\right|_{x=L} + K_2 \left.\frac{d^3y_2}{dx^3}\right|_{x=L} + K_3 \left.\frac{d^3y_3}{dx^3}\right|_{x=L} \\
&\quad + K_4 \left.\frac{d^3y_4}{dx^3}\right|_{x=L}
\end{aligned}\right\} \quad (7.98)$$

This system of equations has to be solved for $K_1$, $K_2$, $K_3$ and $K_4$. However, the first two equations yield the trivial solutions

$$\left.\begin{aligned}
K_1 &= y(0) = 0 \\
K_2 &= \left.\frac{dy(x)}{dx}\right|_{x=0} = 0
\end{aligned}\right\} \quad (7.99)$$

The remaining two equations of the system (7.99) have to solved for $K_3$ and $K_4$. The coefficients of the remaining system,

$$\left.\frac{d^2y_p}{dx^2}\right|_{x=L}, \quad \left.\frac{d^3y_p}{dx^3}\right|_{x=L}, \quad \left.\frac{d^2y_3}{dx^2}\right|_{x=L}, \quad \left.\frac{d^3y_3}{dx^3}\right|_{x=L}, \quad \left.\frac{d^2y_4}{dx^2}\right|_{x=L} \quad \text{and} \quad \left.\frac{d^3y_4}{dx^3}\right|_{x=L}$$

are obtained by three simulation runs yielding solutions $y_p(x)$, $y_3(x)$ and $y_4(x)$ respectively.

The corresponding simulation program is shown in Fig. 7.32. For this study the flexural rigidity and the load intensity have been chosen as the exponential and sinusoidal functions

$$\left.\begin{aligned}
f_1(x) &= 100 \exp\left(-\frac{x}{L}\right) \\
gf_2(x) &= -0.1 \sin\left(\pi \frac{x}{L}\right)
\end{aligned}\right\} \quad (7.100)$$

```
PROGRAM STATIC_CANTILEVER_BEAM
CONSTANT PI=3.14159,L=10
CONSTANT YXX0=-3.18311,YXXX0=0.636623
CONSTANT K=0.1
YXXX=INTEG(-K*SIN(PI*X/L),YXXX0)
YXX=INTEG(YXXX,YXX0)
YX=INTEG(YXX/(100.*EXP(-X/L)),0.)
Y=INTEG(YX,0.)
VARIABLE X=0.
CINTERVAL CI=0.1
TERMT(X.GE.L)
OUTPUT Y,YX,YXX,YXXX
PREPAR Y,YX,YXX,YXXX
END
```

Figure 7.32 Simulation program for the cantilever beam.

respectively, and the beam length has been chosen to be 10. The particular solution is obtained if the statement defining the initial conditions YXX0 and YXXX0 is replaced by

```
CONSTANT YXX0=0.,YXXX0=0
```

which introduces the initial conditions (7.96). Particular solutions for the corresponding homogeneous part are obtained if the load intensity is set to zero by

```
CONSTANT K=0
```

and by setting the initial conditions YXX0, YXXX0 to 1., 0. and 0.,1. for $y_3$ and $y_4$ respectively. Three simulation runs give

$$\left.\frac{d^2 y_p}{dx^2}\right|_{x=L} = -3.183\,11 \qquad \left.\frac{d^3 y_p}{dx^3}\right|_{x=L} = -0.636\,623 \tag{7.101}$$

$$\left.\frac{d^2 y_3}{dx^2}\right|_{x=L} = 1 \qquad \left.\frac{d^3 y_3}{dx^3}\right|_{x=L} = 0 \tag{7.102}$$

$$\left.\frac{d^2 y_4}{dx^2}\right|_{x=L} = 10 \qquad \left.\frac{d^3 y_4}{dx^3}\right|_{x=L} = 1 \tag{7.103}$$

The system to be solved becomes

$$-3.183\,11 + K_3 + 10\,K_4 = 0 \tag{7.104}$$

$$-0.636\,623 + 0\cdot K_3 + K_4 = 0 \tag{7.105}$$

yielding, finally, the initial conditions

$$\left.\frac{d^2 y}{dx^2}\right|_{x=0} = K_3 = -3.183\,11 \tag{7.106}$$

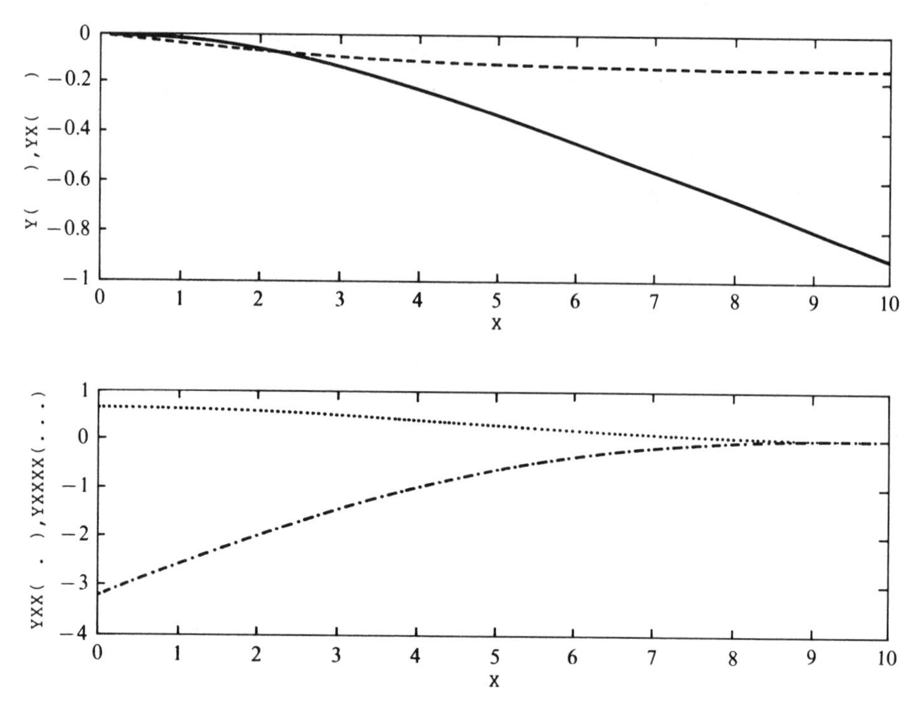


Figure 7.33 Solution of the cantilever beam equation.

and

$$\left.\frac{d^3y}{dy^3}\right|_{x=0} = K_4 = 0.636\,623 \tag{7.107}$$

If initial conditions (7.106) and (7.107) are introduced, all boundary conditions are met as shown in Fig. 7.33, which is obtained by executing the simulation program shown in Fig. 7.32 in its original form.

### 7.4.3 Vibrating Cantilever Beam

Eqn (7.88) is a partial differential equation. Partial differential equations are characterized by the fact that they contain derivatives with respect to more than one independent variable (usually to time and space). It is not our intention here to give a detailed description of partial differential equations and methods for solving them, which are described in books on mathematics for engineering, such as Kreyszig (1988). Nevertheless, some basic concepts of the procedures for solving them and the use of standard simulation tools for ordinary differential equations in these procedures will be given.

The problems of simulation packages involving partial differential equations are stated shortly in Section 3.3.

The methods and tools for the simulation of continuous systems can cope with only one independent variable, so the basic methodologies in continuous simulation, described in Chapter 2, cannot be applied directly to partial differential equations. The following are three well-known approaches for solving partial differential equations:

1. *Separation of variables.* Certain types of *partial differential equations* can be solved by expressing the solution as a product of two or more functions, where each function depends only on one independent variable. In such cases a set of *ordinary differential equations* is obtained, their solutions are evaluated and the result is composed from them. Since the simulation approach is suitable for solving ordinary differential equations it seems on first sight that this is an easy method. But there are obstacles, the *boundary conditions.* In contrast to ordinary differential equations initial and boundary conditions are commonly specified with partial differential equations. In such cases the methods described in the preceding subsection ('Static deflection of the cantilever beam') are applied and the associated problems are the so-called *eigenvalue* problems.
2. *Numerical methods.* In this approach the partial derivatives are replaced by the corresponding difference quotients and the resulting system of difference equations is solved numerically. This method is probably the all-purpose method for solving partial differential equations but since it does not involve the principles of simulation, it will not be treated here. However a special case, the *finite-difference differential equation method,* where all but one partial derivatives are replaced by difference quotients, uses the simulation approach.
3. *The use of the Laplace or Fourier transformation.* Both transformations can be applied only to linear systems, so the application of this approach is limited to a restricted class of problems. Furthermore, it is not simulation oriented and therefore it will not be treated in this book.

## The method of separated variables

The vibrating cantilever beam is described by the partial differential equation (7.88), the boundary conditions (7.89) and (7.90), and the initial conditions (7.91). The general solution of a nonhomogeneous linear partial differential equation is the sum of one particular solution and the general solution of the corresponding homogeneous equation

$$y(x,t) = y_p(x,t) + y_h(x,t) \tag{7.108}$$

Since the static solution obtained in the preceding subsection by the program shown in Fig. 7.32 is a particular solution, which will be denoted here by $y_p(x)$, we are interested only in solving the corresponding homogeneous equation

$$\frac{\partial^2}{\partial x^2}\left(f_1(x)\frac{\partial^2 y_h(x,t)}{\partial x^2}\right)+f_2(x)\frac{\partial^2 y_h(x,t)}{\partial t^2}=0 \tag{7.109}$$

with the same boundary conditions as in Eqns (7.89) and (7.90), but with the following initial condition:

$$y_h(x,0)=-y_p(x) \qquad \left.\frac{\partial y_h(x,t)}{\partial t}\right|_{t=0}=0 \tag{7.110}$$

The method of separating variables yields solutions of Eqn (7.109) of the form

$$y_h(x,t)=g(x)h(t) \tag{7.111}$$

By differentiating this equation with respect to $x$ and $t$, respectively, and inserting the derivatives in Eqn (7.109) we obtain

$$\frac{1}{f_2(x)g(x)}\frac{\mathrm{d}^2}{\mathrm{d}x^2}\left(f_1(x)\frac{\mathrm{d}^2 g(x)}{\mathrm{d}x^2}\right)=-\frac{1}{h(t)}\frac{\mathrm{d}^2 h(t)}{\mathrm{d}t^2} \tag{7.112}$$

The left-hand side of this equation involves functions depending only on $x$, while the right-hand side involves functions depending only on $t$. Hence, both terms must be equal to the same constant, denoted here by $\lambda$. This yields two ordinary differential equations:

$$\frac{d^2}{\mathrm{d}x^2}\left(f_1(x)\frac{\mathrm{d}^2 g(x)}{\mathrm{d}x^2}\right)=\lambda f_2(x)g(x) \tag{7.113}$$

and

$$\frac{\mathrm{d}^2 h(t)}{\mathrm{d}t^2}=-\lambda h(t) \tag{7.114}$$

respectively. The next step in the solution is to determine the boundary conditions for the ordinary differential equation involving $x$. These are obtained from the boundary conditions for $y_h$:

$$g(0)=0 \qquad \left.\frac{\mathrm{d}g(x)}{\mathrm{d}x}\right|_{x=0}=0 \tag{7.115}$$

$$\left.\frac{\mathrm{d}^2 g(x)}{\mathrm{d}x^2}\right|_{x=L}=0 \qquad \left.\frac{\mathrm{d}^3 g(x)}{\mathrm{d}x^3}\right|_{x=L}=0 \tag{7.116}$$

The next step in the solution is to find the solution of $g(x)$ which satisfies the boundary conditions (7.115) and (7.116). Since Eqn (7.113) is linear, the method for solving differential equations with boundary conditions described in the preceding subsection can be applied. With this method, the general solution would be a combination of four particular conditions with orthogonal initial conditions multiplied by four constants. However, due to the initial conditions (7.115) two constants are zero, so

$$g(x) = K_1 g_1(x) + K_2 g_2(x) \tag{7.117}$$

where $g_1(x)$ and $g_2(x)$, which depend on the parameter $\lambda$, are two orthogonal particular solutions of Eqn (7.113) with initial conditions

$$g_1(0) = 0 \quad \left.\frac{dg_1(x)}{dx}\right|_{x=0} = 0 \quad \left.\frac{d^2g_1(x)}{dx^2}\right|_{x=0} = 1 \quad \left.\frac{d^3g_1(x)}{dx^3}\right|_{x=0} = 0 \tag{7.118}$$

and

$$g_2(0) = 0 \quad \left.\frac{dg_2(x)}{dx}\right|_{x=0} = 0 \quad \left.\frac{d^2g_2(x)}{dx^2}\right|_{x=0} = 0 \quad \left.\frac{d^3g_2(x)}{dx^3}\right|_{x=0} = 1 \tag{7.119}$$

respectively. If the boundary conditions (7.116) are inserted in Eqn (7.117), the following two equations for the determination of $K_1$ and $K_2$ are obtained:

$$\left.\begin{aligned} K_1 \left.\frac{d^2g_1(x)}{dx^2}\right|_{x=L} + K_2 \left.\frac{d^2g_2(x)}{dx^2}\right|_{x=L} &= 0 \\ K_1 \left.\frac{d^3g_1(x)}{dx^3}\right|_{x=L} + K_2 \left.\frac{d^3g_2(x)}{dx^3}\right|_{x=L} &= 0 \end{aligned}\right\} \tag{7.120}$$

This system of equations has a nontrivial solution ($K_1 \neq 0$ and $K_2 \neq 0$) if the determinant

$$\begin{vmatrix} \left.\frac{d^2g_1(x)}{dx^2}\right|_{x=L} & \left.\frac{d^2g_2(x)}{dx^2}\right|_{x=L} \\ \left.\frac{d^3g_1(x)}{dx^3}\right|_{x=L} & \left.\frac{d^3g_2(x)}{dx^3}\right|_{x=L} \end{vmatrix} = 0 \tag{7.121}$$

This can be fulfilled only for certain values of $\lambda$. These values are called *eigenvalues* and can be evaluated by the optimization procedure, where the square of the determinant (7.121) is minimized. The corresponding simulation program with $f_1(x) = 100 \cdot \exp(-x/L)$ and $f_2(x) = 0.01 \cdot \sin(\pi x/L)$ is shown in Fig. 7.34. The optimization with respect to KF gives

$$\left.\begin{aligned} \lambda_1 &= 20.7759 \\ \lambda_2 &= 580.777 \\ \lambda_3 &= 4325.42 \\ \lambda_4 &= 16\,516.6 \\ \lambda_5 &= 42\,110.2 \end{aligned}\right\} \tag{7.122}$$

where only the first five eigenvalues were evaluated by setting diversified starting values for the optimized parameter $\lambda$. Note that Eqn (7.113) is unstable and, consequently, numerical problems arise for greater values of $\lambda$. For eigenvalues the

```
PROGRAM CANTILEVER_BEAM_DETERMINATION_OF_EIGENVALUES
CONSTANT PI=3.14159,L=10,LAMBDA=1.
CONSTANT G1XXO=1.,G1XXXO=0.
G1XXX=INTEG(LAMBDA*0.01*SIN(PI*X/L)*G1,G1XXXO)
G1XX=INTEG(G1XXX,G1XXO)
G1X=INTEG(G1XX/(100*EXP(-X/L)),0.)
G1=INTEG(G1X,0.)
CONSTANT G2XXO=0.,G2XXXO=1.
G2XXX=INTEG(LAMBDA*0.01*SIN(PI*X/L)*G2,G2XXXO)
G2XX=INTEG(G2XXX,G2XXO)
G2X=INTEG(G2XX/(100*EXP(-X/L)),0.)
G2=INTEG(G2X,0.)
KF=(G1XX*G2XXX-G2XX*G1XXX)**2
VARIABLE X=0.
CINTERVAL CI=0.1
TERMT(X.GE.L)
OUTPUT G1,G1X,G2,G2X,KF
PREPAR G1,G1X,G2,G2X,KF
END
```

Figure 7.34 Simulation program for the determination of eigenvalues.

determinant (7.121) yields only the ratio of initial conditions $K_1 = \mathrm{d}^2 g(x)/\mathrm{d}x^2|_{x=0}$ and $K_2 = \mathrm{d}^3 g(x)/\mathrm{d}x^3|_{x=0}$, which satisfies the final conditions (7.116):

$$\left.\begin{aligned} \kappa_1 &= \left(\frac{K_2}{K_1}\right)_1 = -0.155\,421 \\ \kappa_2 &= \left(\frac{K_2}{K_1}\right)_2 = -0.457\,300 \\ \kappa_3 &= \left(\frac{K_2}{K_1}\right)_3 = -0.694\,687 \\ \kappa_4 &= \left(\frac{K_2}{K_1}\right)_4 = -0.920\,293 \\ \kappa_5 &= \left(\frac{K_2}{K_1}\right)_5 = -1.116\,52 \end{aligned}\right\} \tag{7.123}$$

Since Eqn (7.113) is linear, every linear combination of solutions is again a solution which satisfies the given boundary conditions. So

$$g(x) = \sum_i A_i \gamma_i(x) \tag{7.124}$$

where

$$\gamma_i(x) = g_1(x) + \kappa_i g_2(x) \tag{7.125}$$

is the solution of Eqn (7.113) for $\lambda_i$, $\mathrm{d}^2 g(x)/\mathrm{d}x^2|_{x=0} = K_1 = 1$ and $\mathrm{d}^3 g(x)/\mathrm{d}x^3|_{x=0} = K_2 = \kappa_i = (K_2/K_1)_i$. Fig. 7.35 depicts the first five solutions $\gamma_i(x)$ for

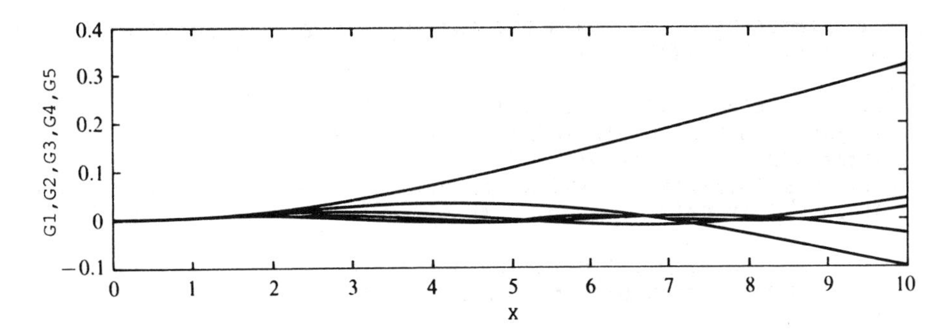


Figure 7.35 First five solutions $\gamma_i(x)$.

$f_1(x) = 100 \cdot \exp(-x/L)$ and $f_2(x) = 0.01 \cdot \sin(\pi x/L)$, which are obtained by the simulation program shown in Fig. 7.36.

The next step is the solution of the ordinary differential equation (7.114) involving $t$. Its general solution for a certain eigenvalue $\lambda_i$ is denoted by $h_i$ and is the weighted sum of two particular solutions

$$h_i(t) = B_1 h_{1i}(t) + B_2 h_{2i}(t) \tag{7.126}$$

where $h_{1i}$ and $h_{2i}$ are the particular solutions for the initial conditions

$$h_{1i}(0) = 1 \qquad \left.\frac{dh_{1i}(t)}{dt}\right|_{t=0} = 0 \tag{7.127}$$

and

$$h_{2i}(0) = 0 \qquad \left.\frac{dh_{2i}(t)}{dt}\right|_{t=0} = 1 \tag{7.128}$$

The solution of the homogeneous partial differential equation is now determined according to Eqns (7.111), (7.124) and (7.126):

$$y_h(x,t) = \sum_i C_i \gamma_i(x) h_{1i}(t) + \sum_i D_i \gamma_i(x) h_{2i}(t) \tag{7.129}$$

where the coefficients $C_i = A_i B_1$ and $D_i = A_i B_2$ have to be determined in such a way that the initial conditions, given by Eqn (7.110), are met. Since $h_{2i}(0) = 0$ and $dh_{1i}/dt|_{t=0} = 0$, it is obvious that the coefficients $C_i$ are determined by the condition $y_h(x,0) = -y_p(x)$ and the coefficients $D_i$ by $\partial y_h(x,t)/\partial t|_{t=0}$ respectively. However, $\partial y_h(x,t)/\partial t|_{t=0}$ and so $D_i = 0$.

The *curve fitting* procedure is used for adjusting the coefficients $C_i$. This procedure is extremely simple if the functions $\gamma_i(x)$ are orthogonal or othogonal with respect to a weight function. The engineering approach to the orthogonality test is simple, i.e. the integrals

$$o_{ij} = \int_0^L \gamma_i(x)\gamma_j(x)\, dx \tag{7.130}$$

```
PROGRAM CANTILEVER_BEAM_SOLUTIONS_FOR_FIRST_FIVE_EIGENVALUES
CONSTANT PI=3.14159,L=10

CONSTANT LAM1=20.7759,G1XX0=1.,G1XXX0=-0.155421
G1XXX=INTEG(LAM1*0.01*SIN(PI*X/L)*G1,G1XXX0)
G1XX=INTEG(G1XXX,G1XX0)
G1X=INTEG(G1XX/(100.*EXP(-X/L)),0.)
G1=INTEG(G1X,0.)

CONSTANT LAM2=580.777,G2XX0=1.,G2XXX0=-0.457300
G2XXX=INTEG(LAM2*0.01*SIN(PI*X/L)*G2,G2XXX0)
G2XX=INTEG(G2XXX,GXXX0)
G2X=INTEG(G2XX/(100.*EXP(-X/L)),0.)
G2=INTEG(G2X,0.)

CONSTANT LAM3=4325.42,G3XX0=1.,G3XXX0=-0.694687
G3XXX=INTEG(LAM3*0.01*SIN(PI*X/L)*G3,G3XXX0)
G3XX=INTEG(G3XXX,G3XX0)
G3X=INTEG(G3XX/(100.*EXP(-X/L)),0.)
G3=INTEG(G3X,0.)

CONSTANT LAM4=16516.6,G4XX0=1.,G4XXX0=-0.929293
G4XXX=INTEG(LAM4*0.01*SIN(PI*X/L)*G4,G4XXX0)
G4XX=INTEG(G4XXX,G4XX0)
G4X=INTEG(G4XX/(100.*EXP(-X/L)),0.)
G4=INTEG(G4X,0.)

CONSTANT LAM5=42110.2,G5XX0=1.,G5XXX0=-1.11652
G5XXX=INTEG(LAM5*0.01*SIN(PI*X/L)*G5,G5XXX0)
G5XX=INTEG(G5XXX,G5XX0)
G5X=INTEG(G5XX/(100.*EXP(-X/L)),0.)
G5=INTEG(G5X,0.)


VARIABLE X=0.
CINTERVAL CI=0.25
NSTEPS NST=10
TERMT(X.GE.L)
OUTPUT G1,G2,G3,G4,G5
PREPAR G1,G2,G3,G4,G5
END
```

Figure 7.36 Simulation program for $\gamma_i(x)$.

are evaluated and if $o_{ij} \ll o_{ii}$, $i \neq j$ for all $i$ and $j$, then the set $\gamma_i$ is orthogonal. If this test is performed in our case, it can be seen that the functions are not orthogonal.

If a nonnegative function $f(x)$ is found such that the orthogonality requirements for the integrals

$$o_{ij} = \int_0^L f(x)\gamma_i(x)\gamma_j(x)\,dx \tag{7.131}$$

are fulfilled, the system $\gamma_i$ is said to be orthogonal with respect to the weight function $f(x)$. A simple test which is performed by adding the following statements:

```
O11=INTEG(SIN(PI*X/L)*G1*G1,0.)
O12=INTEG(SIN(PI*X/L)*G1*G2,0.)
etc.
```

to the program shown in Fig. 7.36 shows that the system $\gamma_i$ is very close to being orthogonal with respect to the weight function $f(x) = 10 \cdot f_2(x) = \sin(\pi x/L)$, as illustrated in Fig. 7.37. A good approximation for the coefficients $C_i$ can therefore be computed using the coefficients of the *generalized Fourier series*

$$C_i = \frac{\int_0^L f(x)\gamma_i(x) y_h(x,0)\,dx}{\int_0^L f(x)\gamma_i^2(x)\,dx} \tag{7.132}$$

If the series $\gamma_i(x)$ were not orthogonal with respect to a weight function, the engineering solution would again be optimization. In this case a criterion function, e.g. the integral of the squared difference of the initial conditions $y_h(x,0) = -y_p(x)$, given by Eqn (7.110), and solution $y_h(x,0)$, given by Eqn (7.129), is minimized. The corresponding part of the simulation program is shown in Fig. 7.38. However,

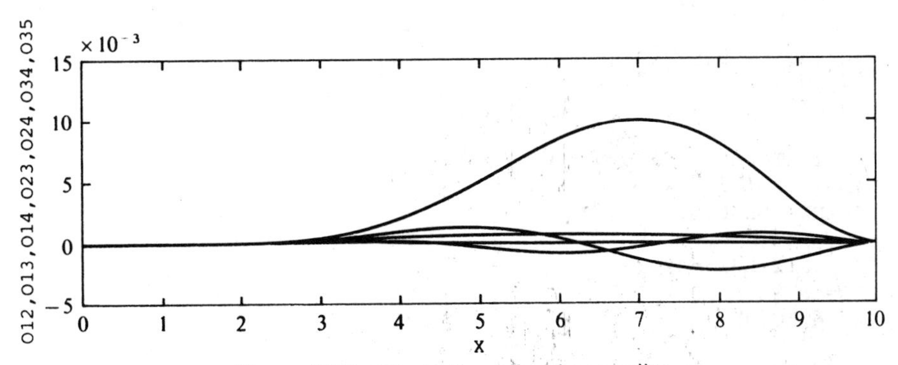


Figure 7.37 Illustration of orthogonality.

```
COMMENT INITIAL CONDITION FUNCTION
CONSTANT YPXXO=-3.18311,YPXXXO=0.636623
YPXXX=INTEG(-0.1*SIN(PI*X/L), YPXXXO)
YPXX=INTEG(YPXXX,YPXXO)
YPX=INTEG(YPXX/(100.*EXP(-X/L)),0.)
YP=INTEG(YPX,0.)
COMMENT OPTIMIZATION
CONSTANT C1=2.83641,C2=0.266414,C3=0.0485939,C4=-0.00262758,...
         C5=-0.00702735
G=C1*G1+C2*G2+C3*G3+C4*G4+C5*G5
KF=INTEG((G-(-YP))*G-(-YP)),0.)
```

Figure 7.38 Part of the program for the optimization of constants $C_i$.

due to the truncation of the series after only a few terms and due to numerical problems while evaluating coefficients $C_i$, it is advisable to perform the optimization in any case. The values obtained by Eqn (7.132) can be used as starting values for the optimization, which in our case yields the following constants $C_i$:

$$\left.\begin{aligned} C_1 &= 2.836\,36 \\ C_2 &= 0.266\,729 \\ C_3 &= 0.049\,633\,6 \\ C_4 &= -0.004\,391\,60 \\ C_5 &= -0.075\,332\,9 \end{aligned}\right\} \tag{7.133}$$

```
PROGRAM CANTILEVER_BEAM_COMPLETE SOLUTION

CONSTANT LAM1=20.7759
H1D=INTEG(-LAM1*H1,0.)
H1=INTEG(H1D,1.)

CONSTANT LAM2=580.777
H2D=INTEG(-LAM2*H2,0.)
H2=INTEG(H2D,1.)

CONSTANT LAM3=4325.42
H3D=INTEG(-LAM3*H3,0.)
H3=INTEG(H3D,1.)

CONSTANT LAM4=16516.6
H4D=INTEG(-LAM4*H4,0.)
H4=INTEG(H4D,1.)

CONSTANT LAM5=42110.2
H5D=INTEG(-LAM5*H5,0.)
H5=INTEG(H5D,1.)

CONSTANT C1=2.83641,C2=0.266414,C3=0.0485939,C4=-0.00262758,...
         C5=-0.00702735
CONSTANT GG1=0.324122,GG2=0.0928195,GG3=0.0426182,...
         GG4=0.0253499,GG5=0.0266068,YP=-0.896445
Y=C1*GG1*H1+C2*GG2*H2+C3*GG3*H3+C4*GG4*H4+C5*GG5*H5+YP

VARIABLE TIME=0.
CONSTANT TFIN=1.5
CINTERVAL CI=0.0375
TERMT(TIME.GE.TFIN)
OUTPUT Y
PREPAR Y
END
```

Figure 7.39 Simulation program for Eqn (7.134).

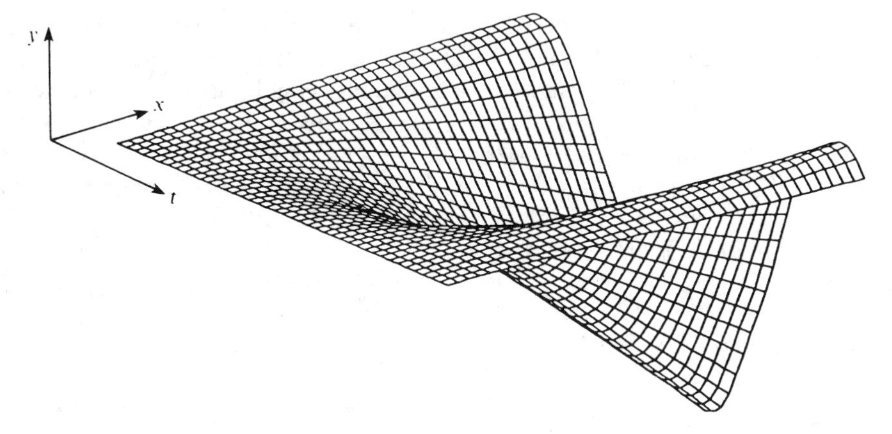


Figure 7.40 Solution of the vibrating cantilever beam.

Finally, the solution of the nonhomogeneous partial differential equation can be written in the following form:

$$y(x,t) = \sum_i C_i \gamma_i(x) h_{1i}(t) + y_p(x) \tag{7.134}$$

Fig. 7.39 shows the simulation program for Eqn (7.134), where the solution for a particular value of $x$ is obtained. The coefficients $C_i$ are given in Eqn (7.133) and the function values $\gamma_i(x)$, which are constant for a constant $x$, are obtained by the simulation run of the program shown in Fig. 7.36. The solution of the nonhomogeneous partial differential equation (7.88), which satisfies all boundary conditions, Eqns (7.89) and (7.90), is shown as a three-dimensional plot in Fig. 7.40.

**The method of finite difference differential equations**

Numerical solutions of partial differential equations are based on the discretization of independent variables. First the increments of the independent variables are chosen and a grid consisting of equidistant lines is built. Then the difference equation approximations of the partial derivatives are used in order to relate the unknown function values in the intersections (also called *mesh* or *lattice points* or *nodes*) to each other and to the given boundary values. This yields a *difference equation*, which is then solved either recursively (if only initial conditions with respect to one independent variable are specified) or as a system of algebraic equations (if boundary conditions are specified).

The finite difference differential equation method uses the same approach for all but one independent variable. So at each discrete value an ordinary differential equation is obtained and the solution of the partial differential equation corresponds to the set of simultaneous solutions of all the ordinary differential equations.

The elementary question in this method is which independent variables are discretized and which one is not? The answer to this question depends upon the boundary conditions. In the case where all boundary conditions with respect to one independent variable apply to the same point, it is advisable to leave this variable continuous and, without loss of generality, to take the boundary conditions as the initial conditions. If this is not the case, the methods for solving boundary value problems must be applied. It is also obvious that the solutions must be of interest only within bounded intervals of discretized variables.

We will not go into the details of finite difference differential equations, so only practical formulae for the finite difference expressions will be given. If $x$ and $t$ are the independent variables and $y(x, t)$ is the dependent variable, then

$$\left.\frac{\partial y(x,t)}{\partial x}\right|_{x=x_i} = \frac{1}{2\,\Delta x}[y(x_i + \Delta x, t) - y(x_i - \Delta x, t)] \tag{7.135}$$

$$\left.\frac{\partial^2 y(x,t)}{\partial x^2}\right|_{x=x_i} = \frac{1}{(\Delta x)^2}[y(x_i - \Delta x, t) - 2y(x_i, t) + y(x_i + \Delta x, t)] \tag{7.136}$$

If the independent variable or its first derivative have to be evaluated between discrete points, the following expressions can be used:

$$y(x_i + \tfrac{1}{2}\,\Delta x, t) = \tfrac{1}{2}[y(x_i, t) + y(x_i + \Delta x, t) \tag{7.137}$$

$$\left.\frac{\partial y(x,t)}{\partial x}\right|_{x=x_i+1/2\,\Delta x} = \frac{1}{\Delta x}[y(x_i + \Delta x, t) - y(x_i, t)] \tag{7.138}$$

The vibrations of a cantilever beam, already studied by the method of separated variables, will now be evaluated by the finite difference differential method. Noting that, in the well-known partial differential equation

$$\frac{\partial^2 y(x,t)}{\partial x^2}\left[f_1(x)\frac{\partial^2 y(x,t)}{\partial x^2}\right] + f_2(x)\frac{\partial^2 y(x,t)}{\partial t^2} = gf_2(x) \tag{7.139}$$

the expression in brackets represents the moment

$$M(x,t) = f_1(x)\frac{\partial^2 y(x,t)}{\partial x^2} \tag{7.140}$$

the boundary conditions can be written as

$$\text{At} \quad x = 0 \qquad y(0,t) = 0 \qquad \left.\frac{\partial y(x,t)}{\partial x}\right|_{x=0} = 0 \tag{7.141}$$

$$\text{At} \quad x = L \qquad M(L,t) = 0 \qquad \left.\frac{\partial M(x,t)}{\partial x}\right|_{x=L} = 0 \tag{7.142}$$

and the initial condition as

$$y(x,0) = 0 \qquad \left.\frac{\partial y(x,t)}{\partial t}\right|_{t=0} = 0 \tag{7.143}$$

Since both conditions with respect to $t$ are specified at the same point ($t = 0$), $t$ remains continuous. The discretization of Eqn (7.139) with respect to $x$ yields

$$\frac{1}{(\Delta x)^2}[M_{i-1} - 2M_i + M_{i+1}] + f_{2i}\frac{d^2 y_i}{dt^2} = gf_{2i} \tag{7.144}$$

where Eqn (7.140) was taken into account and the notation $y_i = y(x_i, t)$, $f_{2i} = f_2(x_i)$, $M_i = M(x_i, t)$, $M_{i+1} = M(x_i + \Delta x, t)$, $M_{i-1} = M(x_i - \Delta x, t)$ was used. Similarly, the discretization of Eqn (7.140) yields

$$M_i = \frac{f_{1i}}{(\Delta x)^2}[y_{i-1} - 2y_i + y_{i+1}] \tag{7.145}$$

Usually, the points are placed equidistantly ($\Delta x$ = constant), dividing the beam into several segments of the same length, so only the following dilemmas have to be solved:

- How to start the grid?
- How to end the grid?
- How many points to use?

The answer to the last question depends not only on the facilities of the simulation tool used but, to a great extent, on the numerics also. If too small a number of points is used, the difference approximations of the derivatives are inaccurate, while too many points lead to stiffness of the system of differential equations and, consequently, to numerical problems. As always, compromises have to be made. There are two possibilities in regard to the first two questions. The grid can be started (ended) either by placing a point at the beginning (end) of the first (last) interval of the independent variable, or by placing two points symmetrical to the boundary. In the first case the interval (and consequently also the segments) begins (ends) at the boundary, while in the second case the boundary is spanned (meaning in the middle of) by the first (last) segment.

As $y(x, t)$ and $M(x, t)$ and their first derivatives are zero at the beginning and at the end, respectively, of the cantilever beam, it is convenient to divide it into segments such that the first and last segments span the beginning and end, respectively, of the beam, as shown in Fig. 7.41. If Eqns (7.137) and (7.138) are now applied to

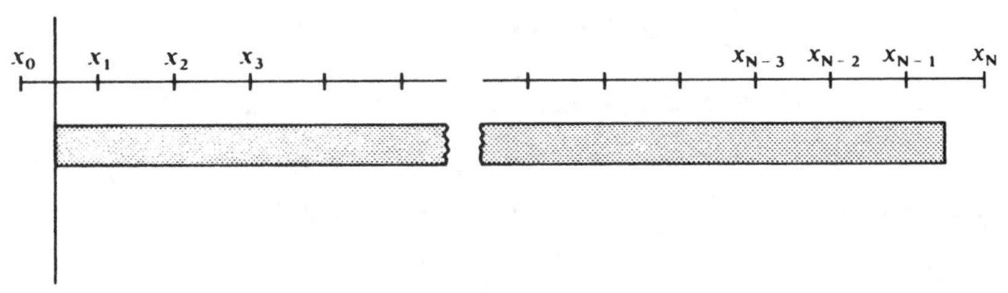


Figure 7.41 The cantilever beam segmentation.

```
program parcial;
const L=10;g=10;N=40;dx=L/N;dt=0.0000375;pi=3.14159;tfin=1.5;nsteps=500;
var y,ydot,yddot,M:array[0..N]of real; t:real;
    i,ncount:integer;f:text;
procedure deriv; (* evaluation of derivatives *)
 begin
   for i:=1 to N-2 do
   M[i]:=100*exp(-(i-0.5)/(N-1))/(dx*dx)*(y[i-1]-2*y[i]+y[i+1]);
   for i:=2 to N-1 do
   yddot[i]:=-g-(M[i-1]-2*M[i]+M[i+1])/(dx*dx*0.01*sin(pi*(i-0.5)/(N-1)));
 end;
procedure output; (* presentation of results *)
 begin
  if ncount<=0
  then
    begin
      writeln(t:5:2,yddot[N-1]:10:5,ydot[N-1]:10:5,y[N-1]:10:5);
       for i:=1 to N-1 do write(f,y[i]:7:3); (* results are written *)
       writeln(f);           (* on the file for graphic presentation *)
       ncount:=nsteps;
     end;
  ncount:=ncount-1;
 end;
procedure integ; (* performs scheduling and integration *)
 begin
  repeat
   begin
    deriv; output; t:=t+dt;
    for i:=2 to N-1 do
     begin
      ydot[i]:=ydot[i]+yddot[i]*dt; (* for simplicity Euler integration *)
      y[i]:=y[i]+ydot[i]*dt;        (* formulae are used *)
     end;
   end
   until t>=tfin;
 end;   (* integ *)
begin   (* main program *)
 for i:=0 to N-1 do
  begin
   y[i]:=0;ydot[i]:=0;  (* set initial conditions *)
  end;
 M[N-1]:=0;M[N]:=0;  (* set boundary conditions for M *)
 ncount:=0; t:=0;  (* initialization *)
 assign(f,'ttt.ttt'); rewrite(f); integ; close(f);
end.
```

Figure 7.42 PASCAL program for the cantilever beam problem.

the boundary conditions, Eqns (7.141) and (7.142), we obtain

$$y_0 + y_1 = 0 \qquad y_1 - y_0 = 0 \tag{7.146}$$

$$M_{N-1} + M_N = 0 \qquad M_N - M_{N-1} = 0 \tag{7.147}$$

yielding

$$y_0 = y_1 = 0 \qquad M_{N-1} = M_N = 0 \tag{7.148}$$

So the difference differential equation (7.144) represents $N - 2$ simultaneous ordinary differential equations for $i = 2, 3, \ldots, N - 1$ and Eqn (7.145) represents $N - 2$ algebraic equations for $i = 1, 2, \ldots, N - 2$. The initial conditions for all ordinary differential equations are zero due to the initial conditions, Eqn (7.143).

The system of ordinary differential and algebraic equations can now be solved using the standard simulation approach since all initial conditions are known. The CSSL standard simulation tools are, however, a little clumsy for tasks of this kind. So a PASCAL program written to the guidelines given in Section 4.4. is shown in Fig. 7.42. Programming in PASCAL, being structured and allowing repeated structures (models), seems to be easy, but it must be noted that the sorting algorithm must be performed 'by hand' and the numerical integration algorithms must first of all be provided by the user. As our intention was to show the principles of the finite differential equation method, the simplest Euler integration alogirthm was used, but due to the stiffness of the problem (especially with a larger number of sections) more sophisticated algorithms must be used in practice. The result of the simulation is the same as that using the method of separated variables shown in Fig. 7.40.

If the cantilever beam is initially curved, so that the initial conditions are

$$y(x, 0) = 0.2\left[1 - \cos\left(2\pi \frac{x}{L}\right)\right] \qquad \left.\frac{\partial y(x,t)}{\partial t}\right|_{t=0} = 0 \tag{7.149}$$

the simulation program shown in Fig. 7.42 has to be modified only at the definition of initial conditions, which are now defined as follows

```
y[i]:=0.2*(1-cos(2*pi*(i-0.5)/(N-1)));ydot[i]:=0;    (* curved beam *)
```

The simulation results are shown in Fig. 7.43.

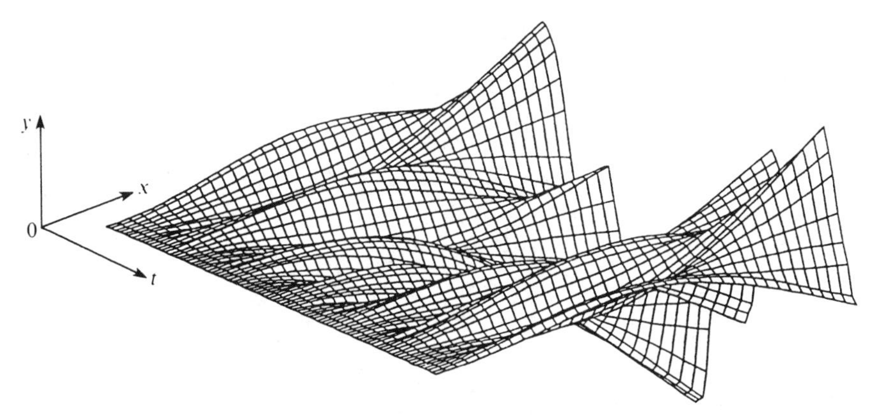


Figure 7.43 Solution of the initially folded cantilever beam problem.

## 7.5 MODEL REFERENCE ADAPTIVE CONTROL

The objective of this case study is to design the model reference adaptive position control of the mechanical load by means of simulation. Here, considerable emphasis will be given to the simulation scheme development and testing procedure. The positioning system is driven by a dc motor and is shown schematically in Fig. 7.44. The input voltage is applied to the armature circuit of the motor, while a fixed voltage is applied to the field winding. The torque developed by the motor is

$$T = k_T i \tag{7.150}$$

where $k_T$ is the motor torque constant and $i$ is the armature current. According to Kirchoff's law the following equation can be written for the armature circuit:

$$u = L\frac{di}{dt} + R_i + k_e\frac{d\theta}{dt} \tag{7.151}$$

where $k_e$ is the back emf constant of the motor and $\theta$ is the angular displacement of the motor shaft. The equation for torque equilibrium is

$$T = J\frac{d^2\theta}{dt^2} + c_f\frac{d\theta}{dt} \tag{7.152}$$

where $J$ is the inertia of the combination of the motor load and gear train referenced to the motor shaft and $c_f$ is the viscous-friction coefficient of the whole system, also referenced to the motor shaft.

The inductivity $L$ is usually small and can be neglected in practice, so the following simplified equation is obtained from Eqns (7.150), (7.151) and (7.152):

$$J\frac{d^2\theta}{dt^2} + c_f\frac{d\theta}{dt} = \frac{k_T}{R}\left(u - k_e\frac{d\theta}{dt}\right) \tag{7.153}$$

The transfer function between the motor shaft displacement $\theta$ and the input voltage

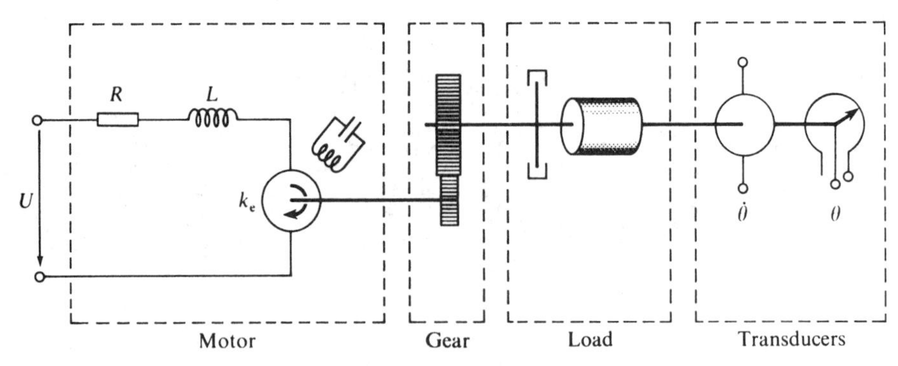


Figure 7.44 Schematic diagram of the positioning system.

$u$ of the simplified system is thus

$$\frac{\Theta(s)}{U(s)} = \frac{k_T}{RJs^2 + (Rc_f + k_T k_e)s} \tag{7.154}$$

Note that the Laplace transform of the angular displacement $\theta(t)$ is denoted as $\Theta(s)$. It is supposed that the motor shaft velocity $d\theta/dt$ can be measured by a tachometer, so the control of the positioning system consists of a proportional-differential term in the feedback path and a constant in the prefilter, as shown in Fig. 7.45. The closed loop transfer function of the system in Fig. 7.45 is

$$G(s) = \frac{\Theta(s)}{W(s)} = \frac{pk_T}{RJs^2 + (Rc_f + k_T k_e + k_T q_0)s + k_T q_1} \tag{7.155}$$

In the terminology of adaptive control systems, the closed loop system in Fig. 7.45 is called the basic control loop.

In practice, the parameters of the positioning system vary with time. In such cases the controller parameters $p$, $q_0$ and $q_1$ must be tuned according to changes of the positioning system parameters, and the resulting control system is called the adaptive control system.

Among several approaches to adaptive control the model reference adaptive control will be used here. The idea of this control scheme is to specify the desired performance of the basic loop, Eqn. (7.155), in the form of a reference model, $G_m(s)$, as shown in Fig. 7.46. The actual output of the basic loop is compared with the reference model output and the parameters of the controller are adjusted in such a

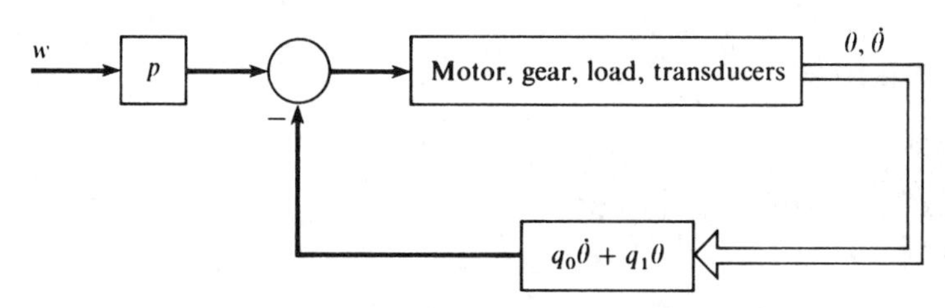


Figure 7.45 Control of the positioning system.

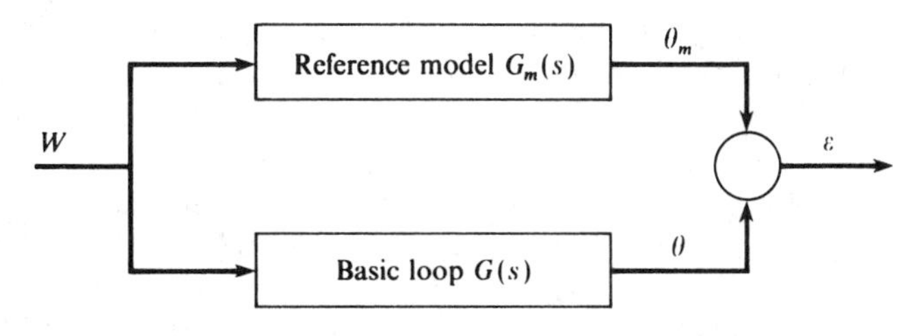


Figure 7.46 Model reference adaptive system.

way that the response of the basic loop follows the response of the reference model. The law for the adjustment of the controller's parameters (the adaptive control law) can be designed in several ways, e.g. by minimization of a criterion function by the gradient technique, or by using Lyapunov's stability theory. In this case study, both approaches will be used and tested by means of simulation. With the gradient method, the error between the desired and the actual output

$$\varepsilon = \theta - \theta_{\mathrm{m}} \tag{7.156}$$

is used in the criterion function

$$Q = \varepsilon^2 \tag{7.157}$$

which is minimized using the steepest descent technique

$$\left.\begin{aligned} \dot{p} &= -\gamma_1 \frac{\partial Q}{\partial p} = -2\gamma_1 \varepsilon \frac{\partial \theta}{\partial p} \\ \dot{q}_0 &= -\gamma_2 \frac{\partial Q}{\partial q_0} = -2\gamma_2 \varepsilon \frac{\partial \theta}{\partial q_0} \\ \dot{q}_1 &= -\gamma_3 \frac{\partial Q}{\partial q_1} = -2\gamma_3 \varepsilon \frac{\partial \theta}{\partial q_1} \end{aligned}\right\} \tag{7.158}$$

where $\gamma_1$, $\gamma_2$ and $\gamma_3$ are positive quantities determining the rate of convergence.

The partial derivatives of the shaft displacement $\theta$ on the controller parameters $p$, $q_0$ and $q_1$, respectively, are obtained using Eqn (7.155):

$$\frac{\partial \Theta}{\partial p} = \frac{1}{p} G(s) W = \frac{1}{p} \Theta \tag{7.159}$$

$$\frac{\partial \Theta}{\partial q_0} = -\frac{s}{p} G^2(s) W = -\frac{s}{p} G(s) \Theta \tag{7.160}$$

$$\frac{\partial \Theta}{\partial q_1} = -\frac{1}{p} G^2(s) W = -\frac{1}{p} G(s) \Theta \tag{7.161}$$

The realization of Eqns (7.160) and (7.161) would require knowledge of the basic loop transfer function $G(s)$, which is, of course, unknown due to the changing parameters of the positioning system. Therefore, an approximation is undertaken and the basic loop transfer function $G(s)$ in Eqns (7.160) and (7.161) is replaced by the reference model transfer function $G_{\mathrm{m}}(s)$. This approximation is justified for the vicinity of the convergence point, which is, of course, $G(s) = G_{\mathrm{m}}(s)$. Denoting

$$\frac{2\gamma_1}{p} = \bar{\gamma}_1 \qquad \frac{2\gamma_2}{p} = \bar{\gamma}_2 \qquad \frac{2\gamma_3}{p} = \bar{\gamma}_3 \tag{7.162}$$

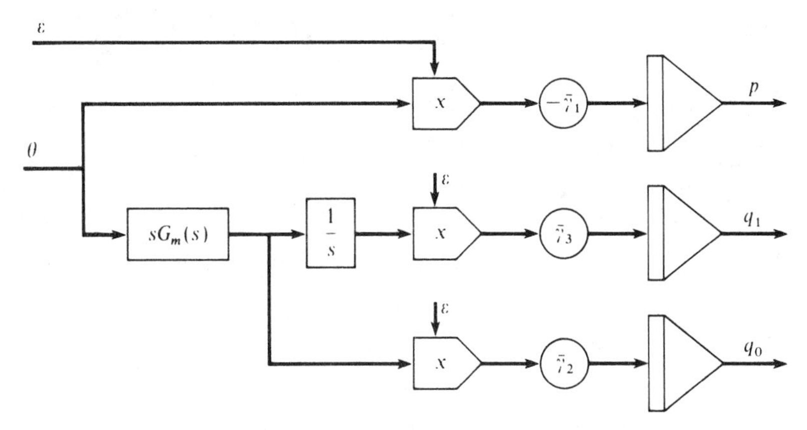


Figure 7.47 Parameter adjustment law for the gradient scheme.

the model reference adaptive control law, Eqn (7.158) becomes

$$\left.\begin{aligned} \dot{p} &= -\bar{\gamma}_1 \varepsilon \theta \\ \dot{q}_0 &= \bar{\gamma}_2 \varepsilon \dot{\theta}_f \\ \dot{q}_1 &= \bar{\gamma}_3 \varepsilon \theta_f \end{aligned}\right\} \tag{7.163}$$

where $\theta_f$ denotes the shaft position filtered by the reference model transfer function

$$\theta_f = G_m(s)\theta \tag{7.164}$$

In practical applications, the quantities $\bar{\gamma}_i$, $i = 1, 2, 3$ are chosen to be positive constants, which results in a variable convergence rate. This is not a serious disadvantage as long as the controller parameter $p$ remains positive. Since the sign of the positioning system gain is supposed to be known (positive) and does not change, the steady state value of $p$ is always positive. During the transient phase, however, the controller parameter $p$ may become negative and convergence problems, discussed later, arise. The scheme of the parameter adjustment law is shown in Fig. 7.47.

The adjustment of the basic loop parameters introduces another feedback into the positioning system, the so-called adaptive loop. The above parameter adjustment scheme is called the MIT rule according to Osburn *et al.* (1961). The main drawback of this scheme is that it does not guarantee the stability of the whole system.

Therefore, another scheme, which is stable due to the design method, will also be investigated. The basic idea of this approach, proposed by Parks (1966), is to introduce a global stability criterion into the design procedure and to choose the adaptive control law in such a way that the requirements of the stability criterion are fulfilled. Perfect matching of the basic control loop and the reference model is required and so the reference model has to be of the same order as the basic control loop, resulting in our case in the transfer function of the reference model:

$$G_m(s) = \frac{\Theta_m(s)}{W(s)} = \frac{\beta}{s^2 + \alpha_1 s + \alpha_2} \tag{7.165}$$

which corresponds to the differential equation

$$\ddot{\theta}_m + \alpha_1\dot{\theta}_m + \alpha_2\theta_m = \beta\omega \tag{7.166}$$

The basic loop (7.155) can also be written as the differential equation

$$\ddot{\theta} + \left(\frac{c_f}{J} + \frac{k_T k_e}{RJ} + \frac{k_T q_0}{RJ}\right)\dot{\theta} + \frac{k_T q_1}{RJ}\theta = \frac{p k_T}{RJ} w \tag{7.167}$$

If the adaptive system error defined by Eqn (7.156) is introduced and Eqn (7.166) is subtracted from the differential equation of the basic loop (7.167), the following error differential equation is obtained:

$$\ddot{\varepsilon} + \alpha_1\dot{\varepsilon} + \alpha_2\varepsilon = \tilde{b}w - \tilde{a}_1\dot{\theta} - \tilde{a}_2\theta \tag{7.168}$$

where the parameter errors $\tilde{b}$, $\tilde{a}_1$ and $\tilde{a}_2$ are defined as follows:

$$\left.\begin{aligned} \tilde{b} &= \frac{p k_T}{RJ} - \beta \\ \tilde{a}_1 &= \frac{c_f}{J} + \frac{k_T k_e}{RJ} + \frac{k_T q_0}{RJ} - \alpha_1 \\ \tilde{a}_2 &= \frac{k_T q_1}{RJ} - \alpha_2 \end{aligned}\right\} \tag{7.169}$$

The adaptive control system is stable if the error differential equation (7.168) is stable. The quantities $\tilde{b}$, $\tilde{a}_1$ and $\tilde{a}_2$ are time variable and we are of course free to determine the law of their adjustment (which is actually the adaptive control law) in a way that ensures the stability of Eqn (7.168).

According to Ljapunov's stability theory a positive definite function

$$V = \alpha_2\varepsilon^2 + \dot{\varepsilon}^2 + \frac{1}{\gamma_4}\tilde{b}^2 + \frac{1}{\gamma_5}\tilde{a}_1^2 + \frac{1}{\gamma_6}\tilde{a}_2^2 \tag{7.170}$$

where $\gamma_4$, $\gamma_5$ and $\gamma_6$ are positive constants, is chosen. It should be noted that $\alpha_2$ is positive as well because the reference model is supposed to be stable. The time derivative of Ljapunov's function, Eqn (7.170), is obtained using Eqn (7.168) in the following form:

$$\begin{aligned} \dot{V} &= 2\alpha_2\varepsilon\dot{\varepsilon} + 2\dot{\varepsilon}(-\alpha_1\dot{\varepsilon} - \alpha_2\varepsilon + \tilde{b}w - \tilde{a}_1\dot{\theta} - \tilde{a}_2\theta) + \frac{2}{\gamma_4}\tilde{b}\dot{\tilde{b}} + \frac{2}{\gamma_5}\tilde{a}_1\dot{\tilde{a}}_1 + \frac{2}{\gamma_6}\tilde{a}_2\dot{\tilde{a}}_2 \\ &= -2\alpha_1\dot{\varepsilon}^2 + 2\tilde{b}\left[\dot{\varepsilon}w + \frac{1}{\gamma_4}\dot{\tilde{b}}\right] + 2\tilde{a}_1\left[-\dot{\varepsilon}\dot{\theta} + \frac{1}{\gamma_5}\dot{\tilde{a}}_1\right] + 2\tilde{a}_2\left[-\dot{\varepsilon}\theta + \frac{1}{\gamma_6}\dot{\tilde{a}}_2\right] \end{aligned} \tag{7.171}$$

and can be made negative semidefinite if the terms in square brackets are made to be zero. Thus

$$\begin{aligned} \dot{\tilde{b}} &= -\gamma_4 \dot{\varepsilon} w \\ \dot{\tilde{a}}_1 &= \gamma_5 \dot{\varepsilon} \dot{\theta} \\ \dot{\tilde{a}}_2 &= \gamma_6 \dot{\varepsilon} \theta \end{aligned} \tag{7.172}$$

If the positioning system parameters $k_T$, $R$ and $J$ are slowly time varying, the following adaptive control laws are obtained using Eqns (7.169) and (7.172):

$$\left.\begin{aligned} p &= -\frac{\gamma_4 RJ}{k_T} \int \dot{\varepsilon} w \, \mathrm{d}t + p(0) = -\bar{\gamma}_4 \int \dot{\varepsilon} w \, \mathrm{d}t + p(0) \\ q_0 &= \frac{\gamma_5 RJ}{k_T} \int \dot{\varepsilon} \dot{\theta} \, \mathrm{d}t + q_0(0) = \bar{\gamma}_5 \int \dot{\varepsilon} \dot{\theta} \, \mathrm{d}t + q_0(0) \\ q_1 &= \frac{\gamma_6 RJ}{k_T} \int \dot{\varepsilon} \theta \, \mathrm{d}t + q_1(0) = \bar{\gamma}_6 \int \dot{\varepsilon} \theta \, \mathrm{d}t + q_1(0) \end{aligned}\right\} \tag{7.173}$$

In practical applications the quantities $\bar{\gamma}_4$, $\bar{\gamma}_5$ and $\bar{\gamma}_6$ are chosen to be positive constants and the adaptive control law can be realized as shown in Fig. 7.48. This scheme is very similar to the scheme in Fig. 7.47. The only differences are that the derivative of the adaptive system error $\dot{\varepsilon}$ is used instead of the error $\varepsilon$ itself, and that the quantities $w$, $\dot{\theta}$ and $\theta$ are not filtered by $G_{\mathrm{m}}(s)$.

As mentioned before, the gradient model reference adaptive scheme may become unstable. One of the reasons for this is the approximation $G(s) = G_{\mathrm{m}}(s)$, which was used in the design procedure. Using high gains $\bar{\gamma}_i$, the behaviour of controller parameters becomes oscillatory and if $p$ becomes negative, the gradient adjustment procedure becomes unstable. Therefore, simulation of the model reference adaptive system is unavoidable before implementation of the scheme. Even if the globally

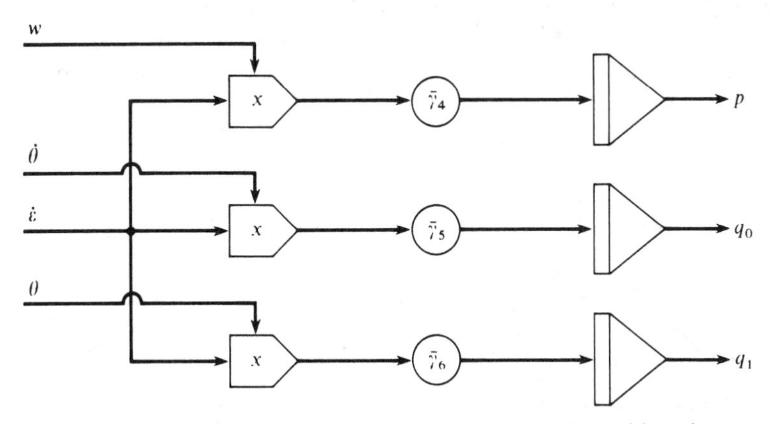


Figure 7.48 Parameter adjustment law for the globally stable scheme.

```
PROGRAM MODEL_REFERENCE_ADAPTIVE_CONTROL_FIRST_TEST
COMMENT POSITIONING SYSTEM
  CONSTANT KT=0.1146,R=0.93,CF=0.001,KE=0.1435
  J=1.5E-4+1.5E-4*STEP(T,0.4)
  THDOT=INTEG((KT*U-(R*CF+KT*KE)*THDOT)/(R*J),0.)
  THETA=INTEG(THDOT,0.)
COMMENT BASIC CONTROL LOOP
  U=P*W-Q0*THDOT-Q1*THETA
COMMENT FIXED CONTROLLER
  CONSTANT P=12.17,Q0=0.0188,Q1=12.17
COMMENT REFERENCE SIGNAL
  W=PULSE(T,0.,0.4,0.2)
COMMENT REFERENCE MODEL
  THMDOT=INTEG(1.E4*W-141.*THMDOT-1.E4*THM,0.)
  THM=INTEG(THMDOT,0.)
COMMENT SIMULATION PARAMETERS
  CINTERVAL CI=0.001
  OUTPUT THETA,THDOT,THM,THMDOT,U
  PREPAR THETA,THDOT,THM,THMDOT,U
  CONSTANT TFIN=0.8
  TERMT(T.GT.TFIN)
END
```

Figure 7.49 Simulation program for basic loop control and the reference model.

stable adaptive scheme is used, simulation of the adaptive system must be performed, since all approximations (e.g. neglecting the inductivity $L$) must be validated.

As mentioned in the first paragraph of this case study, our aim is to describe the simulation scheme development and testing procedure. Usually, the entire simulation scheme, especially if it is extensive and complicated, is not written in one step, but in several successive steps. Such an approach enables intermediate testing of small portions of the simulation scheme.

In our case, it is recommended to simulate first only the basic control loop and the reference model. The corresponding CSSL based simulation program is shown in Fig. 7.49. The parameters of the positioning system were chosen to be $k_T = 0.1146$ Nm/A, $R = 0.93\ \Omega$, $k_e = 0.1435$ V s/rad, $c_f = 0.001$ N ms/rad. The moment of inertia $J$ was supposed to change at $t = 0.4$ s from 0.000 15 kg m$^2$/rad to 0.000 30 kg m$^2$/rad.

The transfer function of the reference model was chosen to be

$$G_m = \frac{10\,000}{s^2 + 141.4s + 10\,000} \tag{7.174}$$

This choice implies that the gain, damping coefficient and natural frequency of the reference model are 1, 0.707 and 100 respectively. The names of the variables in the simulation program in Fig. 7.49 were chosen to elucidate their meaning (`THETA` for $\theta$, `THDOT` for $\dot{\theta}$, `THM` for $\theta_m$, `EPS` for $\varepsilon$, etc.).

A fixed controller was implemented with this first scheme. The controller parameters were chosen to make the basic control loop response with a smaller

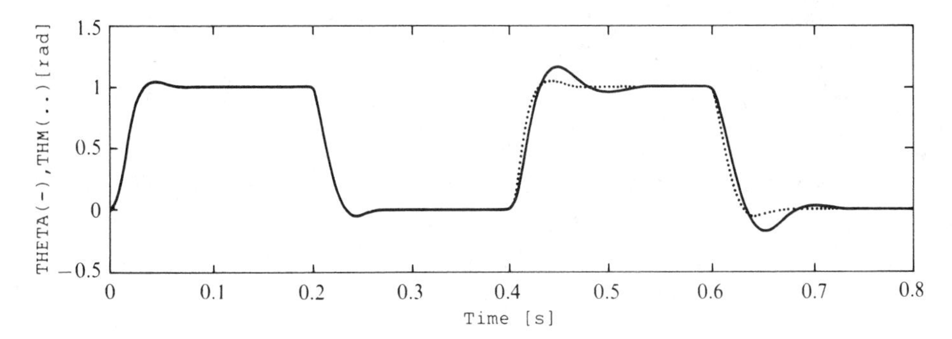


Figure 7.50 Basic loop and reference model response with fixed controller.

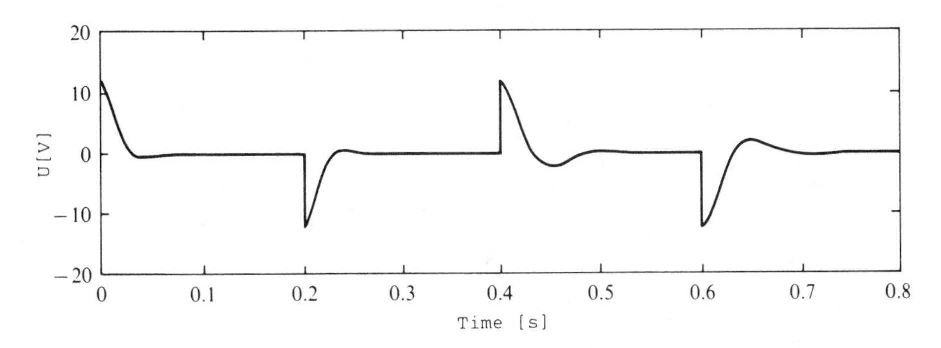


Figure 7.51 Control signal.

moment of inertia ($J = 0.000\,15$ kg m$^2$/rad) equal to the response of the reference model. Figure 7.50 depicts the time response of the basic loop and the reference model response to the pulse reference signal, while Fig. 7.51 shows the corresponding control signal $u$. With a larger moment of inertia, the damping of the basic loop becomes smaller and its overshoot increases.

Our next step in the simulation of the adaptive control system is the implementation of the gradient parameter adjustment law according to Fig. 7.47. The corresponding simulation program is shown in Fig. 7.52. Figure 7.53 shows the corresponding time response of the reference model and the basic loop.

For the realization of the second (globally stable) stable adaptive scheme, only that part of the program which simulates the adaptive controller has to be changed. The gradient control law is replaced by the globally stable control law shown in Fig. 7.48. The corresponding part of the simulation program is represented in Fig. 7.54. The reference model response and the time response of the basic loop for the globally stable adaptive scheme is shown in Fig. 7.55, while Figs 7.56, 7.57 and 7.58 depict the time responses of the controller parameters $p$, $q_0$ and $q_1$ respectively. Comparing the responses of the basic loop with both controllers

```
PROGRAM MODEL REFERENCED ADAPTIVE CONTROL GRADIENT METHOD
COMMENT POSITIONING SYSTEM
  CONSTANT KT=0.1146,R=0.93,CF=0.001,KE=0.1435
  J=1.5E-4+1.5E-4*STEP(T,0.4)
  THDOT=INTEG((KT*U-(R*CF*KT*KE)*THDOT)/(R*J),0.)
  THETA=INTEG(THDOT,0.)
COMMENT BASIC CONTROL LOOP
  U=P*W-Q0*THDOT-Q1*THETA
COMMENT ADAPTIVE CONTROLLER
  CONSTANT GA1BAR=100,GA2BAR=1.,GA3BAR=100.
  CONSTANT P0=12.17,Q00=0.0188,Q10=12.17
  P=INTEG(-GA1BAR*EPS*THETA,PO)
COMMENT FILTERED SHAFT POSITION
  THFDOT=INTEG(1.E4*THETA-141.*THFDOT-1.E4*THF,0.)
  THF=INTEG(THFDOT,0.)
  Q0=INTEG(GA2BAR*EPS*THFDOT,Q00)
  Q1=INTEG(GA3BAR*EPS*THF,Q10)
COMMENT REFERENCE SIGNAL
  W=PULSE(T,0.,0.4,0.2)
COMMENT REFERENCE MODEL
  THMDOT=INTEG(1.E4*W-141.*THMDOT-1.E4*THM,0.)
  THM=INTEG(THMDOT,0.)
COMMENT ADAPTIVE SYSTEM ERROR
  EPS=THETA-THM
  EPSDOT=THDOT-THMDOT
COMMENT SIMULATION PARAMETERS
  CINTERVAL CI=0.002
  NSTEPS NST=40
  OUTPUT THETA,P,Q0,Q1
  PREPAR THETA,THM,U,P,Q0,Q1,J
  CONSTANT TFIN=6.
  TERMT(T.GT.TFIN)
END
```

Figure 7.52 Simulation program for the gradient scheme.

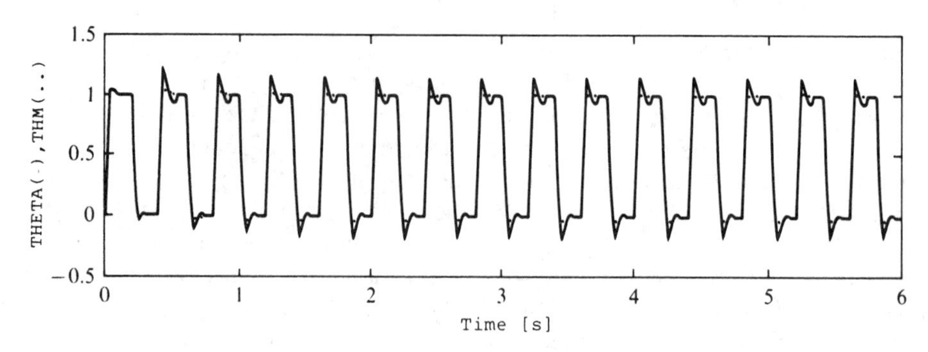


Figure 7.53 Basic loop and reference model response with gradient adaptive controller.

```
COMMENT ADAPTIVE CONTROLLER
  CONSTANT GA4BAR=100,GA5BAR=1.,GA6BAR=100.
  CONSTANT P0=12.17,Q00=0.0188,Q10=12.17
  EPSDOT=THDOT–THMDOT
  P=INTEG(–GA4BAR*EPSDOT*W,P0)
  Q0=INTEG(GA5BAR*EPSDOT*THDOT,Q00)
  Q1=INTEG(GA6BAR*EPSDOT*THETA,Q10)
```

Figure 7.54 Simulation program for the globally stable scheme.

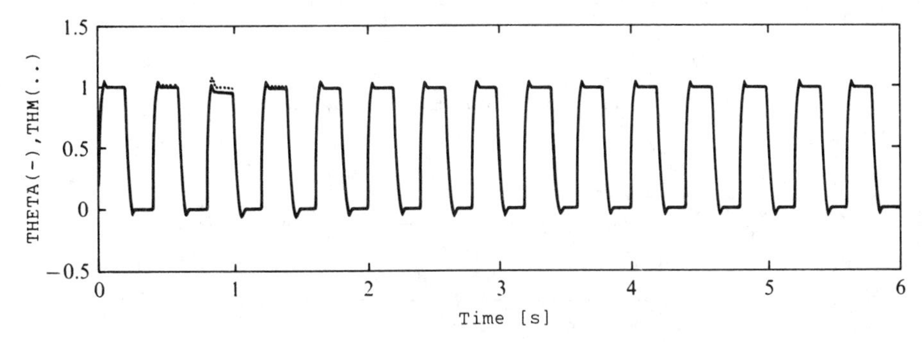


Figure 7.55 Basic loop and reference model response with globally stable controller.

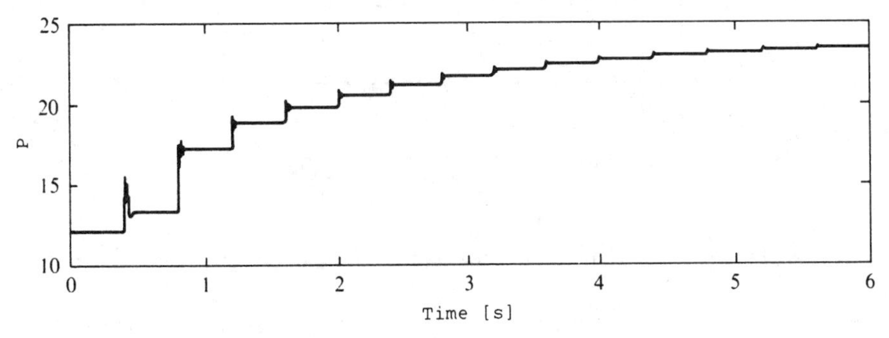


Figure 7.56 Controller parameter $p$.

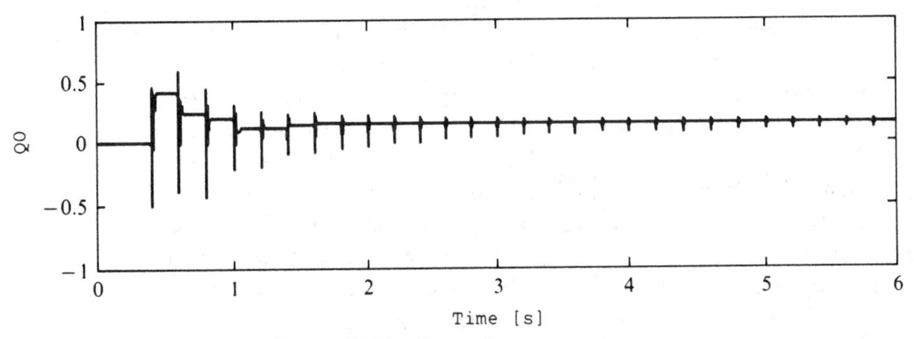


Figure 7.57 Controller parameter $q_0$.

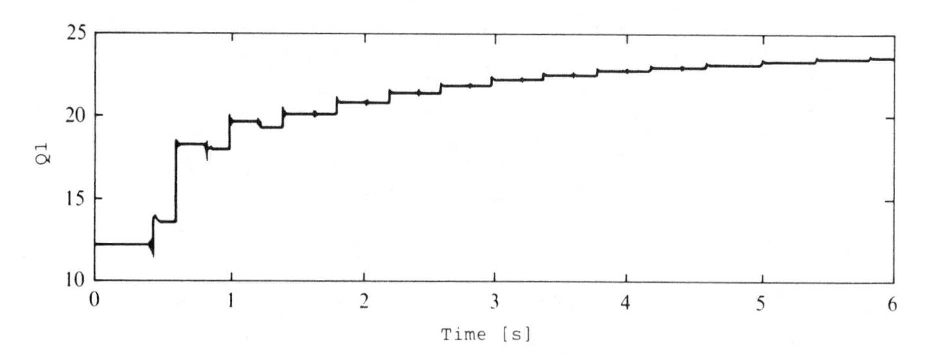


Figure 7.58 Controller parameter $q_1$.

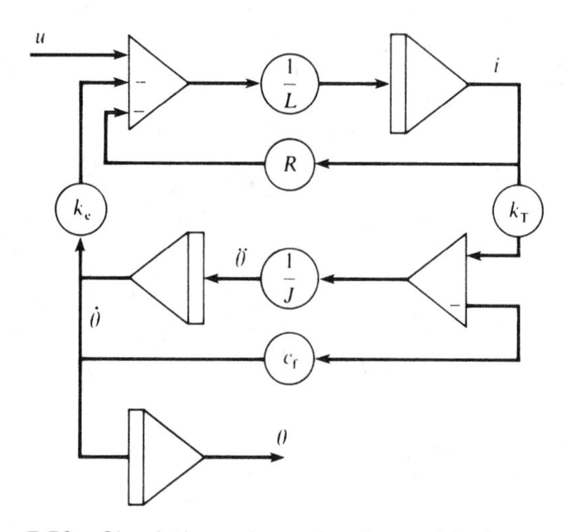


Figure 7.59 Simulation scheme for the positioning system.

(Figs. 7.53 and 7.55) it can be concluded that the globally stable controller is superior to the gradient controller as it ensures smaller overshoot.

The last step in this simulation case study is the investigation of the influence of the unmodelled dynamics on the adaptive system. The unmodelled dynamics in our case are associated with omission of the inductivity $L$. So a new simulation scheme for the positioning system will be developed which takes into account the inductivity $L$, which is supposed to be 100 $\mu$H. This simulation scheme is obtained by the indirect approach using Eqns (7.150), (7.151) and (7.152) and is shown in Fig. 7.59. The corresponding part of the simulation program is shown in Fig. 7.60. Now, that part of the simulation program in Fig. 7.52 which describes the positioning system is replaced by the program in Fig. 7.60. If the simulation is executed, no significant difference from the scheme with unmodelled dynamics is observed, which means that the adaptive control law is robust against the unmodelled dynamics.

```
COMMENT POSITIONING SYSTEM WITH INDUCTIVITY
  CONSTANT KT=0.1146,R=0.93,CF=0.001,KE=0.1435,L=1.E-4
  J=1.5E-4+1.5E-4*STEP(T,0.4)
  I=INTEG((U-R*I-KE*THDOT)/L,0.)
  THDOT=INTEG((KT*I-CF*THDOT)/J,0.)
  THETA=INTEG(THDOT,0.)
```

Figure 7.60 Simulation program for the simulation scheme in Figure 7.59.

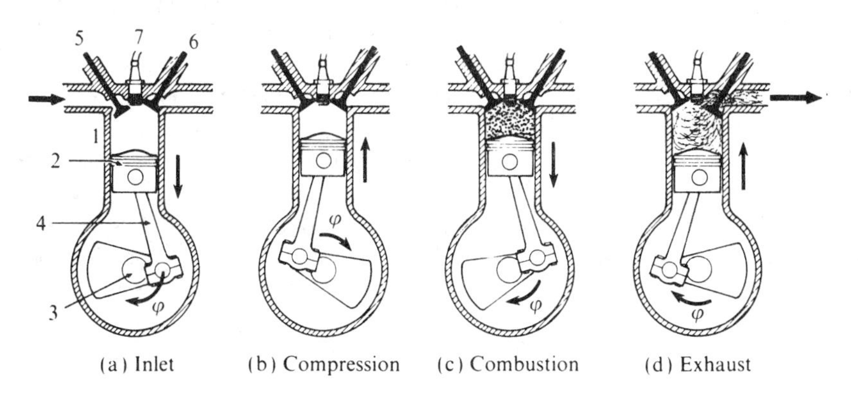


Figure 7.61 Four-stroke process.

## 7.6 SIMULATION OF A DIESEL ENGINE

In a period of increasing consciousness of the ecological and economical problems of motor cars, energy saving and emission reduction are very much the focus of designers of motor car engines. The objective of this case study is to provide a simple model of a one-cylinder four-stroke diesel engine, to simulate it, to perform an optimization of the injection timing and to evaluate the power of the motor as a function of the motor's revolutions per minute. In this case study, the modes controlling the integrators and the optimization, as well as the parameter study options of the simulation language SIMCOS, will be represented.

The theory of combustion engines is described in Ferguson (1986). Here, only a very simplified model will be discussed. A single unit of the combustion engine shown in Fig. 7.61 consists basically of cylinder (1), piston (2), crankshaft (3) and connecting rod (4). The inlet (5) and outlet (6) valves are placed in the cylinder head. By revolutions of the crank shaft the volume of the cylinder is periodically varied by the piston. With four-stroke engines the engine cycle starts at top dead centre (crank angle $\varphi = 0^\circ$) and, as shown in Fig. 7.61, in the first, i.e. the inlet stroke, fresh air is aspirated through the inlet channel. At bottom dead centre ($\varphi = 180^\circ$) the inlet valve is closed and in the following compression stroke the air is compressed. In the case of a diesel engine, the fuel is injected near top dead centre ($\varphi = 360^\circ$) by the injection valve (7). The crank angle difference between the top dead centre and the injection angle is called the injection advance ($\varphi_a = 360° - \varphi_i$). Due to the very high temperature in the cylinder the fuel inflames and produces the combustion heat and pressure which results in the piston moving down and generating a torque on the crankshaft. This (the third) stroke is called the combustion stroke. At bottom dead centre ($\varphi = 540°$) the outlet valve is opened and the exhaust gas is pushed out of the cylinder during the last, i.e. the exhaust, stroke. The engine cycle is completed at $\varphi = 720°$, where the whole procedure is repeated.

Various types of processes (thermodynamic, chemical, hydraulic, mechanical) are involved in combustion engines and thus precise modelling is very difficult. In this case study the model of the diesel engine will be restricted only to those effects which principally influence the outward mechanical behaviour of the engine.

The key process of an engine is the variation of the pressure $p$ in the cylinder. During the exhaust and inlet strokes where a valve is open, the cylinder pressure depends mainly on the shape of the inlet and exhaust pipes and on the opening angles of the valves. This pressure, however, is small compared with the pressure during combustion, so with respect to the resulting torque it will be assumed to be equal to the outside pressure $p_0$:

$$p = p_0 \qquad 0° \leqslant \varphi < 180° \qquad 540° < \varphi \leqslant 720° \tag{7.175}$$

During compression and combustion strokes the cylinder pressure is determined by the thermal equation of the state:

$$pV = mRT \tag{7.176}$$

by the relation of the internal energy of the gas:

$$U = mc_v T \tag{7.177}$$

and by the energy conservation equation:

$$\frac{dU}{dt} = \frac{dQ}{dt} - p\frac{dV}{dt} \tag{7.178}$$

In the above equations, $V$ denotes the cylinder volume; $m$ the gas mass; $R$ the specific gas constant; $T$ the gas temperature; $U$ the gas energy; $c_v$ the specific calorific capacity of the gas; and $Q$ the combustion energy (heat). Using the energy conservation equation (7.178) the heat transfer to the cylinder walls and the mass increase due to the injected fuel are neglected. Inserting Eqn (7.177) into Eqn (7.176), the following equation is obtained:

$$p = \frac{R}{c_v V} U \qquad 180° \leqslant \varphi \leqslant 540° \tag{7.179}$$

The modelling of the combustion process is the most difficult task in the derivation of the entire model. In order to avoid extensive equations of chemical processes, the following descriptive model for the combustion heat release is used, Vibe (1970):

$$\frac{dQ}{d\varphi} = \begin{cases} 6.9\dfrac{m_f H}{\varphi_c}(\mu + 1)\left(\dfrac{\varphi - \varphi_i}{\varphi_c}\right)^{\mu} \exp\left(-6.9\left(\dfrac{\varphi - \varphi_i}{\varphi_c}\right)^{\mu+1}\right) & \varphi_i < \varphi < \varphi_i + \varphi_c \\ 0 \quad \text{otherwise} \end{cases} \tag{7.180}$$

where $m_f$ is the injected fuel mass; $H$ the calorific value of the fuel; $\varphi_i$ the injection angle; $\varphi_c$ the combustion angle (i.e. the angle at which the fuel burns completely); and $\mu$ the combustion shape factor. An important property of the model, Eqn (7.180), is that its integral equals the calorific energy of the fuel:

$$\int_{\varphi_i}^{\varphi_i + \varphi_c} \frac{dQ}{d\varphi} d\varphi = m_f H \tag{7.181}$$

for which the assumption of completely burned fuel is required

In the simulation scheme, the time derivation of the combustion energy is required due to Eqn (7.178), which is obtained by

$$\frac{dQ}{dt} = \frac{dQ}{d\varphi}\frac{d\varphi}{dt} = \frac{dQ}{d\varphi}\omega \tag{7.182}$$

where

$$\omega = \frac{d\varphi}{dt} = \dot{\varphi} \tag{7.183}$$

is the angular crank velocity. The combustion angle $\varphi_c$ is a nonlinear function of the angular crank velocity $\omega$, but in our simple case a linear function

$$\varphi_c = t_c \omega \tag{7.184}$$

will be used. In Eqn (7.184) $t_c$ is the combustion time, which is supposed to be constant.

The volume of the combustion chamber is obtained by kinematic relations and depends on the crank angle $\varphi$, the compression volume $V_c$, the piston surface $A_p$, the crank radius $r_c$ and the length of the connecting rod $l_r$. Denoting the relation of the crank radius and the connecting rod length by $\lambda = r_c/l_r$ the volume of the combustion chamber can be expressed for small $\lambda$ by

$$V(\varphi) = V_c + A_p r_c[1 - \cos\varphi + 0.25\lambda(1 - \cos 2\varphi)] \tag{7.185}$$

In Eqn (7.178) the time derivative of the combustion chamber volume is also required and is obtained by derivation of Eqn (7.185):

$$\frac{dV(\varphi)}{dt} = \dot{V}(\varphi) = A_p r_c \omega(\sin\varphi + 0.5\lambda \sin 2\varphi) \tag{7.186}$$

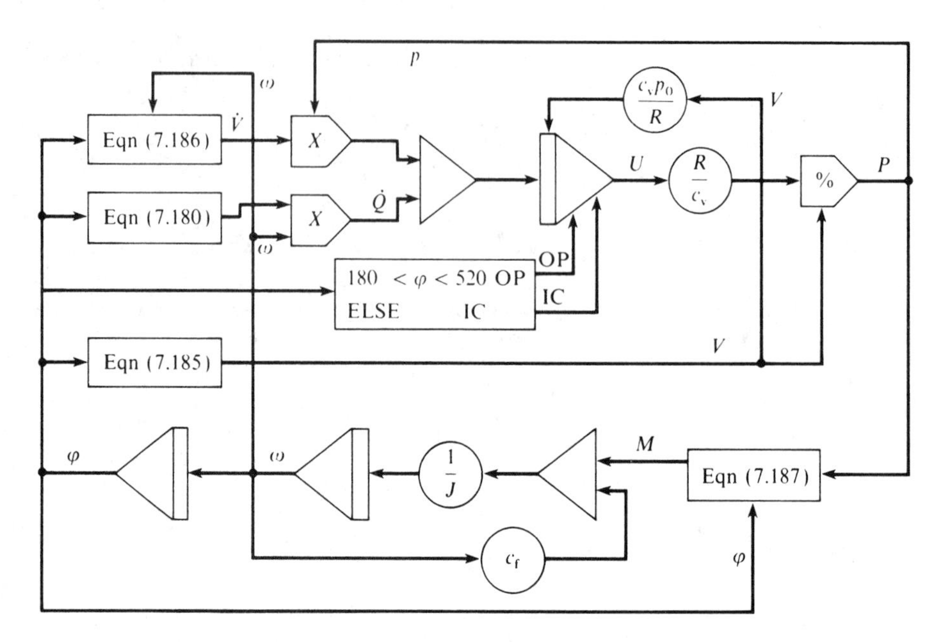


Figure 7.62 Simulation scheme for the diesel engine.

where the angular crank velocity $\omega$ is defined by Eqn (7.183). The cylinder pressure results in the crank shaft torque as follows (valid for small $\lambda$ again):

$$M(\varphi) = r_c(p - p_0)A_p \sin\varphi\left(1 + \frac{\lambda\cos\varphi}{\sqrt{1 - \lambda^2\sin^2\varphi}}\right) \tag{7.187}$$

If the engine is supposed to operate idle and the inertia and friction effects of the piston are neglected, the following equation for the mechanical system is obtained:

$$\frac{d\omega}{dt} = \frac{1}{J}(M - c_f\omega) \tag{7.188}$$

where $J$ is the moment of crankshaft and flywheel inertia and $c_f$ is the friction of the crankshaft bearings.

The model of the diesel engine is now completed and the corresponding simulation scheme and SIMCOS program are shown in Figs. 7.62 and 7.63 respectively.

For better comprehension, the notation in the SIMCOS program in Fig. 7.63 corresponds to the notation in the simulation scheme shown in Fig. 7.62 and to corresponding equations, but there are some details which deserve an explanation. First, the nonstandard part of the program is the periodicity of the crankshaft angle $\varphi$, which is denoted in the program by `PHIR` ($\varphi$ in radians). `PHIR` is then made periodic by the following trick: $\varphi = \arctan\tan\varphi$, and is at the same moment also converted to degrees in the `PROCEDURAL (PHID=PHIR)`. The simulation of both

```
COMMENT CYLINDER; R=8314.4/28.8=288.7
 CONSTANT R=288.7,CV=715.,TO=270.,PO=1.E5$ P=R/(CV*V)*U$ UO=PO*CV*V/R
 PROCEDURAL(IC,OP=PHID)
      IF ((PHID.GE.180.).AND.(PHID.LE.540.)) THEN
         IC=0
         OP=1
      ELSE
         IC=1
         OP=0
      ENDIF
 END
 U=INTEG(QDOT-P*VDOT,UO,IC,OP)$T=TO*IC+U*R*TO/(PO*(VC+2.*AP*RC)*CV)*OP
 CONSTANT MF1=2.693E-5,MF2=1.347E-5,TF1=0.5 $COMMENT FUEL MASS
 MF=MF1+MF2*STEP(TIME,TF1)                    $ COMMENT TO BE INJECTED
COMMENT COMBUSTION
 CONSTANT  PI=3.14159,TC=0.2,PHIA=10.,MU=1.,H=42.E6 $ PHIC=TC*OMEGA
 PROCEDURAL (QDOT=PHID)
      IF ((PHID.GT.360.-PHIA).AND.(PHID.LT.360.-PHIA+PHIC)) THEN
         FAKT1=(PHID-360.+PHIA)/PHIC
         FAKT2=-6.9*FAKT1**(MU+1)
         QDOT=180./PI*OMEGA*6.9*MF*H/PHIC*(MU+1.)*FAKT1**MU*EXP(FAKT2)
      ELSE
         QDOT=0.
      ENDIF
 END
COMMENT KINEMATICS
 CONSTANT VC=5.E-5,RC=0.05,LAMBDA=0.1,AP=0.01
 V=VC+AP*RC*(1-COS(PHIR)+0.25*LAMBDA*(1-COS(2.*PHIR)))
 VDOT=AP*RC*OMEGA*(SIN(PHIR)+0.5*LAMBDA*SIN(2.*PHIR))
 CONSTANT CF=0.3,J=2.,OMEGA0=209.4 $ COMMENT MECHANICS
 M=RC*(P-PO)*AP*SIN(PHIR)*(1.+LAMBDA*COS(PHIR)/...
   SQRT(1.-(LAMBDA*SIN(PHIR))**2))
 OMEGA=INTEG((M-CF*OMEGA)/J,OMEGA0) $ PHIR=INTEG(OMEGA,0.)
 PROCEDURAL (PHID=PHIR)
      PHID=360./PI*ATAN2(SIN(PHIR/2.),COS(PHIR/2.))
      IF(PHID.LT.0.) PHID=PHID+720.
 END
 RPM=60*OMEGA/(2*PI)
 NSTEPS NS=10 $ CINTERVAL CI=0.002 $ COMMENT SIMULATION PARAMETERS
 CONSTANT TFIN=3. $ VARIABLE TIME=0. $ TERMT(TIME.GT.TFIN)
 OUTPUT PHID,RPM,P,V,QDOT $ PREPAR PHID,RPM,P,V,M,QDOT
END
```

Figure 7.63 SIMCOS program for the diesel engine.

régimes for the pressure in the cylinder due to open/closed valves is performed by controlling the mode of the integrator for the internal energy of the gas $U$. With an open valve the integrator is put into initial condition mode and the initial condition of $U$ is calculated using Eqn (7.179) for the pressure $p$ equal to the outside pressure $p_o$, which are denoted by P and PO respectively. With both valves closed the integrator is put into the operate mode and the internal energy of the gas is calculated according

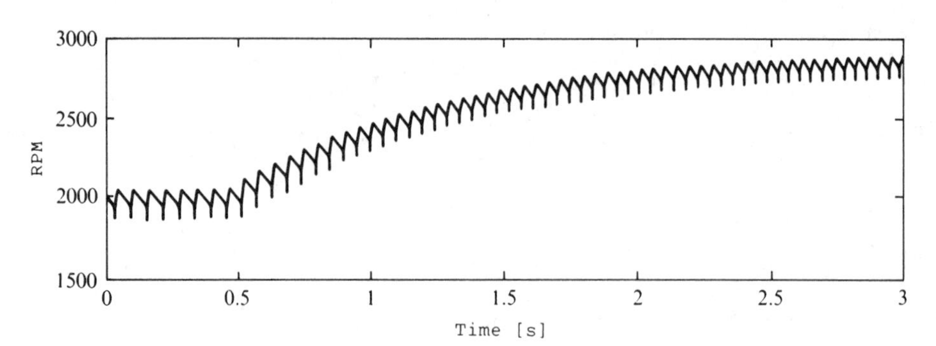


Figure 7.64 Time response of the angular crank velocity.

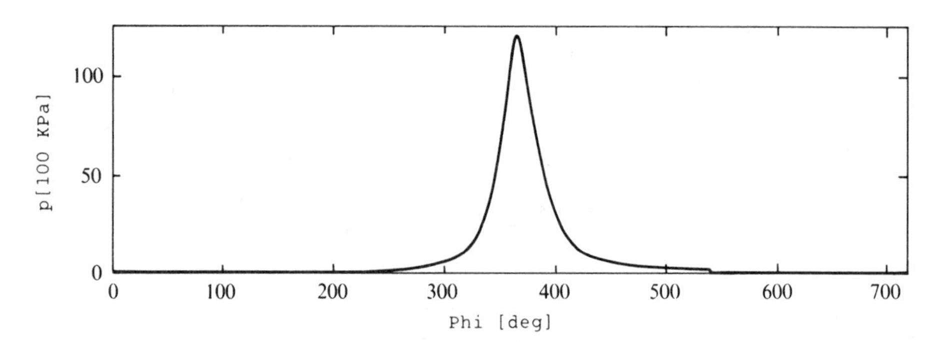


Figure 7.65 Cylinder pressure.

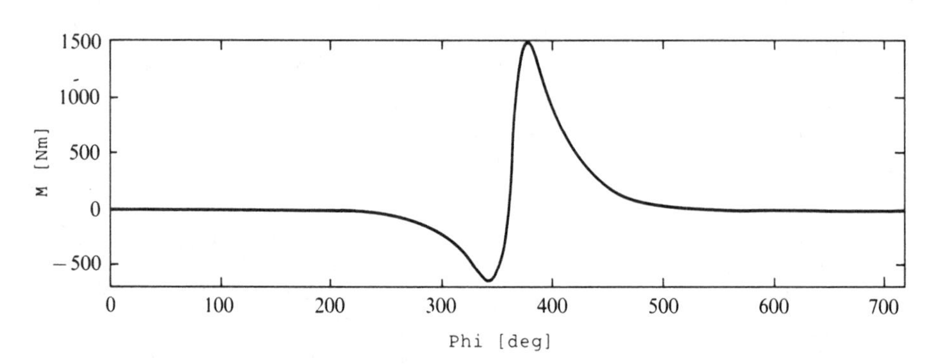


Figure 7.66 Crank shaft torque.

to Eqn (7.178). The signals `IC` and `OP` for the mode control of the integrator are obtained by `PROCEDURAL (IC,OP=PHID)`. Other parts of the program are self-explanatory.

First the diesel engine was simulated for a step change of injected fuel mass $m_f$, which was increased at $t = 0.5$ s from 0.026 93 mg to 0.0404 mg. The combustion shape factor was chosen to be $\mu = 1$ and the injection advance $\varphi_a = 10°$. Figure 7.64

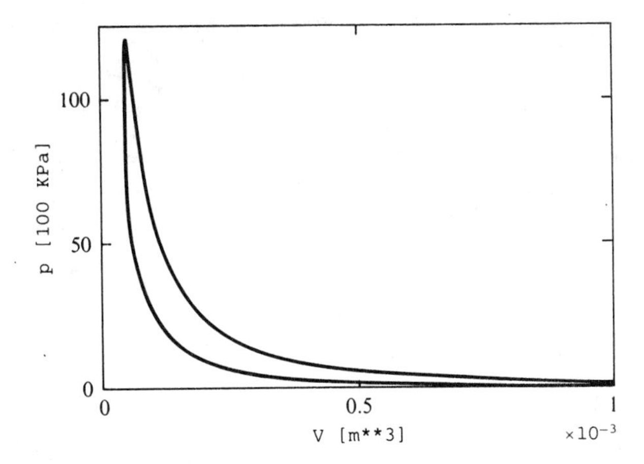


Figure 7.67 Pressure–volume diagram.

```
PROCEDURAL(IC1=PHID)
      IF (PHID.GE.180.) THEN
         IC1=0
      ELSE
         IC1=1
      ENDIF
END
MAVG=INTEG(M*OMEGA/4.*PI),0.,IC1,OP)
POWER=(MAVG–CF*OMEGA)*OMEGA
```

Figure 7.68 DYNAMIC section of the SIMCOS program.

shows the time response of the angular crank velocity in revolutions per minute (RPM), which is obtained from $\omega$ by simple calculation. As we can see from Fig. 7.64, the angular crank velocity is increased from 2000 rpm to 3000 rpm as an approximately first order process with overlapped oscillations due to the four-stroke process. The typical time (crank angle) response of the cylinder pressure and the crank shaft torque are shown in Figs. 7.65 and 7.66, respectively, while Fig. 7.67 depicts the corresponding pressure–volume diagram.

The second objective of this case study is the optimization of the injection advance $\varphi_a$. The criterion function is the average torque which is computed according to the DYNAMIC section of the SIMCOS program shown in Fig. 7.68. The average torque is computed as the integral of the crank shaft torque in the second, third and fourth stroke (the torque in the first and fourth stroke is zero!). The integral is then divided by the time duration of all four strokes ($\omega/4\pi$). The procedure repeats every 720° of the crank shaft angle $\varphi$, which is realized by putting the integrator in the initial condition state in every first stroke. In the DYNAMIC section in Fig. 7.68 the power transmitted to the mechanical load is also computed.

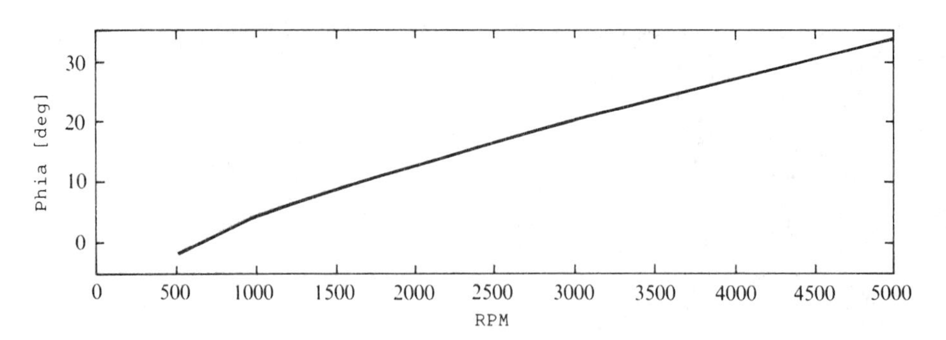


Figure 7.69 Optimal injection advance $\varphi_a$.

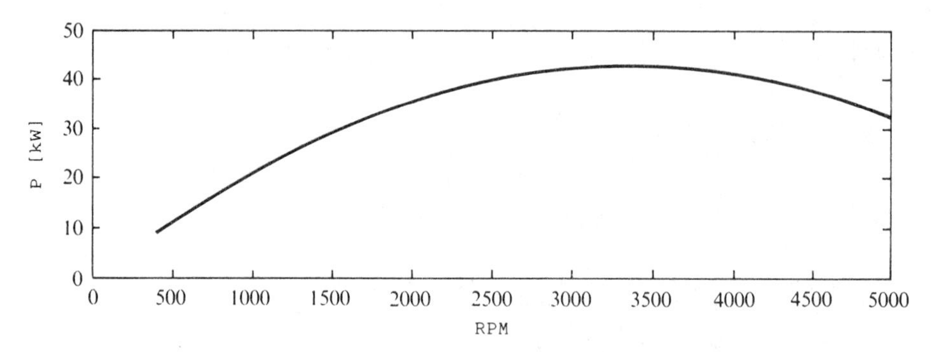


Figure 7.70 Power of the motor as a function of the crank angle velocity.

The optimization of the injection advance was performed for crank shaft velocities from 500 rpm up to 5000 rpm. Figure 7.69 shows the optimal injection advance as a function of the crank shaft velocity. Variable injection advance as a function of the crank shaft velocity was then introduced into the model. The optimal injection advance was approximated by a linear function (obtained by linear regression of the curve in Fig. 7.69) and realized by the SIMCOS statement

```
PHIA=-2.+0.0072*60*OMEGA/(2.*PI)
```

A parameter study was performed as the third objective of the case study. The motor's power was evaluated for crank shaft velocity ranging from 400 rpm up to 5000 rpm in steps of 200 rpm for constant injected fuel mass $m_f = 0.1$ mg. The result of the parameter study is shown in Fig. 7.70. The motor delivers maximum power 43.1 KW at 3400 rpm.

## BIBLIOGRAPHY

AMYOT, J.R. and G. VAN BLOKLAND (1985), 'Parameter optimization with ACSL', *Proceedings of the SCSC '85 Conference*, pp. 63–8.

AMYOT, J.R. and G. VAN BLOKLAND (1987), 'Parameter optimization with ACSL models', *Simulation*, **49** (5), pp. 213–18.

ANDRESON, B.D.O., R.R. BITMEAD, C.R. JOHNSON, P.V. KOTOVIĆ, R.L. KOSUT, I.M.Y. MAREELS, L. PRALY and B.D. RIEDLE (1986), *Stability of Adaptive Systems*, MIT Press, Cambridge, Mass.

ATANASIJEVIĆ, M., R. KARBA and F. BREMŠAK (1985), 'Semibatch distillation modelling and control design', *Proceedings of the 3rd Symposium Simulationstechnik, Simulationstechnik* (D.P.F. Moller, ed.), Springer-Verlag, Berlin, pp. 464–8.

ATANASIJEVIČ, M., R. KARBA, F. BREMŠAK, T. RECELJ, J. GOLOB and L. FELE (1986), 'Comparison of three different approaches to the semibatch distillation column control design', *Preprints of an IFAC Symposium*, Bournemouth, UK, pp. 231–6.

ATANASIJEVIĆ, M., R. KARBA, M. MILANOVIĆ and F. BREMŠAK (1986), 'Computer aided design of semibatch distillation column control', *Workshop on Process Automation*, Ljubljana, Yugoslavia (International Bureau KFA Jülich, ed.), pp. 129–50.

BIRTA, L.G. (1977), 'A parameter optimization module for CSSL-based simulation software', *Simulation*, April, pp. 113–21.

BREITENECKER, F., I. TROCH, R. RUZICKA AND A. SAUBERER (1988), 'GOMA-an optimisation environment for development of automatic control in CSSL-type simulation languages', *Proceedings of the IMACS Conference*, Paris, Vol. 2, pp. 728–30.

BUCKLEY, P.S., W.L. LUYBEN AND J.P. SHUNTA (1985), *Design of Distillation Column Control Systems*, Instrument Society of America, Research Triangle Park, North Carolina.

CELLIER, F.E. (1983), 'Simulation software – today and tomorrow', *Proceedings of the IMACS Conference*, Nantes, France.

DAVIES, W.D.T. (1970), *System Identification for Self-adaptive Control*, Wiley Interscience, NY.

DIXON, L.C.W. (1972), *Nonlinear Optimisation*, The English Universities Press Limited, London.

DRINOVEC, J., A. MRHAR and R. KARBA (1987), 'Evaluation of the influence of haemodialysis on ciprofloxacin pharmacokinetics', *15th International Congress of Chemotherapy*, Vol. 2, Istanbul, pp. 1924–6

EDGART, B. (1979), *Stability of Adaptive Controllers*, Lect. Not. in Cont. and Inf. Sc. Vol., Springer, Berlin.

FERGUSON, C.R. (1986), *Internal Combustion Engines*, Wiley, New York.

GMEHLING, J. and U. ONKEN (1977), 'Vapor–liquid equilibrium data collection, aqueous–organic systems', *Dechema, Chemistry Data Series* (D. Behrens and R. Eckerman, eds), Vol. 1, Part 1, Germany.

GODFREY, K. (1983), *Compartmental Models and their Application*, Academic Press, London.

JACKSON, A.S. (1960), *Analog Computation*, McGraw-Hill, NY.

KARBA, R., F. BREMŠAK, F. KOZJEK, A. MRHAR and D. MATKO (1979), 'Hybrid simulation in drug pharmacokinetics', *Preprints of IMACS Congress Simulation of Systems* (L. Dekker, G. Savastono and G.C. Vansteenkiste, eds), Sorrento, pp. 733–41.

KARBA, R., A. MRHAR, S. PRIMOŽIČ, B. ZUPANČIČ and M. ATANASIJEVIČ-KUNC (1990), 'Modelling and simulation of drugs pharmacokinetics in end stage renal failure', *Preprints of the 11th World Congress IFAC* (V. Utkin and U. Jaaksoo, eds), Vol. 4, Tallinn, Estonia, pp. 51–6.

KARBA, R., B. ZUPANČIČ, F. BREMŠAK, A. MRHAR and S. PRIMOŽIČ (1990), 'Simulation tools in pharmacokinetical modelling', *Acta. Pharm. Jugosl.*, **40**, 247–62.

KLEINERT, W., M. GRÄF, R. KARBA and B. ZUPANČIČ (1988), 'Simulation einer Destillationskolonne Modellierung mit SIMCOS und Vergleich der Ergebnisse von ACSL und SIMSTAR Simulationen', *Proceedings of the 5th Symposium Simulationstechnik, Simulationstechnik* (W. Ameling, ed.), Springer-Verlag, Berlin, pp. 254 9.

KREYSZIG, E. (1988), *Advanced Engineering Mathematics*, John Wiley & Sons, NY.

LANDAU, I.D. (1979), *Adaptive Control – The Model Reference Approach*, Marcel Dekker, NY.

MRHAR, A., F. KOZJEK, S. PRIMOŽIČ and R. KARBA (1987), 'System approach to pharmacokinetical studies for optimal drugs design', *Acta Pharm. Jugosl.*, **37**, pp. 319–29.

OGATA, K. (1990), *Modern Control Engineering*, Prentice Hall, Englewood Cliffs, NJ.

OSBURN, P.V., H.P. WHITAKER and A. KEZER (1961), 'New developments in the design of adaptive control systems', *Inst. of Aeronautical Sciences*, Paper 61—39.

PARKS, P.C. (1966), 'Liapunov redesign of model reference adaptive control systems', *IEEE Tr. A.C.*, **11**, 362–7.

PERRY, R.H. and C.H. CHILTON (1973), *Chemical Engineer's Handbook*, McGraw-Hill, NY.

ROGERS, A.E. and T.W. CONNOLLY (1960), *Analog Computation in Engineering Design*, McGraw-Hill, NY.

ŠEGA, M., S. STRMČNIK, R. KARBA and D. MATKO (1985), 'Control system treatment by program package ANA', *Proceedings of the 11th IMACS World Congress on System Simulation and Scientific Computation*, Oslo, Norway, Vol. 4, pp. 95–8.

ŠEGA, M., MATANASIJEVIĆ, R. KARBA and M. MILANOVIĆ (1986), 'Computer aided design of semibatch distillation column control', *Proceedings of the 2nd European Simulation Congress*, Antwerp, Belgium, pp. 528–34.

SCHUMANN, A., D. MATKO and B. ZUPANČIČ (1991), 'Simulation of a diesel engine using a digital simulation language', *MELECON '91*, Ljubljana.

SHINSKEY, F.G. (1984), *Distillation Control*, McGraw-Hill, NY.

STEPHANOPOULOS, G. (1984), *Chemical Process Control, An Introduction to Theory and Practice*, Prentice Hall, Englewood Cliffs, NJ.

VIBE, I.I. (1970), *Brennverlauf und Kreisprozeß von Verbrennungsmotoren*, VEB Verlag Technik, Berlin.

WEBER, T.W. (1973), *An Introduction to Process Dynamics and Control*, John Wiley, London.

# Index